Bioaccessibility and Digestibility of Lipids from Food

Myriam M.-L. Grundy • Peter J. Wilde
Editors

Bioaccessibility and Digestibility of Lipids from Food

Springer

Editors
Myriam M.-L. Grundy
PEGASE, INRAE
Institut Agro
Saint Gilles, France

Peter J. Wilde
Norwich Research Park
Quadram Institute Bioscience
Norwich, UK

ISBN 978-3-030-56908-2 ISBN 978-3-030-56909-9 (eBook)
https://doi.org/10.1007/978-3-030-56909-9

This Springer imprint is published by the registered company Springer Nature Switzerland AG
The registered company address is: Gewerbestrasse 11, 6330 Cham, Switzerland

Contents

Introduction

Although evidence exists on the health benefits associated with the inclusion of certain lipid-rich foods (e.g. nuts, dairy products and fish) in the diet, the mechanisms that explain the physiological effects and the long-term benefits are not well understood. Lipids in themselves are essential nutrients and have many beneficial health effects: they are a source of energy and essential fatty acids, they are structural components of cell membranes, they are required to solubilise fat soluble compounds, and they serve as precursors of hormones. In general, it is only when dietary lipids are consumed in excess that they begin to have a negative impact on our health. The nutritional quality of our diet has a huge influence on our health and well-being, with a plethora of conditions and diseases being associated with a poor diet. It is becoming increasingly apparent that the nutritional quality of our food is a consequence of not only its nutritional composition but also how its structure influences the rate and extent of nutrient availability. The structure of a food influences the way it is transformed during processing and digestion. This, in turn, has an impact on nutrient bioaccessibility (release) and digestibility, and subsequently on the physiological response and health of the individual who consumes that food. Hence, care needs to be taken when assessing nutritional quality based purely on nutrient composition.

Lipids are generally insoluble in water; therefore, to be digested, lipids are required to be dispersed or emulsified to make them more accessible to the lipases which hydrolyse them into components suitable for absorption, for instance, by being released from the food matrix (from the cells of a plant tissue). The rate and extent of lipid digestion is governed by the way lipids are "presented" to the lipases in the different compartments of the gastrointestinal tract. Lipid digestion can be viewed as a colloidal and interfacial phenomenon where digestive agents (e.g. bile salts and lipases) and food components (e.g. dietary fibres, proteins and phytosterols) interact with the lipid at the lipid–water interface and thereby impact their hydrolysis by lipase(s).

Diets containing foods high in fat are considered to be detrimental to health; however, studies have shown that the consumption of certain foods which are high in fat, such as nuts and certain dairy products, led to beneficial effects on risk factors for cardiovascular disease. A great proportion of lipid in some food matrices, such as almonds, remains undisturbed after mastication and the subsequent digestion processes. Here, the plant cell walls are resistant to digestion in the upper gastrointestinal tract and therefore encapsulate intracellular nutrients, thereby limiting lipid bioaccessibility. In dairy products, the structure of the food matrix can also have an impact on lipid digestibility. Indeed, the composition, microstructure and rheological properties of cheese matrices have been shown to impact the kinetics of the degradation of the cheese and the digestibility of the lipid it contained. Regarding food products where the lipids are readily available, such as milk and plant-based beverages or soups, other parameters can dictate lipid digestibility. These include the size of the droplets and the "quality" of the droplets interface (i.e. surface-active molecules) and the presence of certain compounds that may inhibit lipase activity (e.g. dietary fibres).

Therefore, the overall structure of the food containing the lipids is likely to also play a crucial role, notably by influencing their bioaccessibility and digestibility and ultimately their impact on nutrition and health.

This book aims to cover some of the latest research performed on food structure in the context of lipid digestion. The subjects include lipases, in vitro and in vivo models used to monitor lipid digestion, physiological aspects of lipid digestion and the impact of food structure (plant and animal sources of lipids).

Part I
Digestion of Lipids

Chapter 1
Enzymes Involved in Lipid Digestion

A. Salhi, F. Carriere, Myriam M.-L. Grundy, and A. Aloulou

Lipid digestion is a complex process that takes place at the lipid-water interface and involves various lipolytic enzymes present predominantly in the stomach and the small intestine [34]. These enzymes catalyse the hydrolysis of a variety of dietary lipids from animal and plant sources, such as triacylglycerols (TAGs), phospholipids, galactolipids, cholesterol and vitamin esters. They include gastric lipase, colipase-dependent pancreatic lipase, pancreatic lipase-related proteins 2 (PLRP2), carboxyl ester hydrolase or bile salt-stimulated lipase (CEH, BSSL), and pancreatic phospholipase A2. A debate still exist about the existence of a lingual lipase in human [30, 86, 140], an enzyme that has been demonstrated to be present and active in rat and mice tongue only and which is the product of a gene ortholog [53] to the gene of gastric lipase [24] in humans and many other species. Bakala N'Goma et al. [12] have reviewed the key findings that support the existence of lingual or gastric lipases in several species in term of gene expression, enzyme immunocytolocalization and lipase activity. So far, no supporter of the existence of a lingual lipase in humans has been able to provide similar data.

Contrary to the other major digestive enzymes, i.e. proteases and amylases, lipases act at the lipid-water interface because their substrate is insoluble in water

A. Salhi
Laboratoire de Biochimie et de Genie Enzymatique des Lipases, ENIS, Universite de Sfax, Sfax, Tunisia

Aix Marseille University, CNRS, BIP, UMR7281, Marseille, France

F. Carriere
Aix Marseille University, CNRS, BIP, UMR7281, Marseille, France

M. M.-L. Grundy
PEGASE, INRAE, Institut Agro, Saint Gilles, France
e-mail: myriam.grundy@inrae.fr

A. Aloulou (✉)
Laboratoire de Biochimie et de Genie Enzymatique des Lipases, ENIS, Universite de Sfax, Sfax, Tunisia

M. M.-L. Grundy, P. J. Wilde (eds.), *Bioaccessibility and Digestibility of Lipids from Food*, https://doi.org/10.1007/978-3-030-56909-9_1

[3, 122, 123]. This characteristic makes the development of kinetic models challenging and the usual Michaelis-Menten model no longer applies [115]. As for any enzyme, the rate of lipase activity depends on the initial quantity and type of substrate, the temperature, the pH, but a key parameter is the accessible surface available for lipase adsorption and activity. Indeed, activity increases with the rate of emulsification and the decrease in lipid droplet size, which corresponds to a larger specific surface [17]. The accumulation of lipolysis products and/or surfactants at the interface can however interfere with lipase activity in various ways, including enzyme inhibition or activation. In other words, lipolysis can be affected by environmental conditions and compounds that can alter the structure and/or the integrity of the interface (e.g. surface-active molecules such as proteins, phospholipids and bile salts) [39, 54, 123]. Besides these parameters, the form in which the lipids are delivered to the GI tract, including the food matrix, also impact the rate and extent of lipolysis. This particular topic is reviewed in Chaps. 7, 8, and 9. One important point to consider is that gastrointestinal lipolysis is a dynamic process that starts at the lipid-water interface and progressively proceeds through enzyme reactions on substrates with higher polarity and interaction with water (some lipolysis products like monoacylglycerols (MAGs), phospholipids). The enzymes involved in the steps subsequent to TAG and diacylglycerol (DAG) hydrolysis (carried out by enzymes with high affinity for lipid-water interfaces, the so-called true lipases) have usually a higher affinity for substrates forming mixed micelles and dispersed in the water phase.

This chapter gives a brief overview of the main dietary lipids from both animal and plant origins. Then, the lipolytic enzymes present in the human gastrointestinal (GI) tract and their structure/function relationship are described. Finally, the kinetic models and the sources of enzymes used in vitro models of digestion are discussed.

1.1 Dietary Lipids

Lipids, and in particular TAGs, represent the most condensed form of energy available in biological tissues [71]. Outside of cell membranes, lipids are stored in organelles called lipid bodies, oil bodies or oleosomes in plant seeds, that are present in different cell types. These lipid bodies have a spherical shape and are composed of a core of neutral lipids (TAGs and cholesterol esters) surrounded by a monolayer of phospholipids, cholesterol and proteins [82], as well as other polar lipids in plants and algae. In all organisms, the metabolic pathways responsible for the formation of lipid bodies and their mobilization through hydrolytic reactions have strong similarities. Detailed reviews of lipid bodies in mammals, plants and microorganisms have been published [4, 110, 159].

1.1.1 Membrane Lipids

Eukaryotic cells are compartmentalized into organelles (e.g. endoplasmic reticulum, Golgi apparatus, and mitochondria) that are delimited by membranes with different biological functions.

These membranes are essentially made up of phosphoglycerolipids, with the exception of plant photosynthetic membranes that contain large amounts of galactoglycerolipids. This is the case, for instance, in the plasma membrane which delimits a cell and the endomembrane system which consists of different organelles (e.g. lysosomes and endosomes) dispersed in the cytoplasm of a eukaryotic cell [91]. Phosphoglycerolipids are consisting of two fatty acid residues esterifying a glycerol residue which is itself esterified by a phosphate residue. They are lipids organized around a glycerol-3-phosphate residue, unlike triglycerides which do not contain a phosphate group but are instead formed of three fatty acid residues esterifying a glycerol residue. In contrast to triglycerides, phosphoglycerides are highly amphiphilic molecules, i.e. consist of a hydrophilic polar head (the phosphate group and its substituents, such as choline, ethanolamine, glycerol or inositol) and hydrophobic aliphatic tails (the fatty acid residues).

Other than phosphoglycerides, membranes also contain other lipids such as sphingolipids and sterols, the latter are mainly represented in animals by cholesterol [100]. Biological membranes are organized around a lipid bilayer consisting of two sheets of lipid molecules in which the phosphoglycerides are oriented to present their hydrophilic heads in contact with water and their hydrophobic tails in contact with the other sheet.

Finally, in plants and photosynthetic organisms in general, membranes, such as those of thylakoids present in chloroplast, are mainly composed of galactolipids [5]. The main galactolipids are monoglactosyldiacylglycerol (MGDG), digalactosyldiacylglycerol (DGDG) and sulfoquinovosyldiacylglycerol (SQDG).

In almost all living organisms, the cell membrane consists of a lipid bilayer, as do the membranes surrounding the cell nucleus and organelles [95]. Enveloped viruses also have a membrane around which they surround themselves during exocytosis. The lipid bilayers are impermeable to ions and hydrophilic molecules, allowing the cells to regulate the pH and salinity of their cytosol with the help of transmembrane proteins that act as membrane transporters capable of generating and maintaining a concentration gradient of various chemical species between the cytoplasm and the extracellular medium. These bilayers also tend to close on themselves (budding) to form vesicles. Depending on the nature and composition of the lipids, their interactions can lead to the formation of a distinct phase that separates from water (e.g. oil droplet and emulsions), micelles, and lamellar phases including liposomes or membrane surfaces [87].

1.1.2 Storage Lipids

In mammals, lipids are found in various forms: constitutional lipids which are the primordial cellular constituents of membranes, represent 10% of the dry weight of tissues, serve to protect vital organs (heart, liver, kidneys, spleen, brain and spinal cord), circulating lipids which appear in the blood in the form of lipoproteins, and then disappear at the end of digestion to settle in the liver and other tissues, and reserve lipids which accumulate in adipose tissue and are very important in the production of energy.

Adipose tissue is made up of fat cells called adipocytes, which are animal cells that specialize in energy storage. They contain more than 95% of the body's TAGs [48]. They store lipids up to a critical cell size between 70 and 120 μm. Beyond this maximum size, the adipocyte, is no longer able to store fat, and divides, leading to the formation of a new adipocyte (mitosis process). Once mobilized by lipolysis, the TAGs represent an energy supply for the other organs and in particular the muscle tissue [47]. Adipose tissue is the most important site for storing metabolic energy. In mammals, there are 2 types of adipocytes: white and brown [144].

White adipose tissue represents one of the body's most important energy reserves, it is the main tissue responsible for the storage of TAGs, while brown adipose tissue is involved in the conversion of TAGs into heat (thermogenesis). When the carbohydrate reserves are depleted (in case of fasting, physical exertion, fighting the cold, etc.), or unusable (in case of severe diabetes), the body calls upon the reserves of white adipose tissue to provide the organs with the necessary energy substrates.

Brown fat is involved in thermogenesis through lipolysis of the TAGs of the adipocytes. Unlike white adipocytes, which contain a single lipid droplet per cell, brown adipocytes contain many smaller droplets and a much higher number of mitochondria that contain iron, giving the tissue its brown colour. Brown adipose tissue also contains more capillaries than white adipose tissue because it has a higher oxygen requirement than most other tissues [110].

Small lipid bodies are also detected in different types of tissues, such as liver, muscles, heart, kidneys, small intestine, mammary glands, cultured fibroblasts and macrophages. Lipid droplets rich in cholesterol esters are also detected in steroidogenic cells of the adrenal cortex and in the testes and ovaries [67, 74].

These lipid bodies, rich in cholesterol esters, provide a source of cholesterol that can be used, for example, for the synthesis of steroid hormones [22, 67, 74] or membrane biogenesis [111]. In specific cells such as liver stellate cells, which are rich in retinol esters, lipid bodies contain significant amounts of vitamin A [11].

Lipid globules in breast milk are another form of transient storage and energy transport. The lipid content of breast milk can vary from 0.2% to 61% (w/w). These lipid globules (0.2 to 2 μm in diameter) are formed mainly by a hydrophobic core (triglycerides, esterified cholesterol and fat-soluble vitamins) surrounded by a triple layer membrane made of proteins and phospholipids, and resulting from the exocytosis of lipid droplets during native milk production by the cells of the mammary gland [28]. After processing (pasteurization and homogenization), the phospholipids surrounding milk droplets can be re-organized in monolayers [101, 116].

In vertebrates, lipoproteins are essential for lipid transport to the different tissues of the body. These include chylomicrons (CM) synthesized by the enterocyte during digestion and very low density lipoproteins (VLDL) synthesized by the liver [33, 79, 80, 88, 137]. Lipoproteins are large complexes of proteins and lipids, the outer membrane of which is a monolayer of phospholipids and free cholesterol with one or more proteins called apolipoproteins (e.g., Apo-A, Apo-B, etc.); the core of lipoproteins contains TAGs, cholesterol esters and small amounts of other hydrophobic substances, such as fat-soluble vitamins. CM and VLDL are primarily hydrolyzed by lipoprotein lipase (LPL), while the other lipoproteins, low density lipoproteins (LDL) and high density lipoproteins (HDL), are preferentially hydrolyzed by hepatic lipase (HL) and endothelial lipase (EL) [84, 113, 118].

1.2 Lipolytic Enzymes in the Human GI Tract

1.2.1 True Lipases Acting at the Lipid-Water Interface

These water-soluble enzymes show a high affinity for lipid-water interfaces and are essential for initiation the lipolysis of dietary TAG. In the GI tract, gastric and pancreatic lipases belong to this category of lipolytic enzymes.

1.2.1.1 Gastric Lipase

Human gastric lipase (HGL) is produced in the fundic mucosa of the stomach [107] and is co-located with pepsinogen in the chief cells of the fundic glands [106]. HGL is a highly glycosylated, 50 kDa serine enzyme with an α/β hydrolase fold [37, 126]. Several signals trigger the secretion of gastric lipase such as stomach movements, cholinergic stimuli [40], test meals [36] and the gastrin gastrointestinal hormone gastrin, which also stimulates the secretion of pepsinogen and gastric acid [141, 154]. Gastric lipase is active and stable in the acidic environment of the stomach, where the gastrointestinal lipolysis of dietary fat is initiated [2, 36, 154], although HGL can potentially hydrolyze all esters bonds in a TAG molecule in vitro [39]. Nevertheless, in vivo, MAGs are rarely observed in gastric contents because the gastric lipolysis reaction by HGL is rapidly inhibited by reaction products such as FFA [70, 114]. Gastric lipase has high levels of activity on TAG (Fig. 1.1), to a lesser extent on DAG, and its activity on MAG is generally very low compared to its activity on TAG [62, 64]. Gastric lipase also hydrolyzes polyethylene glycol (PEG) mono- and di-esters and is one of the most active lipases on mixtures of acylglycerols and PEG esters such as the pharmaceutical lipid-based excipients Gelucire®

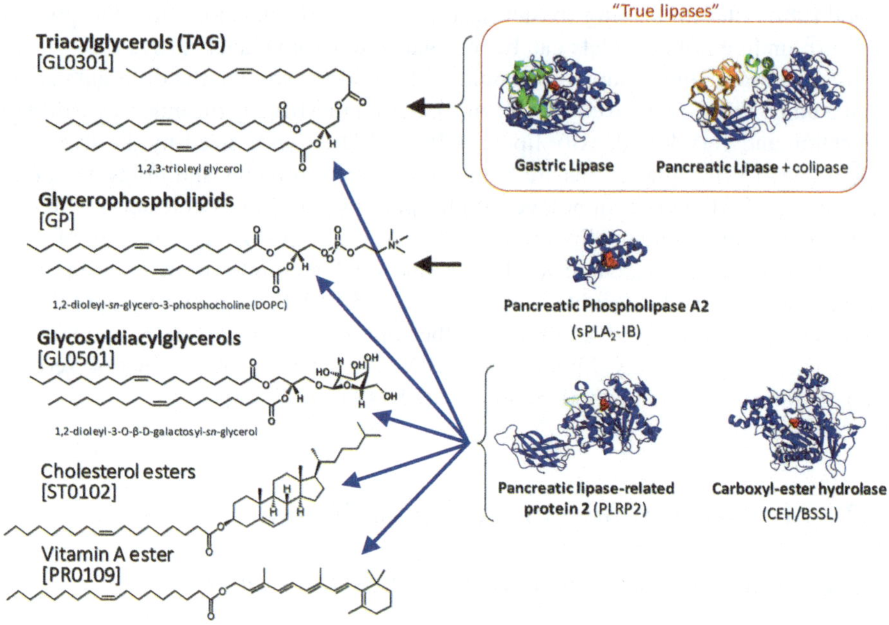

Fig. 1.1 Chemicalstructures of the main lipid classes found in human diet and gastrointestinal lipolytic acting on these lipids. Numbers below the substrate types correspond to the Lipid Maps Classification system (see https://www.lipidmaps.org/) [58, 59]

44/14 and Labrasol® [60–62, 64]. However, HGL does not hydrolyze cholesterol esters or phospholipids (Fig. 1.2) [39].

Gastric lipase shows stereopreference for the hydrolysis of the ester bond at the sn-3 position of TAG, that has been demonstrated both in vitro and in vivo [43]. This enzyme is therefore tailored to release the short and medium chain fatty acids found at this position in milk TAG [125].

In vivo, the level of intragastric lipolysis of TAG ranges from 10% of total TAG acyl chains (normal solid-liquid meal) to 25% (emulsified liquid meal), and DAG and FFAs are primarily generated in the stomach by gastric lipase [36, 38, 42]. During a meal, HGL is always present and stable in the duodenal contents, and can therefore contribute to intestinal lipolysis [36] (Fig. 1.3). Overall, it has been estimated that HGL can release 1 acyl chain out of the four that have to be released for the intestinal absorption of 2 TAG molecules in the form of MAG [36]. This is the reason why it is often mention in literature that HGL can contribute to as much as 25% of fat digestion. Studies in patients with severe exocrine pancreatic insufficiency and no significant levels of pancreatic lipase have confirmed that HGL alone can achieve around 30% of dietary TAG digestion [38].

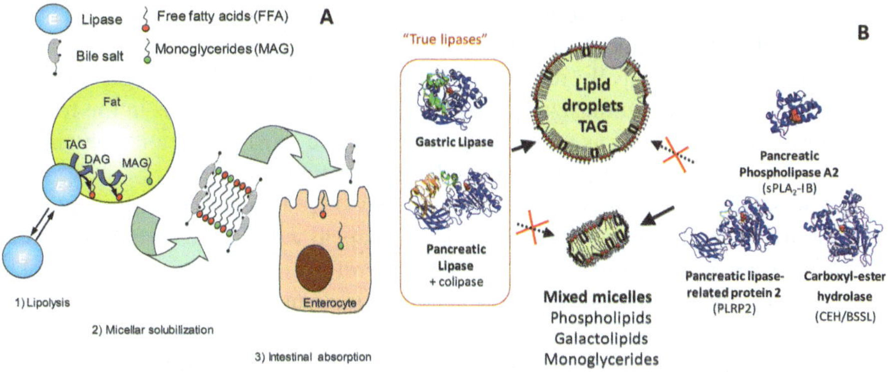

Fig. 1.2 Schematic representation of the essential steps of gastrointestinal lipolysis and intestinal fat absorption (**a**), including (1) the lipolysis reaction after lipase adsorption at the oil-water interface, (2) the micellar solubilisation of lipolysis products, (3) the intestinal absorption of lipolysis products by enterocytes), and substrate specificity of the various lipolytic enzymes found in the GI tract (**b**). Two main categories of lipolytic enzymes can be defined, with the so-called "true lipases" showing a preference for water-insoluble substrates (TAG, DAG) and acting preferentially at the oil-water interface and the enzymes showing a preference for substrates forming mixed micelles with bile salts and dispersed in the aqueous phase like phospholipids, galactolipids and monoglycerides. Adapted from [44, 97]

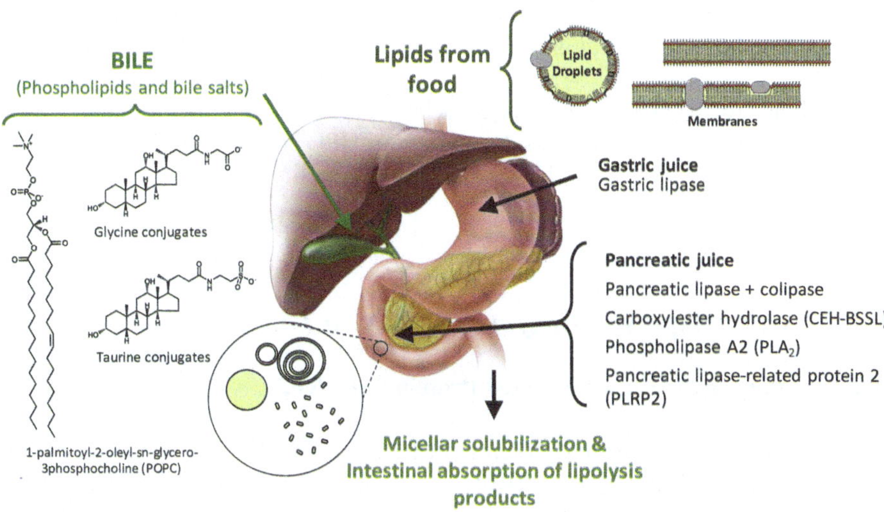

Fig. 1.3 Physiology of gastrointestinal lipolysis

1.2.1.2 Pancreatic Lipase and Colipase

Quantitatively, human pancreatic lipase (HPL) is the major lipase involved in the lipolysis of dietary fat. In the small intestine, HPL acts in the presence of bile with its specific protein cofactor colipase acting as an anchor for the enzyme at the oil-water interface and counteracting the competitive inhibitory effects of bile salts (Fig. 1.3). The biochemical, structural and physiological properties of pancreatic lipase have been extensively reviewed [45, 150]. HPL is a 50.5 kDa glycosylated serine hydrolase, it is synthesized by the acinar cells of the pancreas in the form of an active enzyme and not as an inactive zymogen like most pancreatic enzymes. It is delivered into the intestinal lumen via the pancreatic duct.

In vitro, HPL alone can act on TAG emulsions (Fig. 1.1), but hydrolysis of TAG is inhibited in the presence of bile salts that compete with lipase for adsorption at the lipid-water interface [20]. In vivo, HPL therefore requires its specific10 kDa colipase cofactor, also produced by the pancreas, to counteract the effects of the bile salts and hydrolyze TAGs. Pancreatic lipase also acts on DAG, which actually could be its major substrate after the action of gastric lipase and ultimately produces 2-MAG together with FFA. When DAG serves as an initial substrate for HPL in vitro, the activity measured is usually lower than the activity on TAG [56, 64]. However, when the relative hydrolysis rates of TAG and DAG are estimated in the course of TAG hydrolysis followed by the subsequent hydrolysis of the generated DAG, the rate of DAG hydrolysis was found to be higher than that of TAG [104]. HPL is 1,3-regioselective lipase tailored for efficiently producing 2-MAG from TAG. It is not active on 2-MAG, is weakly active on medium-chain 1(3)-MAGs and is inactive on long-chain 1(3)-MAGs. HPL is not active on phospholipids (Fig. 1.2b) [45, 142], while phospholipase A1 activity has been reported for porcine pancreatic lipase [50]. HPL has very low levels of activity on PEG mono- and di-esters, and is completely inactive on mixtures of these esters and acylglycerols, such as Labrasol® and Gelucire® 44/14 [62, 64]. HPL has no activity on cholesterol esters [1] or galactolipids [6, 10]. Pancreatic lipase has been extensively studied for many years, however, some basic data such as the amount of HPL produced during a meal and its specific activity on test meal TAG had to wait for clinical studies to be obtained [36, 41].

1.2.2 Other Lipolytic Enzymes Present in Pancreatic Secretion

1.2.2.1 Pancreatic Lipase-Related Protein 2 (PLRP2)

Two pancreatic lipase related proteins (PLRP1 and PLRP2) are produced by the pancreas. Human PLRP1 and PLRP2 have 65% and 68% identity respectively with classical pancreatic lipase HPL [72] and a highly conserved catalytic triad (Ser, Asp, His). The three-dimensional structures of these three proteins are superimposable in their structural domains, but differ at the level of the lid that controls access to the active site for PLRP2 and at the level of steric hindrance in the vicinity of the active site for PLRP1 [57, 127, 157]. As a consequence, both native and recombinant HPLRP1 do not exhibit lipolytic activity and its eventual contribution to lipid digestion is still unknown.

PLRP2 exhibits lipolytic activity on various substrates including TAGs, phospholipids (phospholipase A1 activity), galactolipids and cholesterol esters (Fig. 1.1). The lipase activity of PLRP2 is inhibited by micellar concentrations of bile salts and is not reactivated by colipase, suggesting that PLRP2s does not have lipase activity in vivo but may preferentially act as a phospholipase [142] or a galactolipase [136] (Fig. 1.2b). Indeed, PLRP2 also has a high level of galactolipase activity on both MGDG and DGDG [10]. It has been established that human pancreatic juice contains at least two enzymes that effectively hydrolyze DGDG, HPLRP2 and CEH have been identified as these galactolipases [9]. In addition, HPLRP2 is also active on lipophilic vitamin esters [121], as well as on cholesterol esters [8, 130] and is one of the most active monoacylglycerol lipase identified so far [56, 57]. The broad substrate specificity of HPLRP2 is also illustrated by the fact that this enzyme has significant activities on mixtures of acylglycerols and PEG esters such as Gelucire® 44/14 and Labrasol® [62, 64]. Cholesterol esterase activity is the most recent activity attributed to both recombinant human (rHPLRP2) and guinea pig lipase (rGPLRP2) PLRP2 [130] because HPLRP2 has been described as having cholesterol esterase activity [12] (Fig. 1.1). They confirmed a low cholesterol esterase activity for the latter (18.7 ± 1.1 U/mg at pH 6) and a lower activity for GPLRP2 (3.7 ± 0.6 U/mg at pH 6).

1.2.2.2 Carboxyl Ester Hydrolase – Bile Salt-Stimulated Lipase (CEH/BSSL)

Carboxyl ester hydrolase (CEH, EC 3.1.1.1) is a non-specific esterase with different names: bile salt-stimulated lipase (BSSL) [77, 92], cholesterol esterase [108, 109], carboxyl ester lipase (CEL) [129] and lysophospholipase [146]. In humans CEH/BSSL, which is secreted by acinar cells of the exocrine pancreas, is present in pancreatic juice [66, 73]. It is also produced by the mammary glands, present in breast milk and is often referred to as BSSL in this context [23, 77]. Like PLRP2, CEH-BSSL hydrolyzes various substrates in vitro like cholesterol esters, TAG, MAG, vitamin (A, E) esters, and phospholipids [55, 83, 92–94].

CEH-BBSL also hydrolyses carotenoid esters (lutein and capsanthin diesters, esters of β-cryptoxanthin) [29], galactolipids (monogalactosyldiglycerides) [6] and PEG mono- and di-esters [62–64]. Recently, a new potentiometric test has been developed to measure cholesterol esterase activity using cholesterol acetzate as substrate [130]. To measure cholesterol esterase activity, the preparation of the substrate is not straightforward as cholesterol esters are highly apolar and oil-soluble. Other lipids, such as TAGs and phospholipids, have generally be used to allow the dispersion in water of small lipid aggregates containing cholesterol esters. Since most cholesterol esterase enzymes also have lipase and phospholipase Al (PLA1) activities, they can also act on the TAGs or phospholipids present in the reaction mixture, releasing fatty acids from these substrates and it is therefore impossible to estimate the cholesterol esterase activity by titration of the released fatty acids alone. Long-chain cholesterol esters also have a low affinity for bile salt micelles and cannot be dispersed in mixed micelles like phospholipids and galactolipids. This problem was solved by a short-chain cholesterol ester which can be dispersed in water using bile salt micelles. Dispersions of cholesterol acetate (CholA) and sodium taurocholate (NaTC) were found to be good substrate for both pancreatic CEH and PLRP2 [130]. The specific activity of human CEH/BSSL measured under these conditions was found to be the highest recorded with cholesterol esters [17, 27, 112].

CEH-BSSL has no regiospecificity on TAG and can hydrolyse all three ester bonds, whatever their position on the glycerol backbone [155].The physiological role of CEH is still subject to debate [81], but it seems likely that CEH plays a major role in the digestion of cholesterol esters and lipophilic vitamins in adults, and probably in that of acylglycerols in newborn children [76].

Its contribution to the lipolysis of 2-MAG seems to be physiologically relevant only during the lactation period in humans [19, 75]. In vitro bile salt-stimulated lipase alone cannot hydrolyze native milk fat globule TAG, but low rates of hydrolysis by gastric lipase trigger the hydrolysis of milk by pancreatic lipase and bile salt-stimulated lipase. HGL and HPL hydrolyze about two thirds of the total ester bonds present in milk, generating MAG and FFA, and adding bile salt-stimulated lipase results in the hydrolysis of MAG. The concerted action of these three lipases results in the complete digestion of milk TAG, generating free glycerol and FFA as the end-products.

1.2.2.3 Pancreatic Phospholipase A2

Pancreatic phospholipase A2 (PLA2; phosphatide 2-acylhydrolase, EC 3.1.1.4) is involved in the digestion of dietary phospholipids. PLA2 is a stereospecific enzyme which catalyzes the hydrolysis of the sn-2 fatty acyl ester bond of 3-sn-glycerophospholipids, generating FFA and 1-sn-lysophospholipids [50]. Pancreatic PLA2 belongs to Group IB of secretory low molecular weight phospholipases A2 (sPLA2-IB) [52]. It is mainly present in pancreatic secretion, but also in the smooth muscle cells of the lungs and vessels. sPLA2 are involved in many biological functions other than digestion, such as cell proliferation, contraction of smooth muscle in the airways and vessels, and in various inflammatory diseases such as rheumatoid arthritis, endotoxic shock, respiratory distress syndrome and certain types of cancer [85, 145]. On the contrary to HGL, HPL, HPLRP2 and CEH-BSSL that are serine hydrolases with a Ser-His-Asp catalytic triad [57, 126, 157], sPLA2 has no catalytic triad, and the nucleophile required for cleavage of the ester bond is an activated water molecule which is hydrogen-bonded to a His residue, which in turn is hydrogen-bonded to an Asp residue [143].

1.3 Main Structural Features of Gastrointestinal Lipases

1.3.1 Gastric Lipase: Lipase with a Single Domain

HGL is a globular protein consisting of a single polypeptide chain of 379 amino acids [24]. The molecular weight of native HGL, purified from gastric juices, is approximately 50 kDa. This mass corresponds to the mass of the mature 379-amino acid polypeptide plus significant N-glycosylation. There are 4 major isoforms of human gastric lipase with isoelectric points between 6.8 and 7.4 [105]. In 1999, the three-dimensional structure of HGL in its closed form was resolved at 3.0 Å, following the production of the recombinant protein in insect cells using the baculovirus expression system. It revealed several structural elements [126]:

- A main globular domain with an α/β hydrolase fold, organized around a central structure composed of eight β-stands surrounded by six α-helices.
- A flap or lid (residues 210 to 267) composed of 58 amino acids, articulated around two α-helices [126] and controlling the accessibility to the active site. The flap reveals near the active site a large hydrophobic patch of 29 amino acids (residues 215–244), which may be involved in the binding of substrate [102].

1.3.2 Pancreatic Lipases: Lipases with Two Domains

Both HPL and HPLRP2 are composed of two domains: a N-terminal and globular catalytic domain, with an α/β hydrolase fold and a lid controlling the access to the active site as in HGL [57, 96, 147, 149], and a C-terminal domain involved in interactions with colipase and lipids. The C-terminal domain of pancreatic lipases has a sandwich structure containing an exposed hydrophobic loop (loop β5'; residues 405–414) inserted between two strands β5 and β6. This structural domain has an important role in the interaction of lipase with its cofactor, colipase [148, 149]. Structure-function studies have shown that the C-terminal domain also has an important role in the interaction with the lipid interface and has some homologies with C2 lipid binding domains [20, 46]. The β5' loop has a large accessible apolar surface area which is of the order of 877 Å^2. The hydrophobic (H), charged (C) and semi-polar (S) residues of the β5' loop have accessible surfaces of 488 Å^2, 105 Å^2 and 284 Å^2 respectively. The loop β5' is located on the same side of the enzyme as the hydrophobic loops surrounding the active site (lid and β9 loop). In addition, the hydrophobic surface of the β5' loop is not masked after colipase fixation. In HPL, the β5' loop contributes with the open lid, the β9 loop and the hydrophobic tips of colipase to the formation of a large hydrophobic plateau that may parallel the lipid-water interface [96].

HPLRP2 is organized in two domains similar to those found in HPL, except that the lid is found in an open conformation in the absence of surfactants and inhibitors

[57]. This conformation is different from the one observed in the open HPL, which probably explains differences in kinetic properties between HPLRP2 and HPL. Moreover, most of the amino acid residues of HPL involved in the interaction with colipase, both within the lid and the C-terminal domain, are mutated in HPLRP2 [142], which explains why HPLRP2 has only a weak interaction with colipase [136]. Indeed, the lipase activity oh HPLRP2 on TAG is inhibited by bile salts and is not restored by addition of colipase.

1.4 Important Parameters Governing the Activity of Gastrointestinal Lipases

1.4.1 The Role of Bile Salts

The influence of bile salts on the activity of lipases of different origins has been extensively studied in vitro. It has been established that, at low concentrations, bile salts have an activating effect on the hydrolysis of emulsified long-chain TAG by pancreatic [26] and gastric [69] lipases. In interpreting these results, it is generally accepted that bile salts, by lowering the tension at the water-substrate interface, prevent the interfacial denaturation of the enzyme. In the lipolysis reaction, the enzyme partitions between the aqueous phase and the interface [152] (Fig. 1.2a). It has been assumed that the enzyme adsorbed at the interface would exist in a conformational state with an open lid, in equilibrium with an inactive form with the closed lid remaining in the water phase (Fig. 1.2a). It has been shown however that the lid can open in solution in the presence of bile salts with a maximum aperture in the presence of supramicellar concentrations of bile salts [14]. Under physiological conditions, it is therefore expected that the lid opens in solution prior to HPL adsorption at the interface, which is supported by the exposition of a large hydrophobic plateau upon lid opening. The adsorption and activity of pancreatic lipase is however dependent on the presence of colipase, which anchors the lipase at the interface when the concentration of bile salts exceed its micellar concentration [21, 26]. Contrary to HPL, HGL is not inhibited by micellar concentrations of bile salts and does require a cofactor [89].This is due to a higher tensioactivity of HGL, that has a high penetration capacity at the lipid-water interface, even in the presence of phospholipids [51, 131].

Bile salts have a dual impact on the activity of true lipases. On the one hand, they can impair lipase adsorption and activity. On the other hand, they can ensure the continuity of lipolysis by removing the reaction products from the oil-water interface by a process of micellar solubilisation. Concerning the other lipolytic enzymes acting on more polar lipids (phospholipids, galactolipids, MAG), bile salts have an essential role in the dispersion of these substrates in the aqueous phase in the form of mixed micelles (Fig. 1.2) and the bile salt to substrate molar ratio is critical for the activities of HPLRP2, CEH-BSSL and PLA2 [97–99].

1.4.2 Gastric and Duodenal pH Variations

The variations in the gastric and duodenal pH which occur during digestion have been studied in several test meal experiments on healthy humans and patients with exocrine pancreatic insufficiency [36, 38, 42, 124]. After the ingestion of a meal, the pH of the gastric content ranges between 5.5 and 7, depending on the composition of the meal ingested. Then, the gastric content is diluted by the gastric acid secretion and the pH decreases. Gastric emptying also contributes to this pH decrease by removing the meal components from the gastric contents and thus reducing their buffering effects. The intragastric pH value is found in the 4 to 5 pH range at half gastric emptying time (approx. 60 min) in heathy volunteers, which corresponds to the optimum pH conditions for HGL activity on TAG [69, 89].

The duodenal pH variations are less pronounced and are usually restricted to the 5 to 7 pH range, giving a mean value of 6.25 in healthy volunteers. These pH values also correspond to the optimum pH conditions for HPL activity on TAG in the presence of bile salts [25, 89]. Thus, both HGL and HPL are tailored for acting optimally on TAG at the pH values found in the stomach and small intestine, respectively. In the case of HPLRP2 and CEH-BSSL, the optimum pH of activity depends on the type of substrate hydrolyzed and it is usually found in the neutral to slightly alkaline pH range [56]. Both enzymes show however an optimum activity on cholesterol esters at pH 6 [130].

1.4.3 The Effects of Calcium on Lipolysis

Calcium contributes decisively to the catalytic activity of sPLA2-IB, since a calcium ion interacts with the phosphate group of the phospholipid substrate, with the catalytic water molecule and with the tetrahedral intermediate generated during the deacylation reaction (stabilization of the negative charge in the oxyanion hole) [153]. Calcium may also contribute to the interactions of sPLA2 with lipid aggregates. A second calcium ion has been observed near the N-terminal part of the enzyme and the interfacial binding surface (i-face) in the 3D structure of bovine sPLA2-IB [119], as well as in other sPLA2 [135]. Finally, calcium may also contribute to the interfacial organization of the phospholipid substrate itself by condensing the polar heads of negatively charged phospholipids [138].

Therefore, when studying or using PLA2, it is essential to monitor calcium concentration, whereas lipases do not have absolute calcium requirements. A calcium ion is found in the 3D structures of pancreatic lipases, but it is remote from the active site and interfacial binding sites and has probably only structural functions. However, some non-specific effects of calcium may be observed in lipase assays, as calcium can compensate for the electrostatic repulsion that occurs between the enzyme and a charged lipid-water interface. Calcium and FFAs may also form soaps during the lipolysis reaction, and lipase kinetics may be affected by the removal of FFAs from the lipid-water interface [16, 18, 150].

1.5 Kinetic Models of Lipolysis and Enzyme Assays

When we speak of the Michaelis-Menten model, we are talking about homogeneous enzymology where the enzymatic reaction takes place in a homogeneous medium, i.e. in a single phase in which the enzyme and the substrate are soluble. For lipolytic enzymes acting essentially at the lipid-water interface (in a heterogeneous medium), we speak of heterogeneous enzymology and the classical Michaelian model of enzyme kinetics can no longer be applied. One then passes from a three-dimensional or volumetric system to a two-dimensional or surface system. In 1973, an interfacial kinetic model was proposed by Verger and de Haas [151]. According to this model, the enzymatic reaction takes place in two elementary stages:

- An adsorption step of the enzyme at the lipid interface (adsorption or penetration) (E: water soluble enzyme, E*: adsorbed enzyme). This step is governed by two rate constants: a constant of penetration (kp) or adsorption of the enzyme at the lipid-water interface and a constant of desorption (kd) of the enzyme. During this step, the enzyme may or may not undergo a conformational change.
- An actual catalytic step analogous to a Michaelis Menten pseudo-reaction that takes place on the surface and no longer in solution. All concentrations of the species involved in the reaction such as the enzyme (E and E*), the lipid substrate (S), the reaction product (P) and the enzyme/substrate complex (E*S) are expressed in surface density (mole.m-2) instead of volume concentration (mole.m-3). In order to differentiate soluble enzyme forms in the aqueous phase (E) from those present at the interface, the latter are denoted with an asterisk (E*, E*S).

1.5.1 Kinetic Phenomenon of Interfacial Activation

Lipases are water-soluble enzymes, which however have very low activity on water-soluble esters. Their activity becomes very high when the concentration of the substrate exceeds solubility and allows the formation of aggregates such as lipid droplets and thus the formation of a water-lipid interface [65, 134]. This phenomenon is called interfacial activation, which distinguishes lipases from esterases acting only on water-soluble carboxylic esters. Its observation requires the use of non-natural short chain TAG with some solubility in water and its relevance for natural long chain TAG has often been questioned. Nevertheless, it has triggered numerous researches to explain this phenomenon at the enzyme molecular level, first with speculations on conformational changes occurring in the presence of a lipid-water interface, and finally with the obtention of lipase 3D structures showing indeed the presence of a lid controlling the access to the active site through conformational changes in both HPL [147] and gastric lipase [128].

1.5.2 Methods to Measure Lipolytic Enzymes Activity

There are numerous methods for measuring lipase activities and their principles have been reviewed by Beisson et al. [13]. We briefly describe here the methods commonly used for the characterization of digestive lipolytic enzymes.

1.5.3 Measurement of Lipase Activity by the pH-Stat Technique

The pH-stat is a potentiometric method for measuring lipolytic activities at constant pH using emulsified substrates or substrates dispersed in water in the formed of mixed micelles [13, 90, 117]. It is convenient for testing all the lipolytic enzymes of the GI tract using specific substrates and conditions. The principle of this assay is the continuous titration of the FFAs release during substrate hydrolysis by the lipase. The released FFAs are usually titrated by adding 0.1 N sodium hydroxide solution to the reaction medium using an automated burette coupled to a pH-meter and a regulation system in which a pH endpoint can be selected (usually the optimum pH of activity on the selected substrate). The reaction is performed in a thermostated reaction vessel equipped with a mechanical or a magnetic stirrer to ensure fine emulsion and dispersion of the substrate with a large accessible surface. The amount of NaOH added (or fatty acids released) as a function of time corresponds to the enzyme activity expressed in International Units, with 1 IU corresponding to the release of 1 μmole of fatty acid released per minute).

1.5.4 Measurement of Lipase Activity by the Monomolecular Film Technique

The monolayer technique using Langmuir films was adapted to the study lipolytic enzymes and lipid/protein interactions at the air-water interface [68, 120, 150–152]. By analogy with the pH-stat technique, the parameter that is kept constant here is the surface pressure, which characterizes at minima the physical-chemical parameters of the interface. The lipase is acting on a monolayer of substrate (TAG, DAG, phospholipid or galactolipid) spread at the air-water interface, the water phase being used to control the pH and other parameters like salt concentrations. This technique consists of the use of a "zero-order Teflon trough", consisting of two compartments, a reaction compartment, and a reservoir compartment, connected to each other by a small sintered glass channel. Once the two compartments are filled with the reaction solution, the lipid is spread at the surface using few microliters of a solution in chloroform. A monomolecular film of lipids is formed over the entire surface of the trough through surface channels. However, the diffusion of compounds present in the aqueous phase from the reaction compartment to the reservoir compartment is negligible. When the lipolytic enzyme is injected into the reaction compartment, it can partition between the aqueous phase and the monomolecular film and hydrolyze the lipid molecule substrate present in the film. Prior to enzyme injection, surface pressure is measured using a Wilhemy platinium blade connected to an electromicrobalance and can be adjusted using a Teflon mobile barrier to compress the film. The lipids chosen as substrates for these assays are usually medium chain TAG (trioctanoin, [15]), DAG (1,2-dicaprin, [51]), phospholipids (di-lauroyl-PC, [78]) and galactolipids (dilauroyl-MGDG, [136]). While these lipids are insoluble in water and form stable monolayers, their lipolysis products are soluble in water and are transferred from the interface to the bulk, which results in a decrease in molecular packing and surface pressure at the air-water interface. The use of the zero-order trough together with the Teflon mobile barrier controlled by surface pressure changes allows keeping the surface pressure constant as a function of time ("barostat" mode) by compressing the film after injecting the lipase below the monolayer. The enzyme activity is estimated from the surface covered by the mobile barrier as a function of time and its conversion to a number of substrate molecule hydrolyzed based on the molecular area of these molecules deduced from compression isotherms [120]. Thanks to the film compression and the replacement of the hydrolyzed substrate molecules, the reaction can be carried out at a set constant surface pressure and a constant surface concentration of substrate, hence the name "zero order trough" given to this device [151, 152].

1.6 Enzymes Used In Vitro Models of Digestion

In vitro digestion models reproducing the physiological conditions of the digestive tract find nowadays many applications, whether in food [103] or in the field of galenic pharmacy [35]. In particular, they are used to test the digestibility of dietary lipids or the lipolysis of lipid formulations intended for the oral administration of hydrophobic active ingredients [156]. The implementation of these models requires digestive tract enzymes or analogous enzymes obtained from various sources or recombinant. Human digestive lipases are however not commercially available for these applications and the use of digestive juices collected from volunteers are raising ethical concerns. As a result, human enzymes have been replaced by similar enzymes from animal origin. This was easily achieved in the case of pancreatic enzymes since porcine pancreatic extracts (PPEs) or pancreatin are commonly produced as an active pharmaceutical ingredient for the treatment of exocrine pancreatic insufficiency and are commercially available at a reasonable price for this type of application. There are several commercial sources of pepsin for gastric digestion of proteins but it is more difficult to obtain a reliable source of gastric lipase, which resulted in the absence of gastric lipolysis in many protocols or the use of non-relevant microbial lipases [133]. Rabbit gastric extract (RGE) however appeared as a relevant source of both gastric lipase and pepsin [132] and has been used in combination with PPE to replace human enzymes in a two-step in vitro digestion model with a gastric and a duodenal steps [32]. By analogy with in vitro studies on intestinal digestion using porcine pancreatin as a global source of pancreatic enzymes [31, 156], it is important to have access to a global source of gastric enzymes. RGE is such a mixture produced from rabbit stomach as a side product from slaughterhouses. RGE has recently become commercially available and was introduced in the standard INFOGEST static method of in vitro digestion [31].

As a result of these developments, PPEs are increasingly used for in vitro digestion assays. Although the overall content of PPE in lipase, protease and amylase activities are well characterized, the individual enzymes present in these extracts are not well characterized and a limited number of activity assays have been developed. PPEs contain the same lipolytic activities as human pancreatic juice (HPJ) and porcine pancreas pancreatic juice (PPJ), but it was recently shown that their phospholipase, galactolipase and cholesterol esterase activities, relative to the lipase activity, are reduced compared to those found in HPJ [130]. The fact that PPEs from different manufacturers may differ significantly in their lipolytic activities has already been reported. The levels of activity of CEH/BSSL have been shown to be particularly low in some preparations [139]. In addition to differences in the respective levels of porcine and human pancreatic enzymes, comparison of the activity levels of PPL and PPE reveals a significant loss of phospholipase and galactolipase activities in PPE, which is probably due to some proteolysis occurring during the production process of PPE. Indeed, the lipolytic enzymes present in PPE, as well as in PPJ and HPJ, are very sensitive to proteolytic degradation and inactivation by the endogenous proteases present in these enzyme mixtures. The inactivation of the enzymes during the manufacturing process of PPE was also supported by the finding that the activity of CEH/BSSL varies much more than the amount of immunoreactive CEH/BSSL [139]. Galactolipase activity in PPE was the lowest and least stable lipolytic activity, which may explain why it had not been detected previously [6]. This may also explain why PLRP2, the pancreatic enzyme with the highest galactolipase activity [7], has been identified among porcine pancreatic enzymes only recently [130]. The native enzyme has not been isolated yet [49] but characterization of recombinant porcine PLRP2 expressed in mammalian cells has confirmed that this enzyme exhibits galactolipase and phospholipase activities, like HPLRP2, as well as a lipase activity on triolein that is not strongly dependent on the presence of colipase [158]. Thus, porcine PLRP2 may contributes to all lipolytic activities of PPE, making their characterization more complex. Similarly, porcine CEH/BSSL has been identified here in PPE. In addition to lipase, phospholipase A1 and galactolipase activities, CEH/BSSL is believed to be the most important pancreatic enzyme for the digestion of cholesterol esters, as well as esters of vitamins A (retinol) and E (tocopherols and tocotrienols) [92, 93]. The lipolytic activity of PPE on these latter esters, has not been explored in detail yet but it is assumed that it is closely associated with cholesterol esterase activity. The low levels of cholesterol esterase activity in PPE compared to PPJ and HPJ suggest that PPE will be less active on lipophilic vitamin esters than pancreatic juices.

In conclusion, PPE do not allow a complete substitution of the enzymes and lipolytic activities present in human pancreatic juices and one has to be cautious with their use for in vitro digestion, especially for studying the digestibility of cholesterol and vitamin esters.

1.7 Conclusion

Our understanding of the activities of the lipases present in the human GI tract has been growing following decades of study. Thanks to this knowledge, scientists are now able to improve the enzymes they use, either recombinants or extracts obtained from animals. In vitro digestion models are widely employed especially in the pharmaceutical and food areas, for research as well as for the industry. However, in order to acquire accurate and representative lipolysis data, those models ought to include the range of lipases found in vivo.

The digestion models are sometimes coupled with human trials and biophysical methods to validate the results and assess the bioavailability of food compounds, such as for instance carotenoids. Moreover, in vitro models can be used to test the dispersion and solubilization of drugs, and to establish screening tools to design and optimize lipid-based formulations. In the literature, different types of in vitro gastrointestinal digestion models have been developed, ranging from very simple experiments with a mixture of pancreatic enzymes (porcine pancreatin) to complex experimental devices mimicking the entire GI tract. The reader is referred to Chap. 3 to obtain more information on the subject.

References

1. Ali YB, Carriere F, Verger R, Petry S, Muller G, Abousalham A (2005) Continuous monitoring of cholesterol oleate hydrolysis by hormone-sensitive lipase and other cholesterol esterases. J Lipid Res 46(5):994–1000
2. Aloulou A, Carriere F (2008) Gastric lipase: an extremophilic interfacial enzyme with medical applications. Cell Mol Life Sci 65(6):851–854
3. Aloulou A, Rodriguez JA, Fernandez S, van Oosterhout D, Puccinelli D, Carriere F (2006) Exploring the specific features of interfacial enzymology based on lipase studies. Biochim Biophys Acta 1761(9):995–1013
4. Alvarez H, Steinbüchel A (2002) Triacylglycerols in prokaryotic microorganisms. Appl Microbiol Biotechnol 60(4):367–376

5. Amara S, Barouh N, Lecomte J, Lafont D, Robert S, Villeneuve P, De Caro A, Carriere F (2010) Lipolysis of natural long chain and synthetic medium chain galactolipids by pancreatic lipase-related protein 2. Biochim Biophys Acta 1801(4):508–516
6. Amara S, Lafont D, Fiorentino B, Boullanger P, Carriere F, De Caro A (2009) Continuous measurement of galactolipid hydrolysis by pancreatic lipolytic enzymes using the pH-stat technique and a medium chain monogalactosyl diglyceride as substrate. Biochim Biophys Acta 1791(10):983–990
7. Amara S, Lafont D, Parsiegla G, Point V, Chabannes A, Rousset A, Carriere F (2013) The galactolipase activity of some microbial lipases and pancreatic enzymes. Eur J Lipid Sci Technol 115(4):442–451
8. Andersson EL, Hernell O, Bläckberg L, Falť H, Lindquist S (2011) BSSL and PLRP2: key enzymes for lipid digestion in the newborn examined using the Caco-2 cell line. J Lipid Res 52(11):1949–1956
9. Andersson L, Bratt C, Arnoldsson KC, Herslof B, Olsson NU, Sternby B, Nilsson A (1995) Hydrolysis of galactolipids by human pancreatic lipolytic enzymes and duodenal contents. J Lipid Res 36(6):1392–1400
10. Andersson L, Carriere F, Lowe ME, Nilsson A, Verger R (1996) Pancreatic lipase-related protein 2 but not classical pancreatic lipase hydrolyzes galactolipids. Biochim Biophys Acta 1302(3):236–240
11. Azais-Braesco V, Dodeman I, Delpal S, Alexandre-Gouabau MC, Partier A, Borel P, Grolier P (1995) Vitamin A contained in the lipid droplets of rat liver stellate cells is substrate for acid retinyl ester hydrolase. Biochim Biophys Acta 1259(3):271–276
12. Bakala N'Goma JC, Amara S, Dridi K, Jannin V, Carriere F (2012) Understanding the lipid-digestion processes in the GI tract before designing lipid-based drug-delivery systems. Ther Deliv 3(1):105–124
13. Beisson F, Tiss A, Rivière C, Verger R (2000) Methods for lipase detection and assay: a critical review. Eur J Lipid Sci Technol 102(2):133–153
14. Belle V, Fournel A, Woudstra M, Ranaldi S, Prieri F, Thome V, Currault J, Verger R, Guigliarelli B, Carriere F (2007) Probing the opening of the pancreatic lipase lid using site-directed spin labeling and EPR spectroscopy. Biochemistry 46(8):2205–2214
15. Benarouche A, Point V, Carriere F, Cavalier JF (2014) An interfacial and comparative in vitro study of gastrointestinal lipases and Yarrowia lipolytica LIP2 lipase, a candidate for enzyme replacement therapy. Biochimie 102(1):145–153
16. Benzonana G (1968) On role of Ca^{2+} during hydrolysis of insoluble triglycerides by pancreatic lipase in presence of bile salts. Biochim Biophys Acta 151(1):137–146
17. Benzonana G, Desnuelle P (1965) Kinetic study of the action of pancreatic lipase on emulsified triglycerides. Enzymology assay in heterogeneous medium. Biochim Biophys Acta 105(1):121–136
18. Benzonana G, Desnuelle P (1968) Action of some effectors on the hydrolysis of long-chain triglycerides by pancreatic lipase. Biochimica et Biophysica Acta (BBA)-Lipids and Lipid Metabolism 164(1):47–58
19. Bernback S, Blackberg L, Hernell O (1990) The complete digestion of human milk triacylglycerol *in vitro* requires gastric lipase, pancreatic colipase-dependent lipase, and bile salt-stimulated lipase. J Clin Investig 85(4):1221–1226
20. Bezzine S, Carriere F, De Caro J, Verger R, De Caro A (1998) Human pancreatic lipase: an exposed hydrophobic loop from the C-terminal domain may contribute to interfacial binding. Biochemistry 37(34):11846–11855
21. Bezzine S, Ferrato F, Ivanova MG, Lopez V, Verger R, Carriere F (1999) Human pancreatic lipase: colipase dependence and interfacial binding of lid domain mutants. Biochemistry 38(17):5499–5510
22. Bisgaier CL, Chanderbhan R, Hinds RW, Vahouny GV (1985) Adrenal cholesterol esters as substrate source for steroidogenesis. J Steroid Biochem 23(6 PART 1):967–974

23. Blackberg L, Angquist K-A, Hemell O (1987) Bile salt-stimulated lipase in human milk: evidence for its synthesis in the lactating mammary gland. FEBS Lett 217(1):37–41
24. Bodmer MW, Angal S, Yarranton GT, Harris TJR, Lyons A, King DJ, Pieroni G, Riviere C, Verger R, Lowe PA (1987) Molecular cloning of a human gastric lipase and expression of the enzyme in yeast. Biochim Biophys Acta Gene Struct Expr 909(3):237–244
25. Borgstrom B (1975) On the interactions between pancreatic lipase and colipase and the substrate, and the importance of bile salts. J Lipid Res 16(6):411–417
26. Borgstrom B, Erlanson C (1973) Pancreatic lipase and co-lipase. Interactions and effects of bile salts and other detergents. Eur J Biochem 37(1):60–68
27. Borgstrom BAGJ, Dahlqvist A, Lundh G, Sjovall J (1957) Studies of intestinal digestion and absorption in the human. J Clin Invest 36(10):1521–1536
28. Bourlieu C, Michalski MC (2015) Structure-function relationship of the milk fat globule. Curr Opin Clin Nutr Metab Care 18(2):118–127
29. Breithaupt DE, Bamedi A, Wirt U (2002) Carotenol fatty acid esters: easy substrates for digestive enzymes? Comp Biochem Physiol B: Biochem Mol Biol 132(4):721–728
30. Brignot H, Feron G (2019) Oral lipolysis and its association with diet and the perception and digestion of lipids: a systematic literature review. Arch Oral Biol 108:104550
31. Brodkorb A, Egger L, Alminger M, Alvito P, Assuncao R, Ballance S, Bohn T, Bourlieu-Lacanal C, Boutrou R, Carriere F, Clemente A, Corredig M, Dupont D, Dufour C, Edwards C, Golding M, Karakaya S, Kirkhus B, Le Feunteun S, Lesmes U, Macierzanka A, Mackie AR, Martins C, Marze S, McClements DJ, Menard O, Minekus M, Portmann R, Santos CN, Souchon I, Singh RP, Vegarud GE, Wickham MSJ, Weitschies W, Recio I (2019) INFOGEST static in vitro simulation of gastrointestinal food digestion. Nat Protoc 14(4):991–1014
32. Capolino P, Guerin C, Paume J, Giallo J, Ballester J-M, Cavalier J-F, Carriere F (2011) *In vitro* gastrointestinal lipolysis: replacement of human digestive lipases by a combination of rabbit gastric and porcine pancreatic extracts. Food Digestion 2:43–51
33. Cardin AD, Witt KR, Chao J, Margolius HS, Donaldson VH, Jackson RL (1984) Degradation of apolipoprotein B-100 of human plasma low density lipoproteins by tissue and plasma kallikreins. J Biol Chem 259(13):8522–8528
34. Carey MC, Small DM, Bliss CM (1983) Lipid digestion and absorption. Annu Rev Physiol 45:651–677
35. Carriere F (2016) Impact of gastrointestinal lipolysis on oral lipid-based formulations and bioavailability of lipophilic drugs. Biochimie 125:297–305
36. Carriere F, Barrowman JA, Verger R, Laugier R (1993) Secretion and contribution to lipolysis of gastric and pancreatic lipases during a test meal in humans. Gastroenterology 105(3):876–888
37. Carriere F, Gargouri Y, Moreau H, Ransac S, Rogalska E, Verger R (1994) Gastric lipases: cellular, biochemical and kinetic aspects. In: Lipases: their structure, biochemistry and application, pp 181–205
38. Carriere F, Grandval P, Renou C, Palomba A, Prieri F, Giallo J, Henniges F, Sander-Struckmeier S, Laugier R (2005) Quantitative study of digestive enzyme secretion and gastrointestinal lipolysis in chronic pancreatitis. Clin Gastroenterol Hepatol 3(1):28–38
39. Carriere F, Moreau H, Raphel V, Laugier R, Benicourt C, Junien J-L, Verger R (1991) Purification and biochemical characterization of dog gastric lipase. Eur J Biochem 202(1):75–83
40. Carriere F, Raphel V, Moreau H, Bernadac A, Devaux MA, Grimaud R, Barrowman JA, Benicourt C, Junien JL, Laugier R, Verger R (1992) Dog gastric lipase: stimulation of its secretion in vivo and cytolocalization in mucous pit cells. Gastroenterology 102(5):1535–1545
41. Carriere F, Renou C, Lopez V, De Caro J, Ferrato F, Lengsfeld H, De Caro A, Laugier R, Verger R (2000) The specific activities of human digestive lipases measured from the *in vivo* and *in vitro* lipolysis of test meals. Gastroenterology 119(4):949–960
42. Carriere F, Renou C, Ransac S, Lopez V, De Caro J, Ferrato F, De Caro A, Fleury A, Sanwald-Ducray P, Lengsfeld H, Beglinger C, Hadvary P, Verger R, Laugier R (2001) Inhibition of

gastrointestinal lipolysis by Orlistat during digestion of test meals in healthy volunteers. Am J Physiol Gastrointest Liver Physiol 281(1):G16–G28

43. Carriere F, Rogalska E, Cudrey C, Ferrato F, Laugier R, Verger R (1997) In vivo and in vitro studies on the stereoselective hydrolysis of tri-and diglycerides by gastric and pancreatic lipases. Bioorg Med Chem 5(2):429–435
44. Carriere F, Thirstrup K, Hjorth S, Boel E (1994) Cloning of the classical guinea pig pancreatic lipase and comparison with the lipase related protein 2. FEBS Lett 338(1):63–68
45. Carriere F, Withers-Martinez C, van Tilbeurgh H, Roussel A, Cambillau C, Verger R (1998) Structural basis for the substrate selectivity of pancreatic lipases and some related proteins. Biochimica et Biophysica Acta (BBA)-Reviews on Biomembranes 1376(3):417–432
46. Chahinian H, Bezzine S, Ferrato F, Ivanova MG, Perez B, Lowe ME, Carriere F (2002) The β5' loop of the pancreatic lipase C2-like domain plays a critical role in the lipase–lipid interactions. Biochemistry 41(46):13725–13735
47. Coelho M, Oliveira T, Fernandes R (2013) Biochemistry of adipose tissue: an endocrine organ. Arch Med Sci 9(2):191
48. Coppack SW, Jensen MD, Miles JM (1994) In vivo regulation of lipolysis in humans. J Lipid Res 35(2):177–193
49. De Caro J, Eydoux C, Chérif S, Lebrun R, Gargouri Y, Carriere F, De Caro A (2008) Occurrence of pancreatic lipase-related protein-2 in various species and its relationship with herbivore diet. Comp Biochem Physiol B: Biochem Mol Biol 150(1):1–9
50. de Haas GH, Sarda L, Roger J (1965) Positional specific hydrolysis of phospholipids by pancreatic lipase. Biochimica et Biophysica Acta (BBA)/Lipids and Lipid Metabolism 106(3):638–640
51. De La Fournière L, Ivanova MG, Blond JP, Carrière F, Verger R (1994) Surface behaviour of human pancreatic and gastric lipases. Colloids Surf B: Biointerfaces 2(6):585–593
52. Dennis EA (1994) Diversity of group types, regulation, and function of phospholipase A2. J Biol Chem 269(18):13057–13060
53. Docherty AJ, Bodmer MW, Angal S, Verger R, Riviere C, Lowe PA, Lyons A, Emtage JS, Harris TJ (1985) Molecular cloning and nucleotide sequence of rat lingual lipase cDNA. Nucleic Acids Res 13(6):1891–1903
54. Embleton JK, Pouton CW (1997) Structure and function of gastro-intestinal lipases. Adv Drug Deliv Rev 25(1):15–32
55. Erlanson C (1975) Purification, properties, and substrate specificity of a carboxylesterase in pancreatic juice. Scand J Gastroenterol 10(4):401–408
56. Eydoux C, De Caro J, Ferrato F, Boullanger P, Lafont D, Laugier R, Carriere F, De Caro A (2007) Further biochemical characterization of human pancreatic lipase-related protein 2 expressed in yeast cells. J Lipid Res 48(7):1539–1549
57. Eydoux C, Spinelli S, Davis TL, Walker JR, Seitova A, Dhe-Paganon S, De Caro A, Cambillau C, Carriere F (2008) Structure of human pancreatic lipase-related protein 2 with the lid in an open conformation. Biochemistry 47(36):9553–9564
58. Fahy E, Subramaniam S, Brown HA, Glass CK, Merrill AH Jr, Murphy RC, Raetz CRH, Russell DW, Seyama Y, Shaw W (2005) A comprehensive classification system for lipids. Eur J Lipid Sci Technol 107(5):337–364
59. Fahy E, Subramaniam S, Murphy RC, Nishijima M, Raetz CRH, Shimizu T, Spener F, van Meer G, Wakelam MJO, Dennis EA (2009) Update of the LIPID MAPS comprehensive classification system for lipids. J Lipid Res 50(Supplement):S9–S14
60. Fernandez S, Chevrier S, Ritter N, Mahler B, Demarne F, Carriere F, Jannin V (2009) In vitro gastrointestinal lipolysis of four formulations of piroxicam and cinnarizine with the self emulsifying excipients Labrasol and Gelucire 44/14. Pharm Res 26(8):1901–1910
61. Fernandez S, Jannin V, Chevrier S, Chavant Y, Demarne F, Carriere F (2013) In vitro digestion of the self-emulsifying lipid excipient Labrasol((R)) by gastrointestinal lipases and influence of its colloidal structure on lipolysis rate. Pharm Res 30(12):3077–3087

62. Fernandez S, Jannin V, Rodier JD, Ritter N, Mahler B, Carriere F (2007) Comparative study on digestive lipase activities on the self emulsifying excipient Labrasol, medium chain glycerides and PEG esters. Biochim Biophys Acta 1771(5):633–640
63. Fernandez S, Najjar A, Robert S, Rodier J-D, Mahler B, Demarne F, Carriere F, Jannin V (2010) Specific assay of carboxyl ester hydrolase using PEG esters as substrate. Anal Methods 2(8):1013–1019
64. Fernandez S, Rodier J-D, Ritter N, Mahler B, Demarne F, Carriere F, Jannin V (2008) Lipolysis of the semi-solid self-emulsifying excipient Gelucire® 44/14 by digestive lipases. Biochimica et Biophysica Acta (BBA)-Molecular and Cell Biology of Lipids 1781(8):367–375
65. Ferrato F, Carriere F, Sarda L, Verger R (1997) A critical reevaluation of the phenomenon of interfacial activation. In: Methods in enzymology, vol 286. Elsevier, New York, pp 327–347
66. Figarella C (1966) Composition et stabilite de l'equipement enzymatique du pancreas de l'homme et de divers animaux. Bulletin de la Societe de Chimie Biologique 48(1):97
67. Freeman DA, Ascoli M (1982) Studies on the source of cholesterol used for steroid biosynthesis in cultured Leydig tumor cells. J Biol Chem 257(23):14231–14238
68. Gargouri Y, Bensalah A, Douchet I, Verger R (1995) Kinetic-behavior of pancreatic lipase in 5 species using emulsions and monomolecular films of synthetic glycerides. Biochim Biophys Acta Lipids Lipid Metab 1257(3):223–229
69. Gargouri Y, Pieroni G, Riviere C, Lowe PA, Sauniere JF, Sarda L, Verger R (1986) Importance of human gastric lipase for intestinal lipolysis: an *in vitro* study. Biochim Biophys Acta 879(3):419–423
70. Gargouri Y, Pieroni G, Riviere C, Sauniere JF, Lowe PA, Sarda L, Verger R (1986) Kinetic assay of human gastric lipase on short- and long-chain triacylglycerol emulsions. Gastroenterology 91(4):919–925
71. Gibbons GF, Islam K, Pease RJ (2000) Mobilisation of triacylglycerol stores. Biochimica et Biophysica Acta (BBA) - Molecular and Cell Biology of Lipids 1483(1):37–57
72. Giller T, Buchwald P, Blum-Kaelin D, Hunziker W (1992) Two novel human pancreatic lipase related proteins, hPLRP1 and hPLRP2. Differences in colipase dependence and in lipase activity. J Biol Chem 267(23):16509–16516
73. Guy O, Figarella C (1981) The proteins of human pancreatic external secretion. Scand J Gastroenterol 67:59–61
74. Gwynne JT, Strauss JF (1982) The role of lipoproteins in steroidogenesis and cholesterol metabolism in steroidogenic glands. Endocr Rev 3(3):299–329
75. Hernell O, Bläckberg L (1982) Digestion of human milk lipids: physiologic significance of sn-2 monoacylglycerol hydrolysis by bile salt-stimulated lipase. Pediatr Res 16(10):882–885
76. Hernell O, Bläckberg L (1994) Human milk bile salt-stimulated lipase: functional and molecular aspects. J Pediatr 125(5):S56–S61
77. Hernell O, Olivecrona T (1974) Human milk lipases. II. Bile salt-stimulated lipase. Biochim Biophys Acta 369(2):234–244
78. Hjorth A, Carriere F, Cudrey C, Woldike H, Boel E, Lawson DM, Ferrato F, Cambillau C, Dodson GG, Thim L, Verger R (1993) A structural domain (the lid) found in pancreatic lipases is absent in the guinea pig (phospho)lipase. Biochemistry 32(18):4702–4707
79. Hobbs HH, Brown MS, Goldstein JL (1992) Molecular genetics of the LDL receptor gene in familial hypercholesterolemia. Hum Mutat 1(6):445–466
80. Hobbs HH, Brown MS, Russell DW, Davignon J, Goldstein JL (1987) Deletion in the gene for the low-density-lipoprotein receptor in a majority of French Canadians with familial hypercholesterolemia. N Engl J Med 317(12):734–737
81. Howles PN, Carter CP, Hui DY (1996) Dietary free and esterified cholesterol absorption in cholesterol esterase (bile salt-stimulated lipase) gene-targeted mice. J Biol Chem 271(12):7196–7202
82. Huang AHC (1994) Structure of plant seed oil bodies. Curr Opin Struct Biol 4(4):493–498
83. Hyun J, Kothari H, Herm E, Mortenson J, Treadwell CR, Vahouny GV (1969) Purification and properties of pancreatic juice cholesterol esterase. J Biol Chem 244(7):1937–1945

84. Jansen H (2004) Hepatic lipase: friend or foe and under what circumstances? Curr Atheroscler Rep 6(5):343–347
85. Kishino J, Kawamoto K, Ishizaki J, Verheij HM, Ohara O, Arita H (1995) Pancreatic-type phospholipase A2 activates prostaglandin E2 production in rat mesangial cells by receptor binding reaction. The Journal of Biochemistry 117(2):420–424
86. Kulkarni BV, Mattes RD (2014) Lingual lipase activity in the orosensory detection of fat in humans. Am J Physiol Regul Integr Comp Physiol 306(12):R879–R885
87. Le Maire M, Champeil P, Moller JV (2000) Interaction of membrane proteins and lipids with solubilizing detergents. Biochimica et Biophysica Acta (BBA)-Biomembranes 1508(1–2):86–111
88. Leitersdorf E, Tobin EJ, Davignon J, Hobbs HH (1990) Common low-density lipoprotein receptor mutations in the French Canadian population. J Clin Invest 85(4):1014–1023
89. Lengsfeld H, Beaumier-Gallon G, Chahinian H, De Caro A, Verger R, Laugier R, Carriere F (2004) Physiology of gastrointestinal lipolysis and therapeutical use of lipases and digestive lipase inhibitors. In: Müller G, Petry S (eds) Lipases and phospholipases in drug development. Wiley-VCH, Weinheim, pp 195–229
90. Li Y, Hu M, McClements DJ (2011) Factors affecting lipase digestibility of emulsified lipids using an *in vitro* digestion model: proposal for a standardised pH-stat method. Food Chem 126(2):498–505
91. Lipowsky R (1995) Bending of membranes by anchored polymers. Europhysics Letters (EPL) 30(4):197–202
92. Lombardo D, Fauvel J, Guy O (1980) Studies on the substrate specificity of a carboxyl ester hydrolase from human pancreatic juice. I. Action on carboxyl esters, glycerides and phospholipids. Biochim Biophys Acta 611(1):136–146
93. Lombardo D, Guy O (1980) Studies on the substrate specificity of a carboxyl ester hydrolase from human pancreatic juice. II. Action on cholesterol esters and lipid-soluble vitamin esters. Biochimica et Biophysica Acta (BBA)-Enzymology 611(1):147–155
94. Lynn KR, Chuaqui CA, Clevette-Radford NA (1982) Kinetic studies of mammalian and microbial cholesterol esterases in homogeneous aqueous solutions. Bioorg Chem 11(1):19–23
95. MacCallum JL, Tieleman DP (2008) Interactions between small molecules and lipid bilayers. Curr Top Membr 60:227–256
96. Mateos-Diaz E, Amara S, Roussel A, Longhi S, Cambillau C, Carriere F (2017) Probing conformational changes and interfacial recognition site of lipases with surfactants and inhibitors. In: Methods in enzymology, vol 583, pp 279–307
97. Mateos-Diaz E, Bakala N'Goma JC, Byrne D, Robert S, Carrière F, Gaussier H (2018) IR spectroscopy analysis of pancreatic lipase-related protein 2 interaction with phospholipids: 1. Discriminative recognition of mixed micelles versus liposomes. Chem Phys Lipids 211:52–65
98. Mateos-Diaz E, Sutto-Ortiz P, Sahaka M, Byrne D, Gaussier H, Carrière F (2018) IR spectroscopy analysis of pancreatic lipase-related protein 2 interaction with phospholipids: 2. Discriminative recognition of various micellar systems and characterization of PLRP2-DPPC-bile salt complexes. Chem Phys Lipids 211:66–76
99. Mateos-Diaz E, Sutto-Ortiz P, Sahaka M, Rodriguez JA, Carriere F (2018) IR spectroscopy analysis of pancreatic lipase-related protein 2 interaction with phospholipids: 3. Monitoring DPPC lipolysis in mixed micelles. Chem Phys Lipids 211:77–85
100. McMullen TPW, Lewis RNAH, McElhaney RN (2004) Cholesterol–phospholipid interactions, the liquid-ordered phase and lipid rafts in model and biological membranes. Curr Opin Colloid Interface Sci 8(6):459–468
101. Michalski MC (2009) Specific molecular and colloidal structures of milk fat affecting lipolysis, absorption and postprandial lipemia. Eur J Lipid Sci Technol 111(5):413–431
102. Miled N, Canaan S, Dupuis L, Roussel A, Riviere M, Carriere F, de Caro A, Cambillau C, Verger R (2000) Digestive lipases: from three-dimensional structure to physiology. Biochimie 82(11):973–986

103. Minekus M, Alminger M, Alvito P, Ballance S, Bohn T, Bourlieu C, Carriere F, Boutrou R, Corredig M, Dupont D, Dufour C, Egger L, Golding M, Karakaya S, Kirkhus B, Le Feunteun S, Lesmes U, Macierzanka A, Mackie A, Marze S, McClements DJ, Menard O, Recio I, Santos CN, Singh RP, Vegarud GE, Wickham MS, Weitschies W, Brodkorb A (2014) A standardised static *in vitro* digestion method suitable for food - an international consensus. Food and Function 5(6):1113–1124
104. Mitchell DA, Carriere F, Krieger N (2008) An analytical method for determining relative specificities for sequential reactions catalyzed by the same enzyme: general formulation. Biochim Biophys Acta 1784(4):705–715
105. Moreau H, Abergel C, Carriere F, Ferrato F, Fontecilla-Camps JC, Cambillau C, Verger R (1992) Isoform purification of gastric lipases: towards crystallization. J Mol Biol 225(1):147–153
106. Moreau H, Bernadac A, Gargouri Y, Benkouka F, Laugier R, Verger R (1989) Immunocytolocalization of human gastric lipase in chief cells of the fundic mucosa. Histochemistry 91(5):419–423
107. Moreau H, Laugier R, Gargouri Y, Ferrato F, Verger R (1988) Human preduodenal lipase is entirely of gastric fundic origin. Gastroenterology 95(5):1221–1226
108. Mueller JH (1915) The assimilation of cholesterol and its esters. J Biol Chem 15:27
109. Mueller JH (1916) The mechanism of cholesterol absorption. J Biol Chem 27(2):463–480
110. Murphy DJ, Vance J (1999) Mechanisms of lipid-body formation. Trends Biochem Sci 24(3):109–115
111. Nagy L, Freeman DA (1990) Cholesterol movement between the plasma membrane and the cholesteryl ester droplets of cultured Leydig tumour cells. Biochem J 271(3):809–814
112. Nunes CS, Corring T (1979) Pancreatic exocrine secretion in the pig following test meals of different composition and intra-duodenal loads of glucose and maltose. Horm Metab Res 11(05):346–351
113. Nykjaer A, Bengtsson-Olivecrona G, Lookene A, Moestrup SK, Petersen CM, Weber W, Beisiegel U, Gliemann J (1993) The alpha 2-macroglobulin receptor/low density lipoprotein receptor-related protein binds lipoprotein lipase and beta-migrating very low density lipoprotein associated with the lipase. J Biol Chem 268(20):15048–15055
114. Pafumi Y, Lairon D, de la Porte PL, Juhel C, Storch J, Hamosh M, Armand M (2002) Mechanisms of inhibition of triacylglycerol hydrolysis by human gastric lipase. J Biol Chem 277(31):28070–28079
115. Panaiotov I, Verger R (2000) Enzymatic reactions at interfaces: interfacial and temporal organization of enzymatic lipolysis. In: Baszkin A, Norde W (eds) Physical chemistry of biological interfaces. CRC Press, New York, pp 359–400
116. Patton S, Keenan TW (1975) The milk fat globule membrane. Biochim Biophys Acta Rev Biomembr 415(3):273–309
117. Pinsirodom P, Parkin KL (2001) Lipase assays. In: Current protocols in food analytical chemistry. John Wiley & Sons, Hoboken, NJ, pp C3.1.1–C3.1.13
118. Rader DJ, Jaye M (2000) Endothelial lipase: a new member of the triglyceride lipase gene family. Curr Opin Lipidol 11(2):141–147
119. Rajakannan V, Yogavel M, Poi M-J, Jeyaprakash AA, Jeyakanthan J, Velmurugan D, Tsai M-D, Sekar K (2002) Observation of additional calcium ion in the crystal structure of the triple mutant K56, 120,121 M of bovine pancreatic phospholipase A2. J Mol Biol 324(4):755–762
120. Ransac S, Moreau H, Riviere C, Verger R (1991) Monolayer techniques for studying phospholipase kinetics. In: Methods in enzymology, vol 197, pp 49–65
121. Reboul E, Berton A, Moussa M, Kreuzer C, Crenon I, Borel P (2006) Pancreatic lipase and pancreatic lipase-related protein 2, but not pancreatic lipase-related protein 1, hydrolyze retinyl palmitate in physiological conditions. Biochim Biophys Acta 1761(1):4–10
122. Reis P, Holmberg K, Watzke H, Leser ME, Miller R (2009) Lipases at interfaces: a review. Adv Colloid Interf Sci 147-148:237–250

123. Reis P, Watzke H, Leser M, Holmberg K, Miller R (2010) Interfacial mechanism of lipolysis as self-regulated process. Biophys Chem 147(3):93–103
124. Renou C, Carriere F, Ville E, Grandval P, Joubert-Collin M, Laugier R (2001) Effects of lansoprazole on human gastric lipase secretion and intragastric lipolysis in healthy human volunteers. Digestion 63(4):207–213
125. Roman C, Carriere F, Villeneuve P, Pina M, Millet V, Simeoni U, Sarles J (2007) Quantitative and qualitative study of gastric lipolysis in premature infants: do MCT-enriched infant formulas improve fat digestion? Pediatr Res 61(1):83–88
126. Roussel A, Canaan S, Egloff M-P, Rivière M, Dupuis L, Verger R, Cambillau C (1999) Crystal structure of human gastric lipase and model of lysosomal acid lipase, two lipolytic enzymes of medical interest. J Biol Chem 274(24):16995–17002
127. Roussel A, de Caro J, Bezzine S, Gastinel L, de Caro A, Carriere F, Leydier S, Verger R, Cambillau C (1998) Reactivation of the totally inactive pancreatic lipase RP1 by structure-predicted point mutations. Proteins Structure Function and Genetics 32(4):523–531
128. Roussel A, Miled N, Berti-Dupuis L, Riviere M, Spinelli S, Berna P, Gruber V, Verger R, Cambillau C (2002) Crystal structure of the open form of dog gastric lipase in complex with a phosphonate inhibitor. J Biol Chem 277(3):2266–2274
129. Rudd EA (1984) Pancreatic carboxyl ester lipase (cholesterol esterase). Lipases
130. Salhi A, Amara S, Mansuelle P, Puppo R, Lebrun R, Gontero B, Aloulou A, Carriere F (2020) Characterization of all the lipolytic activities in pancreatin and comparison with porcine and human pancreatic juices. Biochimie 169:106–120
131. Sams L, Amara S, Chakroun A, Coudre S, Paume J, Giallo J, Carrière F (2017) Constitutive expression of human gastric lipase in Pichia pastoris and site-directed mutagenesis of key lid-stabilizing residues. Biochim Biophys Acta Mol Cell Biol Lipids 1862(10):1025–1034
132. Sams L, Amara S, Mansuelle P, Puppo R, Lebrun R, Paume J, Giallo J, Carrière F (2018) Characterization of pepsin from rabbit gastric extract, its action on β-casein and the effects of lipids on proteolysis. Food and Function 9(11):5975–5988
133. Sams L, Paume J, Giallo J, Carriere F (2016) Relevant pH and lipase for in vitro models of gastric digestion. Food Funct 7(1):30–45
134. Sarda L, Desnuelle P (1958) Action de la lipase pancreatique sur les esters en emulsion. Biochim Biophys Acta 30(3):513–521
135. Scott DL, White SP, Otwinowski Z, Yuan W, Gelb MH, Sigler PB (1990) Interfacial catalysis: the mechanism of phospholipase A2. Science 250(4987):1541–1546
136. Sias B, Ferrato F, Grandval P, Lafont D, Boullanger P, De Caro A, Leboeuf B, Verger R, Carriere F (2004) Human pancreatic lipase-related protein 2 is a galactolipase. Biochemistry 43(31):10138–10148
137. Simard J, Moorjani S, Vohl MC, Couture P, Torres AL, Gagne C, Despres JP, Labrie F, Lupien PJ (1994) Detection of a novel mutation (stop 468) in exon 10 of the low-density lipoprotein receptor gene causing familial hypercholesterolemia among French Canadians. Hum Mol Genet 3(9):1689–1691
138. Sovago M, Wurpel GWH, Smits M, Müller M, Bonn M (2007) Calcium-induced phospholipid ordering depends on surface pressure. J Am Chem Soc 129(36):11079–11084
139. Sternby B, Nilsson A (1997) Carboxyl ester lipase (bile salt-stimulated lipase), colipase, lipase, and phospholipase A2 levels in pancreatic enzyme supplements. Scand J Gastroenterol 32(3):261–267
140. Stewart JE, Feinle-Bisset C, Golding M, Delahunty C, Clifton PM, Keast RS (2010) Oral sensitivity to fatty acids, food consumption and BMI in human subjects. Br J Nutr 104(1):145–152
141. Szafran Z, Szafran H, Popiela T, Trompeter G (1978) Coupled secretion of gastric lipase and pepsin in man following pentagastrin stimulation. Digestion 18(5–6):310–318
142. Thirstrup K, Verger R, Carriere F (1994) Evidence for a pancreatic lipase subfamily with new kinetic properties. Biochemistry 33(10):2748–2756

143. Thunnissen MM, Eiso AB, Kalk KH, Drenth J, Dijkstra BW, Kuipers OP, Dijkman R, de Haas GH, Verheij HM (1990) X-ray structure of phospholipase A2 complexed with a substrate-derived inhibitor. Nature 347(6294):689–691
144. Tiraby C, Langin D (2003) Conversion from white to brown adipocytes: a strategy for the control of fat mass? Trends in Endocrinology & Metabolism 14(10):439–441
145. Vadas P, Pruzanski W (1986) Role of secretory phospholipases A2 in the pathobiology of disease. Lab Investig 55(4):391–404
146. van den Bosch H, Aarsman AJ, De Jong JGN, Van Deenen LLM (1973) Studies on lysophospholipases: I. purification and some properties of a lysophospholipase from beef pancreas. Biochimica et Biophysica Acta (BBA)-Lipids and Lipid Metabolism 296(1):94–104
147. van Tilbeurgh H, Egloff MP, Martinez C, Rugani N, Verger R, Cambillau C (1993) Interfacial activation of the lipase-procolipase complex by mixed micelles revealed by X-ray crystallography. Nature 362(6423):814–820
148. van Tilbeurgh H, Gargouri Y, Dezan C, Egloff M-P, Nésa M-P, Ruganie N, Sarda L, Verger R, Cambillau C (1993) Crystallization of pancreatic procolipase and of its complex with pancreatic lipase. J Mol Biol 229(2):552–554
149. van Tilbeurgh H, Sarda L, Verger R, Cambillau C (1992) Structure of the pancreatic lipase-procolipase complex. Nature 359(6391):159–162
150. Verger R (1984) Pancreatic lipases. In: Borgström B, Brockman HL (eds) Lipases. Elsevier Science, Amsterdam, pp 83–150
151. Verger R, de Haas GH (1973) Enzyme reactions in a membrane model. 1. A new technique to study enzyme reactions in monolayers. Chem Phys Lipids 10(2):127–136
152. Verger R, de Haas GH (1976) Interfacial enzyme kinetics of lipolysis. Annu Rev Biophys Bioeng 5:77–117
153. Verheij HM, Dijkstra BW (1994) Phospholipase A2: mechanism and structure. In: Petersen SB, Woolley P (eds) Lipases: their structure, biochemistry and application. Cambridge University Press, Cambridge, England, pp 119–138
154. Ville E, Carriere F, Renou C, Laugier R (2002) Physiological study of pH stability and sensitivity to pepsin of human gastric lipase. Digestion 65(2):73–81
155. Wang C-S, Kuksis A, Manganaro F, Myher JJ, Downs D, Bass HB (1983) Studies on the substrate specificity of purified human milk bile salt-activated lipase. J Biol Chem 258(15):9197–9202
156. Williams HD, Sassene P, Kleberg K, Bakala-N'Goma JC, Calderone M, Jannin V, Igonin A, Partheil A, Marchaud D, Jule E, Vertommen J, Maio M, Blundell R, Benameur H, Carriere F, Mullertz A, Porter CJ, Pouton CW (2012) Toward the establishment of standardized in vitro tests for lipid-based formulations, part 1: method parameterization and comparison of in vitro digestion profiles across a range of representative formulations. J Pharm Sci 101(9):3360–3380
157. Winkler FK, Darcy A, Hunziker W (1990) Structure of human pancreatic lipase. Nature 343(6260):771–774
158. Xiao X, Ross LE, Sevilla WA, Wang Y, Lowe ME (2013) Porcine pancreatic lipase related protein 2 has high triglyceride lipase activity in the absence of colipase. Biochim Biophys Acta 1831(9):1435–1441
159. Zweytick D, Athenstaedt K, Daum G (2000) Intracellular lipid particles of eukaryotic cells. Biochim Biophys Acta Rev Biomembr 1469(2):101–120

Chapter 2
Lipid Digestion as a Colloid and Interface Phenomena

David Julian McClements

2.1 Introduction

Colloid and interface science traditionally focuses on the behavior of small particles and surfaces [17, 20]. Many of the most famous physical scientists have made fundamental contributions to this area, including Einstein who developed equations to describe the diffusion of small particles through fluids and to predict the viscosity of colloidal dispersions, Rayleigh who developed a theory to describe the scattering of light from small particles, and Langmuir who made pioneering contributions to surface chemistry. Recently, the fundamental principles of colloidal and interface science have been used to understand the complex processes occurring inside the human body after we eat foods [38, 59]. Many of the most important physicochemical and physiological events occurring during the digestion of foods involve colloids or surfaces. The purpose of this chapter is to focus on the digestion of foods, particularly lipids, from a colloidal science perspective.

Colloidal particles may be present within the foods we eat or they may form within our guts after a food has been consumed. For instance, many foods contain fat droplets, protein particles, or starch granules that enter our gastrointestinal tracts (GITs) through our mouths. Conversely, emulsion droplets, micelles, and vesicles may be formed inside our guts after ingestion of fatty foods due to the mechanical and chemical processes occurring there. In this chapter, the term colloidal particles is used to refer to any type of particle in the gastrointestinal fluids that contains or arises from macronutrients.

D. J. McClements (✉)
Department of Food Science, University of Massachusetts, Amherst, MA, USA
e-mail: mcclements@foodsci.umass.edu

M. M.-L. Grundy, P. J. Wilde (eds.), *Bioaccessibility and Digestibility of Lipids from Food*, https://doi.org/10.1007/978-3-030-56909-9_2

2.2 Overview of Colloidal Events Occurring in GIT

2.2.1 Adsorption and Desorption Phenomena at Particle Surfaces

The adsorption/desorption of substances at interfaces plays an important role in many of the events occurring inside the GIT. The original emulsifier molecules present at the surfaces of lipid droplets may be displaced by other surface-active molecules in the surrounding gastrointestinal fluids, such as bile salts and phospholipids, as a precursor to lipid digestion. Electrically charged biopolymers in the continuous phase, such as mucin or dietary fibers, may adsorb to the surfaces of colloidal particles. Digestive enzymes, such as lipases, proteins, or amylases, have to adsorb to the surfaces of lipid droplets, protein aggregates, or starch granules before they can start hydrolyzing their substrates. Digestion may lead to the generation of surface-active substances, such as free fatty acids or monoacylglycerols, that displace other molecules from the particle surfaces. As a result of these processes, the composition, structure, and physicochemical properties of particle surfaces are modified appreciably as they travel through the human gut.

The coverage of colloidal particle surfaces by surface-active substances depends on their surface-activity and concentration. The *Langmuir adsorption isotherm* provides some valuable insights into the composition of interfacial layers. The fraction of a surface (Θ) covered by a surface-active agent in a solution in contact with it can be estimated using the following equation:

$$\Theta = \frac{\Gamma}{\Gamma_\infty} = \frac{cS}{1+cS} \tag{2.1}$$

Here, S (= $1/c_{½}$) is the surface activity of the surface-active agent, Γ and c are the concentrations of surface-active agent at the interface and in the bulk phase, Γ_∞ represents the concentration of surface-active agent when the interface is saturated, and $c_{½}$ represents the concentration of the surface-active agent in the bulk phase when half the adsorption sites are filled (i.e., $\theta = ½$). The surface activity of a particular substance can be determined from measurements of the interfacial tension versus concentration [32].

In the GIT fluids, there are many different kinds of surface-active agent all competing for the available surfaces of the colloidal particles, such as emulsifiers, enzymes, bile salts, phospholipids, free fatty acids, monoacylglycerols, and peptides. The interfacial composition therefore depends on the relative levels and surface activities of these different substances, which may change throughout the GIT. In the case of a two-component system, the interfacial composition can be related to the levels and surface activities of the two surface-active agents (1 and 2) using the following expressions [45]:

$$\frac{\Gamma_1}{\Gamma_{\text{Total}}} = \frac{S_1 c_1}{S_1 c_1 + S_2 c_2} \tag{2.2a}$$

$$\frac{\Gamma_2}{\Gamma_{\text{Total}}} = \frac{S_2 c_2}{S_1 c_1 + S_2 c_2} \tag{2.2b}$$

A number of simplifying assumptions are made in the derivation of these equations. For instance, it is assumed the two surface-active agents have similar molecular dimensions and form an ideal mixture. In reality, these assumptions are unlikely to be valid. The various kinds of surface-active substances found in the GIT have different molecular dimensions, ranging from small ions to large biopolymers. Moreover, there are often strong interactions between surface-active substances at the interface or in the bulk phase, so the system cannot be assumed to form an ideal mixture. Despite these limitations, the above expressions do provide useful insights into the factors impacting interfacial composition in the presence of mixtures of surface-active substances. For instance, they show that the amount of a substance at an interface increases as its surface activity and concentration increase. For more accurate analysis, theories have been developed to account for molecular dimensions and interactions [14]. These theories require more knowledge about system properties, but provide a more accurate representation of interfacial composition.

2.2.2 *Transport Phenomena*

The movement of molecules and particles from one location to another inside our guts plays an important role in digestion and absorption processes. Macronutrient particles have to be intimately mixed with GIT fluids so they encounter acids, digestive enzymes, bile salts, minerals and the other components that make an important contribution to digestion. If the particles are not thoroughly mixed, then digestion may be inhibited. For instance, digestive enzymes may not easily reach the surfaces of macronutrients trapped inside a gelled phase or a solid matrix. Instead, they may have to diffuse by a tortuous path through the pores in these materials before they reach the macronutrients inside and initiate digestion. Understanding transport phenomena in GIT fluids is therefore critical. The movement of molecules and particles within our guts depends on both the macro-rheology and micro-rheology of our gastrointestinal fluids.

2.2.2.1 Macroscopic Rheology of Intestinal Fluids

The rheology of gastrointestinal fluids may be considerably changed due to the presence of certain food ingredients, particularly dietary fibers, starches, and proteins [13, 43]. These components increase the viscosity by trapping water within their structures or form gels by cross-linking with their neighbors. The macro-rheology of gastrointestinal fluids is usually characterized using conventional shear rheometers. The sample to be analyzed is placed in a temperature-controlled chamber with a cup-and-bob or plate-and-plate measurement device. The instrument then applies a known shear stress to the gastrointestinal fluids and the resulting strain (gels) or rate of strain (fluids) is measured [12]. The modulus (gels) or viscosity (fluids) is calculated from the stress-strain or stress-strain rate profile. A change in the viscosity or gelation-state of the GIT fluids influences the mixing and movement of macronutrients and gastrointestinal components (such as enzymes, bile salts, acids, bases, mucin), which impacts the rate and extent of macronutrient digestion [12, 23].

To a first approximation, the ability of biopolymers, such as dietary fibers or proteins, to increase the shear viscosity (η) of the GIT fluids can be described by the following equation [28]:

$$\frac{\eta}{\eta_1} = \left(1 - \frac{\phi_E}{\phi_C}\right)^{-2} \tag{2.3}$$

Here, η_1 is the shear viscosity of water, ϕ_E is the effective volume fraction of the biopolymers, and ϕ_C is a critical packing parameter (≈0.57). The value of ϕ_E can be estimated using the following expression [28]:

$$\phi_E = \frac{4}{3}\pi r_H^3\left(\frac{cN_A}{M}\right) = \frac{cR_V}{p} \tag{2.4}$$

Here, r_H is the hydration radius of the biopolymer molecules, c is biopolymer concentration, N_A is Avogadro's number, M is the biopolymer molar mass, ρ is biopolymer density, and R_V is the biopolymer volume ratio. This latter parameter is the volume occupied by the hydrated biopolymer divided by the volume occupied by the non-hydrated biopolymer chain. These equations predict that large biopolymers with extended structures, like many dietary fibers, promote a large increase in the viscosity of the gastrointestinal fluids, which can modulate mixing and transport properties [12]. In practice, it should be noted that the viscosity of biopolymer solutions is highly dependent on the shear rate they experience, often exhibiting distinct shear thinning behavior.

Biopolymers that are rapidly hydrolyzed by digestive enzymes, like many starches and proteins, may greatly thicken the gastrointestinal fluids at the beginning of the gut, but have little effect in the latter stages because of the reduction in their molar mass [51]. Moreover, the viscosity of the gastrointestinal fluids may decrease when they are diluted with secretions occurring in the upper part of the GIT [10], or increase when water is extracted in the lower part of the GIT [23]. The macro-rheology of the gastrointestinal fluids may be increased when food matrices are broken down and release biopolymers. Studies have shown that raising the macro-viscosity of the gastrointestinal fluids may retard digestion and absorption of macronutrients [12, 34], which offers a means of controlling post-prandial spikes in blood glucose or lipid levels.

2.2.2.2 Microscopic Rheology of Intestinal Fluids

Knowledge of the macroscopic rheology of the gastrointestinal fluids does not always give a good indication of the movement of molecules and particles at the microscopic level. These entities can move at a very different rate than would be predicted from knowledge of the macro-rheology of the surrounding fluids. The micro-rheology is related to the friction a molecule or particle encounters at the microscopic level as it diffuses through the gastrointestinal fluids [57]. The difference between micro- and macro-rheology is most obvious for systems containing biopolymers. Large biopolymers can cause a large increase in the macroscopic viscosity of water, but molecules and particles can still rapidly diffuse through the large pores between the biopolymer chains. Consequently, the movement of these entities is primarily determined by the viscosity of water, rather than the viscosity of the whole system. The micro-viscosity of a molecule or particle is determined by two main factors: (i) their binding to the biopolymer chains; and, (ii) their diffusion through the pores in the biopolymer network. To a first approximation, the relationship between the diffusion coefficient of a particle through a biopolymer network and the pore size of the network can be described by the expression [4]:

$$D_{\text{Network}} = D_W \exp\left(-\pi\left(\frac{r_N - r_P}{2r_{\text{Pore}} + 2r_P}\right)\right) \tag{2.5}$$

Here, D_W is the diffusion coefficient of the particles in pure water, r_N is the particle radius, r_P is the cross-sectional radius of the biopolymer chains, and r_{Pore} is the radius of the pores in the biopolymer network. This expression predicts that particle diffusion through a biopolymer network is greatly retarded when the particle radius is greater than the pore size. The assumptions underlying the derivation of this expression assumed the particles do not bind to the biopolymer network [2, 7]. More complicated theories are available that take into account these binding interactions [15].

More studies are required to understand the major factors impacting the transport of different substances through the gastrointestinal fluids, including the region of the GIT involved (mouth, stomach, small intestine, or colon), the nature of the foods consumed, and the charactersitics of the molecules or particles involved.

2.2.3 Aggregation Phenomena in GIT Fluids

The aggregation state of macronutrient particles within the GIT impacts the rate of their digestion and transport, thereby altering the kinetics and extent of nutrient and nutraceutical absorption by our bodies. Aggregated particles are often digested more slowly than isolated ones because the surface area exposed to the digestive enzymes is reduced. Aggregated particles also travel more slowly due to diffusion because of their larger dimensions. Moreover, aggregated particles may be unable to penetrate through the pores in the mucus layer coating our epithelium cell, whereas isolated ones may be. It is therefore important to understand the factors impacting the aggregation state of the particles in our guts.

In general, aggregation depends on a delicate balance between attractive and repulsive interactions acting between particles (Fig. 2.1). Particles tend to aggregate when the attractive forces outweigh the repulsive forces, but remain as isolated entities when the repulsive forces dominate [21, 26]. Many kinds of colloidal interactions occur between the particles in GIT fluids. Van der Waals, electrostatic, hydrophobic, depletion, and bridging forces are the most important attractive interactions, whereas steric and electrostatic forces are the most important repulsive ones. To a first approximation, the overall interaction potential between two particles is described by the following equation:

$$w(s) = w_{\mathrm{VDW}}(s) + w_{\mathrm{S}}(s) + w_{\mathrm{ES}}(s) + w_{\mathrm{H}}(s) + w_{\mathrm{D}}(s) + w_{\mathrm{B}}(s) \tag{2.6}$$

Here, $w(s)$ represents the interaction potential at a surface-to-surface separation of s, while the subscripts VDW, S, ES, H, D, and B represent van der Waals, steric, electrostatic, hydrophobic, depletion, and bridging interactions, respectively. The overall interaction potential is governed by the sign, magnitude, and range of the individual interactions. A representative interaction potential for a system where only van der Waals, steric, and electrostatic interactions are important is shown in Fig. 2.1 [30].

The nature of the overall interaction potential depends on the characteristics of the particles, such as their dimensions, morphology, surface hydrophobicity, electrical charge, and interfacial properties, as well as on the properties of the surrounding fluids, such as its pH, ionic strength, and polymer content. Studies have shown that

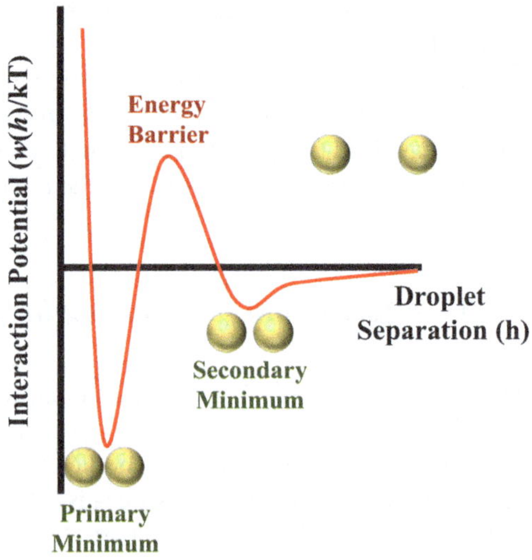

Fig. 2.1 The interaction potential between two droplets in emulsions is the combination of various attractive (van der Waals, hydrophobic, electrostatic, depletion & bridging) and repulsive (steric and electrostatic) forces. The balance of these forces determines emulsion stability

the aggregation state of lipid droplets in the GIT depends on the type of emulsifier used to stabilize them, as well as the nature of any additives in the surrounding liquids [61]. In the remainder of this section, some of the factors impacting the aggregation state of particles in the GIT are outlined.

2.2.3.1 Electrostatic Interactions

The gastrointestinal fluids contain many kinds of charged species, such as salt ions, proteins, peptides, amino acids, polysaccharides, phospholipids, and free fatty acids. The presence of these species modulates the electrostatic interactions between colloidal particles in the GIT in a number of ways.

Ionic strength effects: First, they raise the ionic strength of the GIT fluids, leading to screening of any electrostatic interactions [35]. The impact of ionic strength (I) on the range of electrostatic interactions can be estimated using the following equation: $\kappa^{-1} = 0.304/I^{½}$ nm, where κ^{-1} is the Debye screening length [21]. This value decreases from around 30 nm at low ionic strengths (0.1 mM) to around 0.2 nm at high ionic strengths (1 M). The ionic strength of the gastrointestinal fluids is typically >100 mM and so the range of the electrostatic interactions is usually relatively short. Even so, they may have an important impact on the behavior of macronutrients inside our guts.

Surface potential effects: The electrical properties of colloidal particles are usually characterized by their surface potential, which is usually represented by the ζ-potential, since this parameter is easy to determined experimentally [33]. Analysis of the ζ-potential-pH profile of colloidal particles provides useful insights into how they behave in the mouth, stomach, and small intestine. Conversely, ζ-potential measurements of colloidal particles in the various locations within the GIT provide useful information about the nature of the charged species at their surfaces. Before being ingested, the ζ-potential of food colloids is mainly determined by the type and level of ionized substances at their surfaces. After being ingested, the ζ-potential of the colloidal particles may change for numerous reasons: (i) alterations in pH; (ii) changes in the type and level of ions; (iii) adsorption of ionized substances, including mucin, proteins, phospholipids, or bile salts; (iv) displacement of emulsifiers; (v) production of ionized digestion products at their surfaces, including fatty acids or amino acids.

Alterations in the ζ-potential of the colloid particles suspended in the GIT fluids may alter their aggregation for a number of reasons. When the ζ-potential falls below a certain value, or the ionic strength increases above a certain value, the electrostatic repulsion between the particles may no longer be sufficient to prevent them from coming together. Charged biopolymers in the GIT fluids, such as mucin, may bind to the surfaces of many different droplets, leading to a phenomenon known as bridging flocculation.

2.2.3.2 Steric Interactions

Many of the colloidal particles in the GIT fluids are stabilized against aggregation by steric repulsion [21]. A strong short-range repulsion occurs when the outer layers of two or more colloidal particles overlap with each other [26, 40, 41]. The magnitude and range of this steric repulsion is determined by the type and concentration of the molecules present at the particle surfaces. As a result, anything that alters the composition of the interfacial region can change the stability of the colloidal particles by modulating the steric repulsion. As an example, biopolymer-coated lipid droplets may be stable to aggregation when present inside a food, but become unstable inside the human GIT because the biopolymers are displaced by other surface-active molecules in the gastrointestinal fluids.

2.2.3.3 Bridging Flocculation

Bridging flocculation may occur in situations where adsorbing polymers are present in a suspension of colloidal particles [18]. This phenomenon commonly occurs when the polymers have an opposite charge to the colloidal particles. As an example, protein-coated fat droplets should be positive under acidic gastric conditions, whereas mucin molecules should be negative. A mucin molecule can therefore bind to the surfaces of several protein-coated fat droplets and link them together. The propensity for bridging flocculation to happen is governed by the concentrations of colloidal particles and polymers present, as well as the molecular characteristics of the polymers, such as their molar mass, stiffness, and charge density [29]. Various types of ionized polymers can find their way into the gastrointestinal tract that can capable of promoting bridging flocculation, including mucin, proteins, and polysaccharides [56].

2.2.3.4 Depletion Flocculation

The colloid particles in the GIT can also aggregate due to the presence of non-adsorbing polymers in the gastrointestinal fluids through a *depletion* mechanism. Aggregation usually occurs once the concentration of the non-adsorbing polymers exceeds a critical level [22]. The physicochemical origin of this effect is the fact that the polymer molecules are excluded from a thin zone surrounding the colloidal particle surfaces, which leads to an osmotic force being generated. The strength of this osmotic force increases as the polymer concentration increases. Eventually, it may become strong enough to overcome the repulsive forces acting between the colloidal particles, leading to aggregation. The minimum polymer level required to cause aggregation is known as the critical flocculation concentration or CFC. The value of the CFC depends on colloidal particle size, as well as polymer molar mass and conformation [22].

GIT fluids contain numerous substances capable of promoting flocculation of colloidal particles through a depletion mechanism, such as mucin and dietary fibers [10, 56]. The mucin and dietary fibers concentration change as a food travels through the various regions of the gut, such as the mouth, stomach, small intestine, and colon, which influences the propensity for depletion flocculation to happen.

2.2.3.5 Changes in Aggregation State in GIT

In summary, the gastrointestinal fluids inside humans are structurally and compositionally and complex materials whose properties vary over time. As a result, it is difficult to accurately predict the many factors influencing the aggregation state of colloidal particles within human digestive tracts. Even so, an appreciation of the various forces that can impact their interactions with each other is still helpful for interpreting the behavior of food components in the GIT

2.2.4 Interfacial Reactions

Many of the chemical or biochemical reactions that occur inside the human GIT occur at interfaces. Fat digestion relies on the adsorption of lipase molecules from the gastrointestinal fluids to the surfaces of the fat droplets. The nature of the substances originally present at the fat droplet surfaces therefore influences the rate and extent of fat digestion. The metabolism and chemical degradation of nutraceuticals, such as curcumin and carotenoids, in the gut is often determined by chemical reactions that also happen at fat droplet surfaces. As an example, curcumin degrades faster when it is trapped in small fat droplets than in large ones because the small droplets have a greater specific surface area. As a result, more of the curcumin molecules are in direct contact with hydrophilic components in the water that accelerate degradation. Consequently, there is great interest in understanding the various factors that influence the chemical and biochemical reactions occurring at interfaces within the GIT.

2.3 Application of Colloid Surface Science to Modulate GIT Fate

An understanding of the underlying principles of colloid and interface science is critical for controlling the fate of foods inside the human gut. This knowledge can be employed to modulate the digestion rate of macronutrients such as starch, fat, or proteins, thereby controlling the levels of sugars, lipids, or amino acids in our blood. It can be utilized to boost the bioavailability of micronutrients (such as vitamins and minerals) or nutraceuticals (such as bioactive phytochemicals). Moreover, it can be used to target the release of bioactive substances in specific locations within the human gut, such as the mouth, stomach, small intestine, or colon. A few examples of the application of colloid science for these purposes are given in this section to illustrate its utility for designing healthier functional food products.

2.3.1 Modulation of Digestion Rates

The nutritional and health impacts of ingested macronutrients, such as carbohydrates, fats, and proteins, depend on the rate at which they are digested and absorbed by our bodies. In certain circumstances, food scientists would like to increase the digestion rate of a macronutrient so as to increase its absorption by our bodies or to increase the bioavailability of bioactive agents located inside it. In other circumstances, they would like to reduce the digestion rate so as to avoid a "nutrient spike" in our blood levels, such as having a rapid increase in glucose or lipid levels in our bloodstream. The macronutrient digestion rate can be modulated in various ways, which are summarized below:

- *Surface area*: The digestion of macronutrients is initiated when digestive enzymes adsorb to their surfaces. Thus, the digestion rate can be raised by increasing the specific surface area (A_S) of a macronutrient. Typically, there is an inverse relationship between the specific surface area of an object ant its dimensions. For spheres, $A_S = 6\phi/d_{32}$, where ϕ is the volume fraction of the macronutrient and d_{32} is the surface-weighted mean particle diameter. The rate of macronutrient digestion can therefore be increased by decreasing their particle size. As an example, numerous in vitro studies have shown that the rate of lipid digestion can be increased by decreasing the mean diameter of the droplets in oil-in-water emulsions [11, 47]. Researchers have recently carried out human feeding trials to determine if the size of the droplets in oil-in-water emulsions influenced satiety and energy intake [39]. They found that the initial oil droplet size did not significantly alter satiety – this effect may have been because of changes in the particle size of the emulsions as they passed through the human gut, so there were fewer differences in their digestion rates in the small intestine. This was in contrast to studies where the emulsions had been directly injected into the small intestine of subjects using tubes. This highlights the importance of understanding and controlling the fate of macronutrients throughout the GIT.
- *Aggregation state*: The digestion rate of macronutrient particles often decreases when they are aggregated within the GIT because it is then harder for digestive enzymes to reach their surfaces. Aggregation may involve the particles sticking together but maintaining their individual integrity (as in flocculation or coagulation), the particles completely merging together (as in coalescence), or the particles partly merging together (as in partial coalescence). It is therefore possible to modulate the rate of macronutrient digestion by controlling their aggregation

state within the GIT. Many in vitro studies have shown that macronutrients that are highly aggregated when they enter the stomach and/or small intestine are digested more slowly. For instance, the rate of protein digestion under simulated GIT conditions was decreased as they became more aggregated, due to the reduction in their exposed surface area [52]. Similarly, the rate of lipid digestion in emulsions decreased as they became more highly aggregated [5, 43].

- *Interfacial Engineering*: A macronutrient will only be digested if the digestive enzymes can come into contact with it. It is therefore possible to slow down digestion by covering macronutrients with coatings that limit the access of the digestive enzymes to their substrates. A number of in vitro studies have shown that coating macronutrients with indigestible materials, such as dietary fibers or indigestible nanoparticles, can slow down their digestion [49]. For instance, fat droplets can be coated with dietary fiber layers that limit the ability of lipase molecules to access the lipids below [24, 54]. However, it is important that the particles or fibers remain attached to the lipid droplet surfaces within the stomach and small intestine to have their inhibitory effects [50, 60].
- *Encapsulation*: The rate of macronutrient digestion can also be reduced by encapsulating them within indigestible biopolymer networks, which may be in the form of microgels or macrogels [31]. As a result, the digestive enzymes must penetrate through the biopolymer networks before they can access the macronutrients. The rate of macronutrient digestion can be reduced by increasing the external dimensions of the gelled particles (longer pathlength) or decreasing the size of the pores in the networks (greater restriction of enzyme movement). In vitro GIT studies have shown that encapsulation of proteins [65], starches [46], and lipids [6] in microgel beads can inhibit their digestion rate under simulated stomach and/or small intestine conditions. In this case, it is important that the microgel beads remain intact within the upper GIT, and that the pore size is small enough to inhibit the diffusion of the digestive enzymes inside them.
- *GIT Fluid Rheology*: Macronutrient digestion can also be decreased by thickening or gelling the GIT fluids because this will inhibit the thorough mixing of the gastrointestinal components, as well as retarding the transport of digestive enzymes to the macronutrients and the resulting digestion products to the epithelium cells [9]. In vitro studies have shown that adding thickening or gelling agents to foods prior to subjecting them to a simulated GIT model can retard macronutrient digestion, such as carbohydrates [8, 36], lipids [37], and proteins [62].
- *Binding Interactions*: The digestion of macronutrients in the human GIT depends on numerous different gastrointestinal constituents, including digestive enzymes, bile acids, and minerals [10]. Certain food-grade substances are able to bind to one or more of these constituents, thereby interfering with macronutrient digestion. In vitro studies have shown that cationic chitosan molecules are able to strongly bind to anionic bile acids, which reduces their ability to adsorb to the surfaces of the fat droplets and to solubilize and transport lipid digestion products, thereby inhibiting lipid digestion [1, 16, 53]. Other studies have shown that

cellulose ethers [55], pectin [3], alginate [19], and xanthan [44]can also bind to gastrointestinal components, and so have the potential to modulate macronutrient digestion.

2.3.2 *Enhancement of Bioavailability*

An understanding of the colloidal and interfacial phenomenon occurring inside the human GIT is also useful for increasing the bioavailability and bioactivity of nutraceuticals, micronutrients, and macronutrients. In this section, the focus will be on increasing the bioavailability of hydrophobic bioactive substances, such as oil-soluble vitamins, omega-3 oils, and non-polar nutraceuticals. Usually, these substances are originally located within the oil phase of foods, and so their bioavailability depends on this phase being digested before they are released into the GIT fluids. Consequently, anything that increases the rate and degree of lipid digestion increases their bioavailability. The bioavailability of hydrophobic bioactives can therefore be increased by dissolving them in small, non-aggregated, and non-encapsulated lipid droplets that have interfaces that are not resistant to lipase adsorption.

The bioavailability of hydrophobic bioactives is also dependent on the characteristics of the oil phase surrounding them, as this determines the nature of the mixed micelles formed during digestion. Mixed micelles are examples of natural colloidal dispersions found in the human body after ingestion of a meal, which consist of relatively small micelles (typically <10 nm) and vesicles (typically 100–1000 nm) [58]. The amounts and structures of the mixed micelles generated during the lipid digestion process governs the bioaccessibility of hydrophobic bioactives [42]. Mixed micelles have hydrophobic domains inside them with dimensions that are governed by the chain length and unsaturation degree of the fatty acid and monoacylglycerol chains arising from digestion of the lipid phase. The dimensions of these hydrophobic domains are around twice the length of the fatty acid chains [27]. As a result, the dimensions of the hydrophobic domains in the mixed micelles increase as the fatty acid chains become longer and less saturated. A relatively small hydrophobic bioactive will easily fit into these hydrophobic domains and will therefore have a high bioaccessibility. Conversely, a relatively large hydrophobic bioactive will be unable to be accommodated and so it will have a low bioaccessibility. The impact of lipid type on the bioaccessibility of hydrophobic bioactive substances has been demonstrated for carotenoids (such as β-carotene, lycopene, or lutein): these long hydrophobic molecules are easily solubilized within mixed micelles formed from long-chain fatty acids (such as corn oil), but not from those formed from short- or medium-chain fatty acids (such as MCT) [42, 48].

2.3.3 Targeted or Controlled Delivery

Finally, a knowledge of colloid and interface science is critical for the development of delivery systems that can encapsulate, retain, and protect bioactive substances within some regions of the GIT, but they release them at a controlled rate or in response to an environmental trigger in another region of the GIT. These delivery systems may be designed to release bioactive substances within the mouth, stomach, small intestine, or colon by manipulating their compositions and structures [31]. Delivery systems fabricated from starches begin to breakdown in the mouth and small intestine and so can be used for oral delivery. Those constructed from proteins and lipids tend to breakdown mainly in the stomach and small intestine and so can be used for gastric or intestinal delivery. Conversely, those constructed from dietary fibers tend to be resistant to digestion in the upper GIT but are fermented in the colon by gut microbes, which makes them more suitable for delivery of substances to the large intestine. Knowledge of the behavior of different food components within the GIT can therefore be utilized to create food-grade delivery systems that breakdown and release bioactive substances in these different locations within the GIT. Biopolymer microgels are one of the most widely explored delivery systems for the targeted or controlled release of bioactive agents within the GIT. These systems consist of small particles that are fabricated from food-grade biopolymers, such as proteins and/or polysaccharides [63].

Biopolymer microgels can be designed to stay intact in some locations of the GIT, but then dissociate in other locations, thereby releasing any entrapped bioactive agents. This can be achieved by controlling the attractive interactions between the biopolymer molecules that make up the hydrogel network, or by enzymatically or chemically degrading them. Alternatively, microgels can be designed so that they strongly bind the bioactive agents in some regions of the GIT, but then release them in other regions, without actually breaking down. This can be achieved by controlling the attractive interactions between the bioactive agents and the microgel network. Consequently, it is possible to control the release rate and location within the GIT by controlling how the microgel responds to environmental triggers, such as pH, ionic strength, or enzyme activity [63]. Studies have shown that biopolymer microgels can be created that dissociate in the mouth because of alterations in pH [64] or temperature [66]. The pH-triggered release was achieved by controlling he electrostatic interactions between the biopolymers inside the microgels, whereas the temperature-triggered release was achieved by utilizing a helix-to-coil transition of one of the biopolymers used to fabricate the microgels.

Biopolymer microgels can also be created that are able to release bioactive substances within other locations inside the GIT. For example, alginate microgels have been shown to retard lipid digestion in the mouth, stomach, and small intestine, but then release them inside the colon [25]. Food scientists have a variety of different strategies that they can use to create food-grade delivery systems to control the release of bioactive substances in the GIT, which is an active area of research.

2.4 Conclusions

An understanding of the underlying principles of colloid and interface science is crucial for understanding and controlling the complicated events that happen inside the human GIT. The digestion and absorption of macronutrients involves many colloidal and interfacial phenomenon, such as adsorption/desorption, mass transport, particle aggregation, and surface chemistry. Food researchers are utilizing their knowledge of these phenomena to design functional foods with improved nutritional profiles to enhance human health and wellness. They are doing this by controlling macronutrient digestion rates, bioactive bioavailability, and bioactive release site. This is a highly active area that is likely to lead to innovative new products that may tackle the increasing rise of chronic diseases, such as diabetes, obesity, hypertension, and cancer. Nevertheless, more studies are required to test new forms of delivery systems in human feeding trials and to make economically viable products that consumers want to eat.

Acknowledgements This material was partly based upon work supported by the National Institute of Food and Agriculture, USDA, Massachusetts Agricultural Experiment Station (MAS00491) and USDA, AFRI Grants (2014-67021, 2016-25147, 2016-08782).

References

1. Beysseriat M, Decker EA, McClements DJ (2006) Preliminary study of the influence of dietary fiber on the properties of oil-in-water emulsions passing through an in vitro human digestion model. Food Hydrocoll 20:800–809
2. Boegh M, Nielsen HM (2015) Mucus as a barrier to drug delivery - understanding and mimicking the barrier properties. Basic Clin Pharmacol Toxicol 116:179–186
3. Cervantes-Paz B, Ornelas-Paz JD, Perez-Martinez JD, Reyes-Hernandez J, Zamudio-Flores PB, Rios-Velasco C, Ibarra-Junquera V, Ruiz-Cruz S (2016) Effect of pectin concentration and properties on digestive events involved on micellarization of free and esterified carotenoids. Food Hydrocoll 60:580–588
4. Chan AW, Neufeld RJ (2009) Modeling the controllable pH-responsive swelling and pore size of networked alginate based biomaterials. Biomaterials 30:6119–6129
5. Chang YG, McClements DJ (2016) Influence of emulsifier type on the in vitro digestion of fish oil-in-water emulsions in the presence of an anionic marine polysaccharide (fucoidan): caseinate, whey protein, lecithin, or Tween 80. Food Hydrocoll 61:92–101
6. Chen F, Deng ZY, Zhang ZP, Zhang RJ, Xu QY, Fan GQ, Luo T, McClements DJ (2018) Controlling lipid digestion profiles using mixtures of different types of microgel: alginate beads and carrageenan beads. J Food Eng 238:156–163
7. Crater JS, Carrier RL (2010) Barrier properties of gastrointestinal mucus to nanoparticle transport. Macromol Biosci 10:1473–1483
8. Dartois A, Singh J, Kaur L, Singh H (2010) Influence of guar gum on the in vitro starch digestibility-rheological and microstructural characteristics. Food Biophysics 5:149–160
9. Dikeman CL, Murphy MR, Fahey GC (2006) Dietary fibers affect viscosity of solutions and simulated human gastric and small intestinal digesta. J Nutr 136:913–919

10. Espinal-Ruiz M, Parada-Alfonso F, Restrepo-Sanchez LP, Narvaez-Cuenca CE, McClements DJ (2014) Impact of dietary fibers methyl cellulose, chitosan, and pectin on digestion of lipids under simulated gastrointestinal conditions. Food Funct 5:3083–3095
11. Golding M, Wooster TJ (2010) The influence of emulsion structure and stability on lipid digestion. Curr Opin Colloid Interface Sci 15:90–101
12. Gouseti O, Jaime-Fonseca MR, Fryer PJ, Mills C, Wickham MSJ, Bakalis S (2014) Hydrocolloids in human digestion: dynamic in-vitro assessment of the effect of food formulation on mass transfer. Food Hydrocoll 42:378–385
13. Grundy MML, Edwards CH, Mackie AR, Gidley MJ, Butterworth PJ, Ellis PR (2016) Re-evaluation of the mechanisms of dietary fibre and implications for macronutrient bioaccessibility, digestion and postprandial metabolism. Br J Nutr 116:816–833
14. Guzman E, Llamas S, Maestro A, Fernandez-Pena L, Akanno A, Miller R, Ortega F, Rubio RG (2016) Polymer-surfactant systems in bulk and at fluid interfaces. Adv Colloid Interf Sci 233:38–64
15. Hansing J, Ciemer C, Kim WK, Zhang XL, DeRouchey JE, Netz RR (2016) Nanoparticle filtering in charged hydrogels: effects of particle size, charge asymmetry and salt concentration. Eur Phys J E 39(5):53
16. Helgason T, Weiss J, McClements DJ, Gislason J, Einarsson JM, Thormodsson FR, Kristbergsson K (2008) Examination of the interaction of chitosan and oil-in-water emulsions under conditions simulating the digestive system using confocal microscopy. J Aqua Food Prod Technol 17:216–233
17. Hiemenz PC, Rajagopalan R (1997) Principles of colloid and surface chemistry. Marcel Dekker, New York
18. Hogg R (2013) Bridging flocculation by polymers. Kona Powder Part J 30:3–14
19. Hu M, Li Y, Decker EA, McClements DJ (2010) Role of calcium and calcium-binding agents on the lipase digestibility of emulsified lipids using an in vitro digestion model. Food Hydrocoll 24:719–725
20. Hunter RJ (1994) Introduction to modern colloid science. Oxford University Press, Oxford, UK
21. Israelachvili J (2011) Intermolecular and surface forces, 3rd edn. Academic Press, London
22. Jenkins P, Snowden M (1996) Depletion flocculation in colloidal dispersions. Adv Colloid Interf Sci 68:57–96
23. Lentle RG, Janssen PWM (2011) The physical processes of digestion. Springer Science, New York
24. Li Y, Hu M, Xiao H, Du YM, Decker EA, McClements DJ (2010) Controlling the functional performance of emulsion-based delivery systems using multi-component biopolymer coatings. Eur J Pharm Biopharm 76:38–47
25. Li Y, Hu M, Du Y, Xiao H, McClements DJ (2011) Control of lipase digestibility of emulsified lipids by encapsulation within calcium alginate beads. Food Hydrocoll 25:122–130
26. Lu K (2008) Theoretical analysis of colloidal interaction energy in nanoparticle suspensions. Ceram Int 34:1353–1360
27. Marze S (2017) Bioavailability of nutrients and micronutrients: advances in modeling and in vitro approaches. In: Doyle MP, Klaenhammer TR (eds) Annual review of food science and technology, vol 8
28. McClements DJ (2000) Comments on viscosity enhancement and depletion flocculation by polysaccharides. Food Hydrocoll 14:173–177
29. McClements DJ (2005) Theoretical analysis of factors affecting the formation and stability of multilayered colloidal dispersions. Langmuir 21:9777–9785
30. McClements DJ (2014) Nanoparticle- and microparticle-based delivery systems. CRC Press, Boca Raton
31. McClements DJ (2017) Designing biopolymer microgels to encapsulate, protect and deliver bioactive components: physicochemical aspects. Adv Colloid Interf Sci 240:31–59
32. McClements DJ (2015) Food emulsions: principles, practices, and techniques. CRC Press, Boca Raton

33. McClements DJ, DeLoid G, Pyrgiotakis G, Shatkin JA, Xiao H, Demokritou P (2016) The role of the food matrix and gastrointestinal tract in the assessment of biological properties of ingested engineered nanomaterials (iENMs): state of the science and knowledge gaps. NanoImpact 3:47–57
34. Moxon TE, Gouseti O, Bakalis S (2016) In silico modelling of mass transfer & absorption in the human gut. J Food Eng 176:110–120
35. Nokhodchi A, Asare-Addo K (2014) Drug release from matrix tablets: physiological parameters and the effect of food. Expert Opin Drug Deliv 11:1401–1418
36. Ou SY, Kwok KC, Li Y, Fu L (2001) In vitro study of possible role of dietary fiber in lowering postprandial serum glucose. J Agric Food Chem 49:1026–1029
37. Pasquier B, Armand M, Castelain C, Guillon F, Borel P, Lafont H, Lairon D (1996) Emulsification and lipolysis of triacylglycerols are altered by viscous soluble dietary fibres in acidic gastric medium in vitro. Biochem J 314:269–275
38. Phan S, Salentinig S, Prestidge CA, Boyd B (2014) Self-assembled structures formed during lipid digestion: characterization and implications for oral lipid-based drug delivery systems. Drug Deliv Transl Res 4:275–294
39. Poppitt SD, Budgett SC, MacGibbon AK, Quek SY, Kindleysides S, Wiessing KR (2018) Effects of lipid emulsion particle size on satiety and energy intake: a randomised cross-over trial. Eur J Clin Nutr 72:349–357
40. Pyrgiotakis G, Blattmann CO, Demokritou P (2014) Real-time nanoparticle-cell interactions in physiological media by atomic force microscopy. ACS Sustain Chem Eng 2:1681–1690
41. Pyrgiotakis G, Blattmann CO, Pratsinis S, Demokritou P (2013) Nanoparticle-nanoparticle interactions in biological media by atomic force microscopy. Langmuir 29:11385–11395
42. Qian C, Decker EA, Xiao H, McClements DJ (2012) Nanoemulsion delivery systems: influence of carrier oil on beta-carotene bioaccessibility. Food Chem 135:1440–1447
43. Qin DK, Yang XJ, Gao SR, Yao JH, McClements DJ (2016) Influence of hydrocolloids (dietary fibers) on lipid digestion of protein-stabilized emulsions: comparison of neutral, anionic, and cationic polysaccharides. J Food Sci 81:C1636–C1C45
44. Qiu CY, Zhao MM, Decker EA, McClements DJ (2015) Influence of anionic dietary fibers (xanthan gum and pectin) on oxidative stability and lipid digestibility of wheat protein-stabilized fish oil-in-water emulsion. Food Res Int 74:131–139
45. Razumovsky L, Damodaran S (2001) Incompatibility of mixing of proteins in adsorbed binary protein films at the air-water interface. J Agric Food Chem 49:3080–3086
46. Rose DJ, Keshavarzian A, Patterson JA, Venkatachalam M, Gillevet P, Hamaker BR (2009) Starch-entrapped microspheres extend in vitro fecal fermentation, increase butyrate production, and influence microbiota pattern. Mol Nutr Food Res 53:S121–SS30
47. Salvia-Trujillo L, Qian C, Martin-Belloso O, McClements DJ (2013) Influence of particle size on lipid digestion and beta-carotene bioaccessibility in emulsions and nanoemulsions. Food Chem 141:1472–1480
48. Salvia-Trujillo L, Qian C, Martín-Belloso O, McClements DJ (2013) Modulating beta-carotene bioaccessibility by controlling oil composition and concentration in edible nanoemulsions. Food Chem 139:878–884
49. Sarkar A, Zhang SN, Murray B, Russell JA, Boxal S (2017) Modulating in vitro gastric digestion of emulsions using composite whey protein-cellulose nanocrystal interfaces. Colloids and Surfaces B-Biointerfaces 158:137–146
50. Silva HD, Poejo J, Pinheiro AC, Donsi F, Serra AT, Duarte CMM, Ferrari G, Cerqueira MA, Vicente AA (2018) Evaluating the behaviour of curcumin nanoemulsions and multilayer nanoemulsions during dynamic in vitro digestion. J Funct Foods 48:605–613
51. Singh J, Kaur L, Singh H (2013) Food microstructure and starch digestion. In: Henry J (ed) Advances in food and nutrition research, vol 70
52. Singh TK, Oiseth SK, Lundin L, Day L (2014) Influence of heat and shear induced protein aggregation on the in vitro digestion rate of whey proteins. Food Funct 5:2686–2698

53. Thongngam M, McClements DJ (2005) Isothermal titration calorimetry study of the interactions between chitosan and a bile salt (sodium taurocholate). Food Hydrocoll 19:813–819
54. Tokle T, Lesmes U, Decker EA, McClements DJ (2012) Impact of dietary fiber coatings on behavior of protein-stabilized lipid droplets under simulated gastrointestinal conditions. Food Funct 3:58–66
55. Torcello-Gomez A, Fraguas CF, Ridout MJ, Woodward NC, Wilde PJ, Foster TJ (2015) Effect of substituent pattern and molecular weight of cellulose ethers on interactions with different bile salts. Food Funct 6:730–739
56. Vingerhoeds MH, Blijdenstein TBJ, Zoet FD, van Aken GA (2005) Emulsion flocculation induced by saliva and mucin. Food Hydrocoll 19:915–922
57. Waigh TA (2016) Advances in the microrheology of complex fluids. Rep Prog Phys 79:074601
58. Williams HD, Trevaskis NL, Yeap YY, Anby MU, Pouton CW, Porter CJH (2013) Lipid-based formulations and drug supersaturation: harnessing the unique benefits of the lipid digestion/absorption pathway. Pharm Res 30:2976–2992
59. Yao M, Xiao H, McClements DJ (2014) Delivery of lipophilic bioactives: assembly, disassembly, and reassembly of lipid nanoparticles. Annu Rev Food Sci Technol 5(5):53–81
60. Zeeb B, Lopez-Pena CL, Weiss J, McClements DJ (2015) Controlling lipid digestion using enzyme-induced crosslinking of biopolymer interfacial layers in multilayer emulsions. Food Hydrocoll 46:125–133
61. Zhang RJ, Zhang ZP, Zhang H, Decker EA, McClements DJ (2015) Influence of emulsifier type on gastrointestinal fate of oil-in-water emulsions containing anionic dietary fiber (pectin). Food Hydrocoll 45:175–185
62. Zhang S, Vardhanabhuti B (2014) Intragastric gelation of whey protein-pectin alters the digestibility of whey protein during in vitro pepsin digestion. Food Funct 5:102–110
63. Zhang ZP, Zhang RJ, Chen L, Tong QY, McClements DJ (2015) Designing hydrogel particles for controlled or targeted release of lipophilic bioactive agents in the gastrointestinal tract. Eur Polym J 72:698–716
64. Zhang ZP, Zhang RJ, Decker EA, McClements DJ (2015) Development of food-grade filled hydrogels for oral delivery of lipophilic active ingredients: pH-triggered release. Food Hydrocoll 44:345–352
65. Zhang ZP, Zhang RJ, McClements DJ (2017) Control of protein digestion under simulated gastrointestinal conditions using biopolymer microgels. Food Res Int 100:86–94
66. Zhang ZP, Zhang RJ, Tong QY, Decker EA, McClements DJ (2015) Food-grade filled hydrogels for oral delivery of lipophilic active ingredients: temperature-triggered release microgels. Food Res Int 69:274–280

Chapter 3
Lipid Digestion: *In Vitro* and *In Vivo* Models and Insights

Dafna Meshulam Pascoviche and Uri Lesmes

3.1 Introduction

Dietary lipids are an important source of calories, nutrients, micronutrients and other bioactive compounds introduced into the human body via the alimentary canal. By definition, nutritional lipids are defined as food components that dissolve in organic solvents but are insoluble in water. In fact, lipids are a vast family of substances that include triacylglycercols (TAGs), diacylglycercols (DAGs), monoacylglycercols (MAGs), free fatty acids (FFAs), phospholipids, sphingolipids, sterols, waxes, essential and flavor oils as well as caretonoids and insoluble vitamins, such as vitamin D (Fig. 3.1).

The predominant chemical groups of food lipids are the TAGs (can be up to 40–50% of an oil or a fat) and they naturally exist as complex mixtures of acylglycerols with other lipid fractions. Thus, food lipids have high chemical and physical diversity and techno-functionalities and are vastly consumed as water-lipid mixtures, i.e. emulsions. This section will discuss qualitative and quantitative insights pertinent to the digestive fate of lipid intake.

Adopting a pharmacological approach, lipid digestion involves lipid liberation from the oral intake, its absorption, distribution, metabolism and excretion (LADME). All are a complex series of spatio-temporal events occurring in controlled stirred compartments of the human gastro-intestine (GI), from mouth to anus. In fact, lipid bioaccessibility and bioavailability are detrimental bottleneck processes delineating the subsequent lipid bio-efficacy in modulating on biological functions and human health. To this end, bioaccessibility is defined as the substance

D. M. Pascoviche • U. Lesmes (✉)
Department of Biotechnology and Food Engineering, Technion – Israel Institute of Technology, Haifa, Israel
e-mail: lesmesu@bfe.technion.ac.il

M. M.-L. Grundy, P. J. Wilde (eds.), *Bioaccessibility and Digestibility of Lipids from Food*, https://doi.org/10.1007/978-3-030-56909-9_3

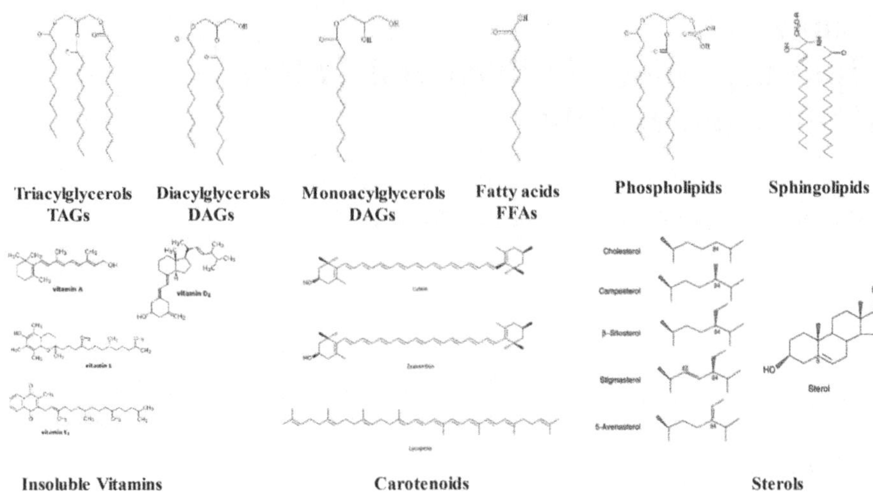

Fig. 3.1 Illustration of central lipid groups found in foodstuffs

fraction released from the food matrix into the GI lumen and becomes available for absorption. Similarly, bioavailability is the substance fraction released into the body circulation and rendered available to the body tissues for further uptake and effect on biological functions.

Being distinctly different from hydrophilic compounds, lipid systems, their behavior during digestion as well as bioaccessibility/bioavailability have been extensively studied and shown to involve complex interfacial and colloidal phenomena directed by the anatomical and physiological conditions of the human gut [1–4]. Moreover, studies show that lipid digestion can also be engineered through rational modulation of lipid chemistry, structure and organization to exhibit targeted and controlled breakdown in terms of rate, extent and loci of lipid digestion. Thus, the end products of lipid digestion as well as digestion dynamics hold valuable information for food and health professionals.

3.2 What Are Lipids Exposed to from Ingestion to Defecation?

The human GI has evolved over human history to handle intake of foods and control the molecular entities that enter into the circulation from the luminal space, which per se is an external space to the body. This digestive canal comprises of four key digestive organs: the mouth, stomach, small intestine and colon as well as various accessory organs, i.e. the salivary glands, lingual glands, esophagus, liver, gallbladder and exocrine pancreas. As any oral intake, lipids undergo a series of detailed physicochemical processes in the four different organs with varying times of retention and transit between the processing organs.

Broadly, digestive physical processes include chewing, swallowing, mixing, peristalsis, diffusion towards the gut epithelium as well as flow into subsequent organs. The biochemical events of luminal digestion entail: lipid release, enzymatic degradation and self-assembly into various structures, such as micelles and lamellae. The digestive breakdown of lipids is predominantly due to the action of numerous enzymes (Table 3.1) secreted along the GI that act on lipids at the water-lipid interface [1, 3, 8, 15]. Further, ingested but unabsorbed lipids reaching the colon can be fermented by the colon microflora. Studies show that the key determinants of lipid digestion are formulation characteristics, e.g. bulk or emulsified lipid, and consumer physiological and psychological status, e.g. muscle tone, levels of enzymes, luminal composition and mental status/mood.

Table 3.1 Central lipolytic actuators secreted along the human gastro-intestinal tract

Location of secretion (Source)	Enzymes	Action	References
Mouth (salivary glands)	Lingual lipase	Arguable existence in humans. Assumed to liberate FFA in oral cavity	Feher [5] N'goma [6]
Stomach (chief cells of fundic mucosae)	Gastric lipase	Hydrolysis of 5–40% of consumed TAGs, preferentially at sn-3 position	Sams [7] Armand [8]
Small intestine (forms in acinar cells of the exocrine pancreas)	Lipase (requires the presence of Colipase)	Hydrolysis of 40–70% of consumed TAGs, preferentially at sn-1 and sn-3 position	Armand [8]
	Pancreatic lipase related protein 2 (PLRP2)	Broad substrate specificity, e.g. phospholipids, galactolipids, lipophilic vitamin esters, acylglycerols and PEG esters	N'goma [6] Lowe [9]
	Phospholipase A2 (PLA2)	Liberate the *sn*-2 fatty acyl chains from phospholipids to yield non-esterified fatty acids and lysophospholipids	Hui [10] Murakami [11]
Small intestine and also secreted by mammary glands (often is referred to as bile salt stimulated lipase, (BSSL), or pancreatic bile salt-dependent lipase (BSDL) in that context)	Cholesterol esterase (also: carboxyl ester hydrolase, CEH)	Non-specific esterase, that hydrolyzes various substrates e.g. acylglycerols, phospholipids, cholesterol, lipophilic vitamin esters, carotenoid esters, galactolipids, and PEG di- and mono-esters	N'goma [6] Armand [8]
Colon (bacteria)	–	Adsorption of medium chain fatty acids with negligible adsorption of long chain fatty acids	Mu and Hoy [12]
	–	Implicated in phytosterol metabolism and bioavailability	Bohn et al. [13] Danneskiold-Samsøe et al. [14]

Studying lipid digestion is not just challenging from the biological perspective but also a formidable analytical challenge. To this end, a wide range of analytical methods have been coupled to *in vitro* and *in vivo* digestion models, these include both destructive and nondestructive methods (such as GC, GC-MS, NMR, laser scattering, analytical centrifugation, ultrasound, X-ray diffraction, synchrotron X-ray diffraction and even small angle neutron scattering). For example, various compositional and structural aspects can be examined via microscopic methods e.g. cross polarized light microscopy (CPLM), cryogenic-transmission electron microscopy (cryo-TEM), cryogenic field emission scanning electron microscopy (cryo-FESEM), through scattering techniques like polarized and depolarized dynamic light scattering (DLS/DDLS), small and wide-angle X-ray scattering (SAXS/WAXS also referred to as X-ray diffraction, XRD), small angle neutron scattering (SANS), and also using spectroscopic techniques such as nuclear magnetic resonance spectroscopy (NMR), Raman spectroscopy and multiple coherent anti-Stokes Raman scattering micro-spectroscopy (CARS) and electron paramagnetic resonance spectroscopy (EPR). All these were thoroughly reviewed by Phan et al. (2014) [16]. Such studies can take place in various models: *in vitro*, *ex vivo*, *in vivo* as well as investigate isolated distinct events of digestion, i.e. in the mouth, stomach, small intestine and colon.

3.2.1 Oral Processing of Lipids

The mouth is where lipids are introduced into the gastrointestinal tract via bolus formation. This involves initial mechanical breakdown of foods and mixing with saliva to decrease its size, improve its flow behavior and nutrient release by increasing available surface area for further digestion. Although liquid foods/beverages spend only a short while in the oral cavity, they are also affected by interactions with salivary enzymes, proteins, mucins as well as other foodstuff fragments being processed [4]. In fact, saliva exposes foods to various ions and mucin that have been shown to induce droplet flocculation or colloidal aggregation via depletion or bridging flocculation [17–20]. Bolus formation and properties are also elemental in texture perception and flavor release [21, 22]. Subsequent lubricated bolus mixture passes along the digestion through the esophagus into the stomach.

3.2.2 Gastric Processing of Lipids

In the stomach, the bolus is transformed into chyme due to exposure to the acidic environment, enzymatic secretions and peristaltic motion. These conditions are time dependent and promote further breakdown of the food matrix, liberation of entrapped lipids and even emulsification of the released lipids [4]. While in the stomach, emulsified lipids tend to be physically destabilized via flocculation and/or coalescence as well as biochemically broken down by gastric lipase [1, 2, 8, 23–26]. Gastric proteolysis may also affect the physical stability of emulsions stabilized by proteins [27, 28]. Thus, lipid droplet sizes may increase or decrease due to the competitive nature of these processes [29, 30]. Overall, droplet surface area accessible to lipases changes in the stomach affecting the properties of the effluent poured into the small intestine.

3.2.3 Small Intestinal Processing of Lipids

Digesta effluents passing through the pyloric sphincter flow into the small intestine, which is the key locus for digestion and absorption of nutrients. Therein, chyme is subjected to a milieu of changes: elevation of pH (to pH 6–7.5), mixing with various salts, pancreatic enzymes, coenzymes and bile components, namely bile salts and phospholipids [31, 32]. Moreover, gastric lipase may remain active in the intestine to some extent [33]. Thus, lipid breakdown in the intestine is a complex process of luminal and intestinal breakdown as well as various events of absorption and uptake. Inevitably, this is accompanied by numerous reactions and interactions with other nutritional and physiological components. In general, lipid droplets undergo biochemical breakdown via interfacial action of lipases, mediated by co-lipase, Ca + 2 levels and emulsifier displacement by bile acids [1, 15, 31, 34, 35]. These events lead to the generation of various hydrolytic products with acidic nature, such as surface active monoacylglycerols and free fatty acids [24, 36]. Such breakdown products are incorporated into structures of mixed micelles consisting of phospholipids or bile forming architectures of micelles, multi- or uni-lamellar vesicles, before being absorbed into circulation of lymphatic system [16, 25, 37]. Undigested and unabsorbed lipids may persist into the colon.

3.2.4 Colonic Processing of Lipids

Intestinal effluents transit through the ileocecal valve into the large intestine, where reabsorption of water and minerals as well as fermentation by the gut microflora occur [38]. Evidence suggest that high fat effluents may stimulate undesirable fatty diarrhea (Steatorrhea), as shown to be the case in the use of some lipase inhibitors [1]. Most of the relevant data is related to the fermentability and metabolism of lipophilic compound's, such as lipid-soluble vitamins and antioxidants. To date, the impact of lipids on the colon microbiota composition and metabolism is not well-established [14].

Overall, there is still a need to further elucidate the complex processes of lipid digestion. Thus it is no surprise that there are numerous platforms to interrogate the digestibility of lipids [21, 39–45]. The underlying hypotheses of such studies rely on a pharmacological LADME approach with recent efforts focused on the first bottle-necks for lipid digestion. These are liberation or bioaccessibility (defined as the amount of a consumed nutrient that is available for absorption) and bioavailability (defined as the rate and extent to which the active substance is absorbed and becomes available at the site of action) [46, 47]. To this end, the recent scientific literature is strewn with *in vitro* and *in vivo* studies.

3.3 Which *In Vitro* Models Are Used to Study Digestive Lipolysis?

Numerous *in vitro* studies have sought to underpin the governing principles directing emulsion behavior and breakdown in the human mouth, stomach and small intestine [1, 27, 28, 43, 48]. Such studies have applied different *in vitro* methods, from simple isolated methods that can be done using beakers to highly advanced models using bioreactors, tissue cultures, *ex-vivo* samples (e.g. Ussing chambers) and coupling them to advanced analytics [25, 30, 41, 49].

In vitro digestion systems can be roughly classified according to the physiological compartment simulated (e.g. gastric or gastro-intestinal models) or according to their mode of operation, i.e. static, semi-dynamic and dynamic models [39–41, 49, 50]. All of these models recreate various aspects and events of digestion, e.g. ionic composition, dynamics of pH and mixing, transit between organs and various physiological secretions. Since intestinal lipolysis is pivotal in lipid digestion, the literature is strewn with pH-stat methods aiming to study luminal lipolysis in the intestine via monitoring of the generation of free fatty acids [51]. This kind of study can be performed with a computer controlled auto-titrator maintaining the pH of digesting content, while lipase is hydrolyzing the sample. The bioreactor should also contain co-lipase, calcium and bile salts in order to allow process perpetuation. Tested sample is initially inserted into a bio- reactor containing suitable concentrations of digestive components in a neutral gut- typical pH. During lipolysis, free fatty acids are released, lowering solution pH which is quickly being re-adjusted with alkali as its needed volume is being recorded over reaction time. The output of such set-up will be the volume of titrated alkali (usually NaOH at a known and calibrated concentration) and that can be further translated into the released FFA [52, 53]. With this well-accepted procedure one can study the breakdown rate and extent of lipid digestion released fatty acids from the triacylglycerols [34, 51, 54]. Moreover, recent collaborative efforts done by the international INFOGEST network of scientists have published a harmonized protocol for *in vitro* simulation of digestion [55, 56].

Concomitantly, lipid uptake through epithelial barrier and the adsorption into the circulation have been studied using various cell cultures, organoids or *ex-vivo* samples. The fundamental model is commonly accomplished with monolayer Caco-2 cells, a human epithelial colorectal adenocarcinoma cell line. These cell cultures mimic the human intestinal epithelium and can be used to interrogate the absorption of lipid digestion products across the intestinal epithelium using trans-well set ups [21]. Today, attempts are made to improve the bio-relevance of such models by growing the cells in a three-dimensional enviroment [57] or by co-cultivation of various cell lines such as Caco-2/HT29-MTX (human intestinal goblet-type) [58], Caco-2/murine Peyer's patch lymphocytes [59] and Caco-2/THP-1 (human leukemia monocytic cell line) [60]. More examples as well as in depth explanations are available on in recent publications [61].Another model for intestinal absorption is the intestinal crypt organoids model, being a long-term primary multi-cellular cluster structures gastrointestinal organoids (or enteroids). Other studies depict additional ways to study digestive lipolysis including the everted gut sac, trans-mucosal perfusion through isolated mucosa, intestinal segment models and Ussing chambers [62–64]. More progress in recreating digestive lipolysis includes efforts to develop advanced *in silico* and "human-gut-on-a-chip" micro-devices [44, 64, 65].

Modeling the dynamic aspects of digestive events is increasingly described in the scientific literature. Historically, such experimental recreation of complex digestion events was initiated during the 1990s [45, 50]. Today, semi-dynamic and dynamic models can be found. The first are models recreating some of the dynamic aspects of digestion, for example, mastication events or gastric pH changes. The later are more advanced dynamic models that recreate the dynamics of more than a single physiological parameter, e.g. gastric pH changes, pulsatile mixing and gastric emptying. Such models are increasingly used by specialized labs and include a variety of models, such as the TNO models (TIM-1, tiny TIM and TIMagc) [66, 67], commercial Applikon bioreactor models [68, 69], the DIDGI® System [70], the dynamic gastric model (DGM) [71] and the human gastric simulator (HGS) [72]. All of these technological advancements enable better simulation of gastro-intestinal digestion and improved ability to underpin the determinants of bioaccessibility and bioavailability as well as characterize the properties of chyme. To date, there is sporadic information about oral digestion of lipids, despite the recognized importance. Recreation of oral digestion can be achieved using real saliva from volunteers [73–75] or the use a simulated salivary fluids (SSF) [55, 76] and the dynamic simulation of mastication is broadly reviewed by others [76–78].

Inevitably, *in vitro* models suffer from various drawbacks that limit their biorelevance. To this end, recreation of the complex biological feedback mechanisms, the involvement of the immune and endocrine systems as well as the adequate simulation of physiological secretions still require improvement. Particularly, imitation of gastric lipolysis is an open challenge. Review of the recent literature shows various commercial substitutes, e.g. rabbit gastric lipase, dog gastric lipase, and a recombinant human gastric lipase (rHGL) that could potentially help bridge the gap in our understanding [7, 79–81]. This stimulated attempts to evaluate, compare and use such enzymes to study gastric lipolysis *in vitro* [30, 33, 80–83].

To date, static *in vitro* digestion modes are frequently used as robust, high throughput and ethics-free screening tools. Interrogation of *in vitro in vivo* correlations show good agreement of findings at experimental "end-points" but sub-optimal correlation in respect to the kinetics of digestion events. The existing data torrents have also stimulated on-going efforts to generate *in silico* computational models with the aim of generating predictive models of digestion [44, 45, 64, 84]. On the other hand, dynamic *in vitro* digestion systems are more accurate and bio-relevant [85], yet, they are complex to operate, costly, offer lower throughput and are still not fully reliable replacements to clinical or *in vivo* trials.

3.4 What Have *In Vitro* Studies Taught Us?

Adapting an interfacial engineering approach, numerous studies have identified the droplet interface as a pivot to emulsion stabilization during shelf life as well as during digestion. Such studies reported the use of a multitude of emulsifiers, nanoparticles and structural design strategies to modulate droplet interfaces and their subsequent *in vitro* susceptibility to digestive events [86–91]. Therefore, recent studies shed light on the importance of lipid structure and processing technologies on the performance of lipids, e.g. pasteurized milk, infant formula and oil bodies [83, 92, 93].

In respect to oral behavior, studies highlight the important role of mastication and salivary mucins in release of lipids from the food matrix and emulsion flocculation [17, 20, 28, 94]. Mucin-induced flocculation has been demonstrated for various protein-stabilized emulsions and it is believed to stem from bridging flocculation of lipid droplets by the glycated mucins [19, 28, 94]. In fact, the reversibility of emulsion flocculation has even been linked to the sensory perception of emulsions [95]. In respect to biochemical breakdown, some scientists believe a lingual lipase may act on lipid intake in the mouth; yet, the topic is highly controversial, as no such enzyme has been successfully isolated from humans. Moreover, identified breakdown of lipids in mouth of neonates has been attributed to maternal lipases in mother's milk.

In respect to the gastric digestion of lipids, *in vitro* studies have explored the gastric behavior and breakdown of emulsions, mainly from the structural perspective rather than the chemical one (i.e. neglecting gastric lipolysis) [28, 41, 89, 96–98]. Overall, these studies highlight that gastric digestion of emulsions involves increase in particle sizes attributed to two competing mechanisms: droplet coalescence and/or flocculation (See illustration in Fig. 3.2).

On one hand, the dynamics of gastric acid secretion can affect droplet-droplet interactions, reduce electrostatic repulsion and in turn induce flocculation. On the other hand, biochemical degradation of emulsifiers, namely proteolysis by pepsin, can diminish levels of emulsifiers available for droplet stabilization. Consequently, studies have found that gastric conditions can accelerate droplet coalescence of protein-stabilized emulsions. In life, both of these competing mechanisms also compete with physiological gastric emptying which will expel emulsions from the stomach into the small intestine where other environmental stressors (e.g. pancreatic trypsin and bile) will continue the lipid digestion.

In respect to intestinal lipolysis, various models have been described in the literature with the INFOGEST digestion model [55] and the pH-stat method [51, 52] being the predominant ones in the scientific literature. Such methods have enabled elucidation of the various factors governing intestinal lipolysis, i.e. effects of droplet

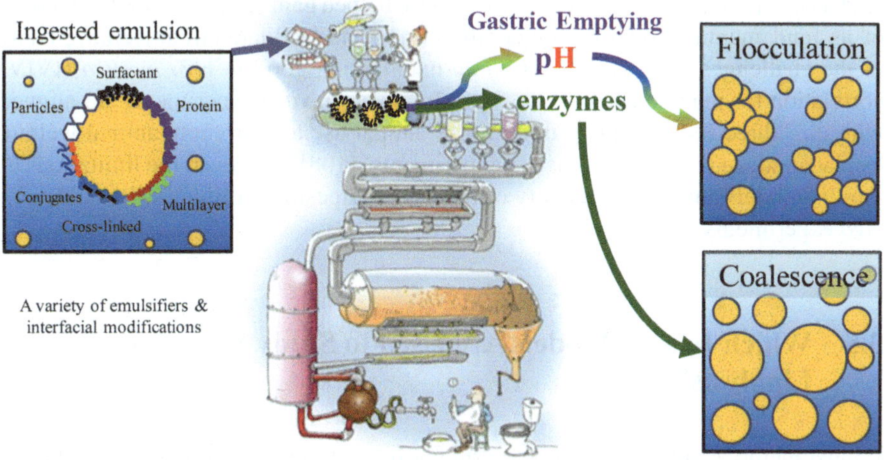

Fig. 3.2 Illustrative summary of *in vitro* insights of gastric behavior of lipid-based formulations

size, calcium, free fatty acids, glyco-conjugates, nano-laminates, interfacial cross-linking and even some food grade nano-particles [51, 53, 88, 90, 91, 96, 99–102]. Other studies also show the importance of co-lipase and bile acids in intestinal lipolysis [1, 103, 104]. Such studies have even enabled identification of natural compounds, like galactolipids, that may interfere with intestinal lipolysis and by altering the activity of pancreatic lipases at the water-lipid interface [105].

Persistence of lipids through the small intestine may lead to their colonic degradation. Thus, various *in vitro* methods of colonic fermentations [40] have been used to study the effects of lipids on colonic microbiota composition and metabolism [14]. To date, the role of lipids in colonic fermentation and health is less well established, although, some lipids have been found to elicit changes in the colon microbiota. For example, one study showed that lipid type leads to changes in *Bacteroides*, *Turicibactor*, and *Bilophila* while a corresponding fish oil-based diet enriched for *Bifidobacterium, Adlercreutzia, Lactobacillus, Streptococcus* and *Akkermansia muciniphila* [106]. Thus, further studies are needed to fully elucidate the ramifications of dietary lipids on the composition and functional capacity of the gut microbiota.

The key limitation of all the *in vitro* studies outlined herein is that they tend to neglect the action of various aspects of human digestion, be it endocrine feedbacks, immunological processes or other physiological capabilities of the human gut. This limits the bio-relevance of findings, however, without significantly compromising their correlation with *in vivo* data in regards to macronutrients [54]. One identified limitation of gastric model is the lack of gastric lipase. Therefore, recent studies have also explored the importance of this stereo-selective lipase to digestive lipolysis and its ramifications to downstream digestive lipolysis in the small intestine. Such studies already depict experimental setups to simulate the action of gastric lipase and the screening of viable commercial alternatives and methods for future research [7, 8, 33, 107].

Thus, *in vitro* studies of lipid digestion have helped gain significant understanding of its fundamentals. Such highly controlled models facilitate circumventing many of the challenges of *in vivo* trials while maintaining good *in vivo-in vitro* (IVIV) correlation for macronutrients [41, 54]. Moreover, *in vitro* experiments are aligned with the "three Rs" initiative to minimize animal use and promote animal welfare (replace/reduce refine animal use in experiments) [108]. However, the limitations of *in vitro* models and the complexity of the *in vivo* digestion of lipids still mandate *in vivo* experiments.

3.5 Which *In Vivo* Models Are Used to Study Digestive Lipolysis?

In vivo methods are irreplaceable, particularly as it pertains to clinical and health claims. In addition, *in vivo* experiments support the ability to produce more accurate *in vitro* protocols and extend the physiological conditions that can be duplicated, e.g. those found in the GI of infants or the elderly [109]. Thus, it is no surprise that rodents, dogs, pigs and various human patients have been part of studies into lipid digestion. Also, numerous methods have been applied in such studies, for example NMR, magnetic resonance imaging (MRI), Capsule endoscopy to invasive/destructive cases of animal cannulation and sacrifice. Despite their high bio-relevance, *in vivo* methods have their own limitations; one of key concerns is the poor correlation between animal and human experiments. For example, in rodents, oral lipid digestion is appreciable due to significant activity of salivary lipases [110]. However, human lingual lipases have not been isolated successfully so far, and their existence is controversial. Pharmaceutical research of lipid-based oral formulations is often performed in rodents and dogs. In that field, rats are frequently used for proof-of-concept studies and dog are more preferred as a comparable animal model. In respect to food, dogs digestion is considered to be well correlated to that in humans [111], however, recent studies on foods and lipid metabolism have been done in pigs and mini-pigs, due to various practical considerations [112, 113].

One of the recent efforts in developing *in vivo* models is focused on infants. Several animals have been proposed lately as comparable systems to newborns e.g. rats, mice, rabbits and guinea pigs, however, pigs were established as most suitable to this end [114, 115]. An important aspect of infant digestion is the minor production of pancreatic lipase that can result in mal-digestion and malabsorption of fat. Based on this ground, several investigations were made concerning neonate nutrition and Infant formula [115]. Despite the similarities in anatomical, physiological and metabolic features between newborn pig and newborn human, some claim that a piglet pancreas is capable of producing BSDL/BSSL as opposed to human infant who lacks of such enzymes and receives them from his mother's milk. As a result, another model was suggested of an exocrine pancreatic insufficient (EPI) piglet. EPI pigs also serve as a model to cystic fibrosis patients [114].

Contrary, human trials are the golden standard for nutritional research, however, these are extremely challenging ethically, scientifically, logistically and financially. Some relevant studies have been performed on metabolic pathogeneses, such as cystic fibrosis, pancreatic insufficiency and ileostomy patients; however, their relevance to healthy individuals is highly questionable. Research in healthy individuals has also been described in the literature, however, to a more limited extent. Altogether, human trials pose an analytical challenge and highly limit the analytical arsenal that can be used. Yet, relevant studies describe various methods, including monitoring of body indices (weight, BMI, fat mass, etc.), the use of isotopes as biomarkers (e.g. C^{13} breath testing), collection of bodily secretions (e.g. urine), drawing of blood and non-invasive platforms such as capsule endoscopy, ultrasound and MRI or even questioners.

Overall, *in vivo* animal studies are highly bio-relevant but they are ethically constrained' poorly reproducible and correlated to the human case [42]. Human trials are undoubtedly the best platform for investigation of lipid digestion, yet, humans are also the most challenging case and have the lowest throughput; thus, justifying the continued use of *in vitro* and animal models in research and development efforts.

3.6 What Have Human Trials Taught Us?

Age-related differences in lipolysis products generated from emulsions and the lipid oxidation rates of various FAs have been demonstrated in adults and children [34, 116]. More in depth insights have been generated via the collection of aspirates through naso-gastric intubations. These have enabled analysis of the content of various elements in human GIT juices, even in aspirates collected from the duodenum (the first section of the small intestine). More recently, MRI and computational tools have been used to understand gastric behavior of test emulsions, gastric emptying, secretion and motility dynamics of blood flow as well as physically localize lipids and other ingredients [117–119]. The improved ability to visualize and to image digestion processes generated new insights i.e. that gastric emptying rate was accelerated for phase separated emulsions in comparison to stable emulsions. Ultrasound has been used to study gastric emptying effects, pyloric flow, pyloric-duodenal interaction and various behavior patterns of the GI organs [120]. For instance, gallbladder contractions measured via ultrasound showed to be affected by Orlistat, a commercial lipase inhibitor [34].

Capsule endoscopy (e.g. Pillcam™) has allowed visualizing the gastrointestinal tract lining even in distal areas, with high detail and spatio-temporal resolution. In fact, its power has increased since it has become a gold standard in diagnostics of small bowl disorders. More development is still required to minimize capsule size to allow its use in children, allow external independent locomotion of the device [121] or chemical analysis of the pill's surroundings (e.g. by near IR sensors). Another application of this concept is the wireless motility and pH capsule (WMC). This smart pill is capable of measuring various features such as gastric emptying time, small bowel transit time, and colon transit time at a single study [122, 123]. Such a research scheme has been successfully applied to investigate protein digestion, e.g. in the case of alpha-lactalbumin [124]. Yet, no current research could be found on lipids or emulsions, perhaps due to practical consideration of camera obscuration. Studies performed in ileostomy patients seek to adapt a controlled clinical model for humans, however, evidence show such patients to have reduced gastric emptying rate and a higher terminal ileum bacterial counts [125, 126]. In fact, the ileostomy model was used to study resistant starch [127] and for modeling lipid bioavailability [75]. Yet, the extrapolation of these insights to healthy individuals should be done with great care.

3.7 What Does the Future Hold?

All in all, human trials are still the golden standard for lipid research, yet, they are very expensive, have strict regulations, ethical constraints and show significant variations between individuals as well as difficulties in compliance and controlling confounders, such as subject's diet and lifestyle. Due to these many limitations, *in vitro* models offer great potential to advance our fundamental understanding of lipid digestion. Careful and meticulous efforts are needed to correlate *in vitro* and *in vivo* findings. These are likely to facilitate prospective rational design of lipid-based systems and minimize the need for animal and human trials.

References

1. Wilde PJ, Chu BS (2011) Interfacial & colloidal aspects of lipid digestion. Adv Colloid Interf Sci 165:14–22
2. Sarkar A, Goh KKT, Singh H (2010) Properties of oil-in-water emulsions stabilized by β-lactoglobulin in simulated gastric fluid as influenced by ionic strength and presence of mucin. Food Hydrocoll 24:534–541
3. Carrière F (2016) Impact of gastrointestinal lipolysis on oral lipid-based formulations and bioavailability of lipophilic drugs. Biochimie 125:297–305
4. McClements DJ (2018) Recent developments in encapsulation and release of functional food ingredients: delivery by design. Curr Opin Food Sci 21:1–6
5. Feher J (2017) Quantitative human physiology, pp 821–833
6. N'Goma J-CB, Amara S, Dridi K, Jannin V, Carrière F (2012) Understanding the lipid-digestion processes in the GI tract before designing lipid-based drug-delivery systems. Ther Deliv 3:105–124
7. Sams L, Paume J, Giallo J, Carrière F (2016) Relevant pH and lipase for in vitro models of gastric digestion. Food Funct 7:30–45
8. Armand M (2007) Lipases and lipolysis in the human digestive tract: where do we stand? Curr Opin Clin Nutr Metab Care 10:156–164
9. Lowe ME (2000) Properties and function of pancreatic lipase related protein 2. Biochimie 82:997–1004

10. Hui DY (2012) Phospholipase A2 enzymes in metabolic and cardiovascular diseases. Curr Opin Lipidol 23:235–240
11. Murakami M, Taketomi Y, Sato H, Yamamoto K (2011) Secreted phospholipase A_2 revisited. J Biochem 150:233–255
12. Mu H, Hoy CE, Høy CE (2004) Lipid-based formulations for oral administration of poorly water-soluble drugs. Prog Lipid Res 43:105–133
13. Bohn T, McDougall GJ, Alegría A, Alminger M, Arrigoni E, Aura A-M, Brito C, Cilla A, El SN, Karakaya S, Martínez-Cuesta MC, Santos CN (2015) Mind the gap-deficits in our knowledge of aspects impacting the bioavailability of phytochemicals and their metabolites-a position paper focusing on carotenoids and polyphenols. Mol Nutr Food Res 59:1307–1323
14. Danneskiold-Samsøe NB, Barros HD, Santos R, Bicas JL, Cazarin CB, Madsen L, Kristiansen K, Pastore GM, Brix S, Junior MR (2019) Interplay between food and gut microbiota in health and disease. Food Res Int 115:23–31
15. Reis P, Holmberg K, Watzke H, Leser ME, Miller R (2009) Lipases at interfaces: a review. Adv Colloid Interf Sci 147–148:237–250
16. Phan S, Salentinig S, Prestidge CA, Boyd BJ (2014) Self-assembled structures formed during lipid digestion: characterization and implications for oral lipid-based drug delivery systems. Drug Deliv Transl Res 4:275–294
17. Vingerhoeds MH, Blijdenstein TBJ, Zoet FD, van Aken GA (2005) Emulsion flocculation induced by saliva and mucin. Food Hydrocoll 19:915–922
18. Carpenter GH (2013) The secretion, components, and properties of saliva. Annu Rev Food Sci Technol 4:267–276
19. Sarkar A, Goh KKT, Singh H (2009) Colloidal stability and interactions of milk-protein-stabilized emulsions in an artificial saliva. Food Hydrocoll 23:1270–1278
20. Vingerhoeds MH, Silletti E, de Groot J, Schipper RG, van Aken GA (2009) Relating the effect of saliva-induced emulsion flocculation on rheological properties and retention on the tongue surface with sensory perception. Food Hydrocoll 23:773–785
21. Marze S (2017) Bioavailability of nutrients and micronutrients: advances in modeling and in vitro approaches. Annu Rev Food Sci Technol 8:35–55
22. Van Vliet T (2010) Colloidal aspects of texture perception. Food Eng Ingredients 35:38–42
23. Mackie A, Macierzanka A (2010) Colloidal aspects of protein digestion. Curr Opin Colloid Interface Sci 15:102–108
24. McClements DJ, Decker EA (2007) Controlling lipid bioavailability using emulsion-based delivery systems. In: McClements DJ (ed) Understanding and controlling the microstructure of complex foods. Woodhead Publishing Limited, New York, pp 483–500
25. Vors C, Capolino P, Guérin C, Meugnier E, Pesenti S, Chauvin M-A, Monteil J, Peretti N, Cansell M, Carrière F, Michalski M-C (2012) Coupling in vitro gastrointestinal lipolysis and Caco-2 cell cultures for testing the absorption of different food emulsions. Food Funct 3:537
26. Chahinian H, Snabe T, Attias C, Fojan P, Petersen SB, Carrière F (2006) How gastric lipase, an interfacial enzyme with a Ser-His-Asp catalytic triad, acts optimally at acidic pH. Biochemistry 45:993–1001
27. Singh H, Ye A (2013) Structural and biochemical factors affecting the digestion of protein-stabilized emulsions. Curr Opin Colloid Interface Sci 18:360–370
28. Singh H, Sarkar A (2011) Behaviour of protein-stabilised emulsions under various physiological conditions. Adv Colloid Interf Sci 165:47–57
29. Golding M, Wooster TJ, Day L, Xu M, Lundin L, Keogh J, Clifton P, Cliftonx P (2011) Impact of gastric structuring on the lipolysis of emulsified lipids. Soft Matter 7:3513
30. Levi CS, Goldstein N, Portmann R, Lesmes U (2017) Emulsion and protein degradation in the elderly: qualitative insights from a study coupling a dynamic in vitro digestion model with proteomic analyses. Food Hydrocoll 69:393–401
31. Singh H, Ye AQ, Horne D (2009) Structuring food emulsions in the gastrointestinal tract to modify lipid digestion. Prog Lipid Res 48:92–100
32. McClements DJ, Decker EA (2009) In: Designing functional foods. Elsevier, pp 502–546

33. Bakala-N'Goma JC, Williams HD, Sassene PJ, Kleberg K, Calderone M, Jannin V, Igonin A, Partheil A, Marchaud D, Jule E, Vertommen J, Maio M, Blundell R, Benameur H, Müllertz A, Pouton CW, Porter CJH, Carrière F (2015) Toward the establishment of standardized in vitro tests for lipid-based formulations. 5. Lipolysis of representative formulations by gastric lipase. Pharm Res 32:1279–1287
34. Golding M, Wooster TJ (2010) The influence of emulsion structure and stability on lipid digestion. Curr Opin Colloid Interface Sci 15:90–101
35. Armand M, Pasquier B, Andre M, Borel P, Senft M, Peyrot J, Salducci J, Portugal H, Jaussan V, Lairon D (1999) Digestion and absorption of 2 fat emulsions with different droplet sizes in the human digestive tract. Am J Clin Nutr 70:1096–1106
36. Kalantzi L, Goumas K, Kalioras V, Abrahamsson B, Dressman JB, Reppas C (2006) Characterization of the human upper gastrointestinal contents under conditions simulating bioavailability/bioequivalence studies. Pharm Res 23:165–176
37. Bauer E, Jakob S, Mosenthin R (2005) Principles of physiology of lipid digestion. Asian-Australasian J Anim Sci 18:282–295
38. Yao M, Xiao H, Mcclements DJ (2014) Delivery of lipophilic bioactives: assembly, disassembly, and reassembly of lipid nanoparticles. Annu Rev Food Sci Technol 5:53–81
39. Guerra A, Etienne-Mesmin L, Livrelli V, Denis S, Blanquet-Diot S, Alric M (2012) Relevance and challenges in modeling human gastric and small intestinal digestion. Trends Biotechnol 30:591–600
40. Payne AN, Zihler A, Chassard C, Lacroix C (2012) Advances and perspectives in in vitro human gut fermentation modeling. Trends Biotechnol 30:17–25
41. Marze S, Choimet M (2012) In vitro digestion of emulsions: mechanistic and experimental models. Soft Matter 8:10982–10993
42. Hur SJ, Lim BO, Decker EA, McClements DJ (2011) In vitro human digestion models for food applications. Food Chem 125:1–12
43. Marze S (2015) Bioaccessibility of lipophilic micro-constituents from a lipid emulsion. Food Funct 6:3218–3227
44. Marze S (2015) Refining in silico simulation to study digestion parameters affecting the bioaccessibility of lipophilic nutrients and micronutrients. Food Funct 6:114–123
45. Muttakin S, Moxon TE, Gouseti O (2019) In vivo, in vitro, and in silico studies of the GI tract. In: Interdisciplinary approaches to food digestion. Springer International Publishing, Cham, pp 29–67
46. Lacatusu I, Badea N, Niculae G, Bordei N, Stan R, Meghea A (2014) Lipid nanocarriers based on natural compounds: an evolving role in plant extract delivery. Eur J Lipid Sci Technol 116:1708–1717
47. Parada J, Aguilera JM (2007) Food microstructure affects the bioavailability of several nutrients. J Food Sci 72:R21–R32
48. Walker R, Decker EA, McClements DJ (2015) Development of food-grade nanoemulsions and emulsions for delivery of omega-3 fatty acids: opportunities and obstacles in the food industry. Food Funct 6:41–54
49. McClements DJ, Li Y (2010) Review of in vitro digestion models for rapid screening of emulsion-based systems. Food Funct 1:32–59
50. Yoo JY, Chen XD. Int J Food Eng
51. Li Y, Hu M, McClements DJ (2011) Factors affecting lipase digestibility of emulsified lipids using an in vitro digestion model: proposal for a standardised pH-stat method. Food Chem 126:498–505
52. Li Y, Mcclements DJ (2010) New mathematical model for interpreting pH-stat digestion profiles: impact of lipid droplet characteristics on in vitro digestibility. J Agric Food Chem 58:8085–8092
53. Meshulam D, Lesmes U (2013) Responsiveness of emulsions stabilized by lactoferrin nanoparticles to simulated intestinal conditions. Food Funct 5:65–73
54. Bohn T et al (2017) Crit Rev Food Sci Nutr:1–23

55. Minekus M, Alminger M, Alvito P, Ballance S, Bohn TO, Bourlieu C, Carriere F, Boutrou R, Corredig M, Dupont D, Dufour C (2014) A standardised static in vitro digestion method suitable for food–an international consensus. Food Funct 5:1113
56. Brodkorb A, Egger L, Alminger M, Alvito P, Assunção R, Ballance S, Bohn T, Bourlieu-Lacanal C, Boutrou R, Carrière F, Clemente A (2019) INFOGEST static in vitro simulation of gastrointestinal food digestion. Nat Protocol 14:991–1014
57. Kim HJ, Ingber DE (2013) Gut-on-a-Chip microenvironment induces human intestinal cells to undergo villus differentiation. Integr Biol 5:1130
58. Hilgendorf C, Spahn-Langguth H, Regårdh CG, Lipka E, Amidon GL, Langguth P (2000) Caco-2 versus caco-2/HT29-MTX co-cultured cell lines: permeabilities via diffusion, inside- and outside-directed carrier-mediated transport. J Pharm Sci 89:63–75
59. Tyrer P, Ruth Foxwell A, Kyd J, Harvey M, Sizer P, Cripps A (2002) Validation and quantitation of an in vitro M-cell model. Biochem Biophys Res Commun 299:377–383
60. Chanput W, Mes JJ, Wichers HJ (2014) THP-1 cell line: an in vitro cell model for immune modulation approach. Int Immunopharmacol 23:37–45
61. Verhoeckx K, Cotter P, López-Expósito I, Kleiveland C, Lea T, Mackie A, Requena T, Swiatecka D, Wichers H (2015) The impact of food bioactives on health: in vitro and ex vivo models. Springer Nature, Cham
62. Luo Z, Liu Y, Zhao B, Tang M, Dong H, Zhang L, Lv B, Wei L (2013) Ex vivo and in situ approaches used to study intestinal absorption. J Pharmacol Toxicol Methods 68:208–216
63. Jain AS, Dhawan VV, Sarmento B, Nagarsenker MS (2016) In vitro and ex vivo evaluations of lipid anti-cancer nanoformulations: insights and assessment of bioavailability enhancement. AAPS PharmSciTech 17:553–571
64. Lefebvre DE, Venema K, Gombau L, Valerio LG Jr, Raju J, Bondy GS, Bouwmeester H, Paul Singh R, Clippinger AJ, Collnot E-M, Mehta R, Stone V (2015) Utility of models of the gastrointestinal tract for assessment of the digestion and absorption of engineered nanomaterials released from food matrices. Nanotoxicology 9:523–542
65. Kim HJ, Huh D, Hamilton G, Ingber DE (2012) Human gut-on-a-chip inhabited by microbial flora that experiences intestinal peristalsis-like motions and flow. Lab Chip 12:2165
66. Minekus M, Marteau P, Havenaar R, Huisintveld JHJ (1995) Multicompartmental dynamic computer-controlled model simulating the stomach and small intestine. ATLA 23:197–209
67. Bellmann S, Lelieveld J, Gorissen T, Minekus M, Havenaar R (2016) Development of an advanced in vitro model of the stomach and its evaluation versus human gastric physiology. Food Res Int 88:191–198
68. Elashoff JD, Reedy TJ, Meyer JH (1982) Analysis of gastric emptying data. Gastroenterology 83:1306–1312
69. Levi CS, Lesmes U (2014) Bi-compartmental elderly or adult dynamic digestion models applied to interrogate protein digestibility. Food Funct 5:2402–2409
70. Ménard O, Picque D, Dupont D (2015) In: The impact of food bioactives on health. Springer International Publishing, Cham, pp 73–81
71. Vardakou M, Mercuri A, Barker SA, Craig DQM, Faulks RM, Wickham MSJ (2011) Achieving antral grinding forces in biorelevant in vitro models: comparing the USP dissolution apparatus II and the dynamic gastric model with human in vivo data. AAPS PharmSciTech 12:620–626
72. Kong FB, Singh RP (2010) A human gastric simulator (HGS) to study food digestion in human stomach. J Food Sci 75:E627–E635
73. Chen C, Zhang B, Fu X, You L-J, Abbasi AM, Liu RH (2016) The digestibility of mulberry fruit polysaccharides and its impact on lipolysis under simulated saliva, gastric and intestinal conditions. Food Hydrocoll 58:171–178
74. Grundy MML, Grassby T, Mandalari G, Waldron KW, Butterworth PJ, Berry SEE, Ellis PR (2015) Effect of mastication on lipid bioaccessibility of almonds in a randomized human study and its implications for digestion kinetics, metabolizable energy, and postprandial lipemia. Am J Clin Nutr 101:25–33. https://doi.org/10.3945/ajcn.114.088328

75. Grassby T, Mandalari G, Grundy MML, Edwards CH, Bisignano C, Trombetta D, Smeriglio A, Chessa S, Ray S, Sanderson J, Berry SE, Ellis PR, Waldron KW (2017) In vitro and in vivo modeling of lipid bioaccessibility and digestion from almond muffins: the importance of the cell-wall barrier mechanism. J Funct Foods 37:263–271. https://doi.org/10.1016/j.jff.2017.07.046
76. Morell P, Hernando I, Fiszman SM (2014) Understanding the relevance of in-mouth food processing. A review of in vitro techniques. Trends Food Sci Technol 35:18–31
77. Woda A, Mishellany-Dutour A, Batier L, François O, Meunier J-P, Reynaud B, Alric M, Peyron M-A (2010) Development and validation of a mastication simulator. J Biomech 43:1667–1673
78. van Aken GA, Vingerhoeds MH, de Hoog EHA (2007) Food colloids under oral conditions. Curr Opin Colloid Interface Sci 12:251–262
79. Ele Ekouna J-P, Boitel-Conti M, Lerouge P, Bardor M, Guerineau F (2017) Enhanced production of recombinant human gastric lipase in turnip hairy roots. Plant Cell Tissue Organ Cult 1:601–610
80. Capolino P, Guérin C, Paume J, Giallo J, Ballester J-M, Cavalier J-F, Carrière F (2011) In vitro gastrointestinal lipolysis: replacement of human digestive lipases by a combination of rabbit gastric and porcine pancreatic extracts. Food Dig 2:43–51
81. Sassene PJ, Fanø M, Mu H, Rades T, Aquistapace S, Schmitt B, Cruz-Hernandez C, Wooster TJ, Müllertz A (2016) Comparison of lipases for in vitro models of gastric digestion: lipolysis using two infant formulas as model substrates. Food Funct 7:3989–3998
82. Chatzidaki MD, Mateos-Diaz E, Leal-Calderon F, Xenakis A, Carrière F (2016) Water-in-oil microemulsions versus emulsions as carriers of hydroxytyrosol: an in vitro gastrointestinal lipolysis study using the pHstat technique. Food Funct 7:2258–2269
83. Bourlieu C, Ménard O, De La Chevasnerie A, Sams L, Rousseau F, Madec MN, Robert B, Deglaire A, Pezennec S, Bouhallab S, Carrière F, Dupont D (2015) The structure of infant formulas impacts their lipolysis, proteolysis and disintegration during in vitro gastric digestion. Food Chem 182:224–235
84. Giang TM, Gaucel S, Brestaz P, Anton M, Meynier A, Trelea IC, Le Feunteun S (2016) Dynamic modeling of *in vitro* lipid digestion: individual fatty acid release and bioaccessibility kinetics. Food Chem 194:1180–1188
85. Dupont D, Alric M, Blanquet-Diot S, Bornhorst G, Cueva C, Deglaire A, Denis S, Ferrua M, Havenaar R, Lelieveld J, Mackie AR, Marzorati M, Menard O, Minekus M, Miralles B, Recio I, Van den Abbeele P (2019) Can dynamic in vitro digestion systems mimic the physiological reality? Crit Rev Food Sci Nutr 59:1546–1562
86. Sarkar A, Ademuyiwa V, Stubley S, Esa NH, Goycoolea FM, Qin X, Gonzalez F, Olvera C (2018) Pickering emulsions co-stabilized by composite protein/polysaccharide particle-particle interfaces: impact on in vitro gastric stability. Food Hydrocoll 84:282–291
87. Marku D, Wahlgren M, Rayner M, Sjöö M, Timgren A (2012) Characterization of starch Pickering emulsions for potential applications in topical formulations. Int J Pharm 428:1–7
88. Shimoni G, Shani Levi C, Levi Tal S, Lesmes U (2013) Emulsions stabilization by lactoferrin nano-particles under in vitro digestion conditions. Food Hydrocoll 33:264–272
89. Zhang R, Zhang Z, Zhang H, Decker EA, McClements DJ (2015) Influence of emulsifier type on gastrointestinal fate of oil-in-water emulsions containing anionic dietary fiber (pectin). Food Hydrocoll 45:175–185
90. Tzoumaki MV, Moschakis T, Scholten E, Biliaderis CG (2013) In vitro lipid digestion of chitin nanocrystal stabilized o/w emulsions. Food Funct 4:121–129
91. Tokle T, Lesmes U, Decker EA, McClements DJ (2012) Impact of dietary fiber coatings on behavior of protein-stabilized lipid droplets under simulated gastrointestinal conditions. Food Funct 3:58–66
92. de Oliveira SC, Bourlieu C, Ménard O, Bellanger A, Henry G, Rousseau F, Dirson E, Carrière F, Dupont D, Deglaire A (2016) Impact of pasteurization of human milk on preterm newborn in vitro digestion: gastrointestinal disintegration, lipolysis and proteolysis. Food Chem 211:171–179

93. Gallier S, Acton D, Garg M, Singh H (2017) Natural and processed milk and oil body emulsions: bioavailability, bioaccessibility and functionality. Food Struct 13:13–23
94. Chang YG, McClements DJ (2016) Characterization of mucin–lipid droplet interactions: influence on potential fate of fish oil-in-water emulsions under simulated gastrointestinal conditions. Food Hydrocoll 56:425–433
95. Vingerhoeds MH, de Wijk RA, van Aken GA (2011) Textural perception of liquid emulsions: role of oil content, oil viscosity and emulsion viscosity. Food Hydrocoll 25:789–796
96. McClements DJ (2018) Enhanced delivery of lipophilic bioactives using emulsions: a review of major factors affecting vitamin, nutraceutical, and lipid bioaccessibility. Food Funct 9:22–41
97. Meyer JH, Elashoff JD, Lake R (1999) Gastric emptying of indigestible versus digestible oils and solid fats in normal humans. Dig Dis Sci 44:1076–1082
98. Moreau H, Bernadac A, Gargouri Y, Benkouka F, Laugier R, Verger R (1989) Immunocytolocalization of human gastric lipase in chief cells of the fundic mucosa. Histochemistry 91:419–423
99. Janssen RH, Vincken JP, Van Den Broek LAM, Fogliano V, Lakemond CMM, Agric J (2017) Nitrogen-to-protein conversion factors for three edible insects: Tenebrio molitor, Alphitobius diaperinus, and Hermetia illucens. Food Chem 65:2275–2278
100. Lesmes U, McClements DJ (2012) Controlling lipid digestibility: response of lipid droplets coated by β-lactoglobulin-dextran Maillard conjugates to simulated gastrointestinal conditions. Food Hydrocoll 26:221–230
101. Mun S, Decker EA, Park Y, Weiss J, McClements DJ (2006) Influence of interfacial composition on in vitro digestibility of emulsified lipids: potential mechanism for chitosan's ability to inhibit fat digestion. Food Biophys 1:21–29
102. Meshulam D, Slavuter J, Lesmes U (2014) Behavior of emulsions stabilized by a hydrophobically modified inulin under bio-relevant conditions of the human gastro-intestine. Food Biophys 9:416–423
103. Chu BS, Gunning AP, Rich GT, Ridout MJ, Faulks RM, Wickham MSJ, Morris VJ, Wilde PJ (2010) Adsorption of bile salts and pancreatic colipase and lipase onto digalactosyldiacylglycerol and dipalmitoylphosphatidylcholine monolayers. Langmuir 26:9782–9793
104. Salentinig S, Sagalowicz L, Leser ME, Tedeschi C, Glatter O (2011) Transitions in the internal structure of lipid droplets during fat digestion. Soft Matter 7:650–661
105. Chu BS, Rich GT, Ridout MJ, Faulks RM, Wickham MSJ, Wilde PJ (2009) Modulating pancreatic lipase activity with galactolipids: effects of emulsion interfacial composition. Langmuir 25:9352–9360
106. Caesar R, Tremaroli V, Kovatcheva-Datchary P, Cani PD, Bäckhed F Crosstalk between gut microbiota and dietary lipids aggravates WAT inflammation through TLR signaling. Cell Metab. https://doi.org/10.1016/j.cmet.2015.07.026
107. Carriere F, Renou C, Lopez V, De Caro J, Ferrato F, Lengsfeld H, De Caro A, Laugier R, Verger R (2000) The specific activities of human digestive lipases measured from the in vivo and in vitro lipolysis of test meals. Gastroenterology 119:949–960
108. Hooijmans CR, Leenaars M, Ritskes-Hoitinga M (2010) A gold standard publication checklist to improve the quality of animal studies, to fully integrate the three Rs, and to make systematic reviews more feasible. Altern Lab Anim 38:167–182
109. Shani-Levi C, Alvito P, Andrés A, Assunção R, Barberá R, Blanquet-Diot S, Bourlieu C, Brodkorb A, Cilla A, Deglaire A, Denis S (2017) Extending in vitro digestion models to specific human populations: perspectives, practical tools and bio-relevant information. Trends Food Sci Technol 60:52–63
110. Qi K, Seo T, Jiang Z, Carpentier YA, Deckelbaum RJ (2006) Triglycerides in fish oil affect the blood clearance of lipid emulsions containing long-and medium-chain triglycerides in mice. J Nutr 136:2766–2772
111. Hauss DJ (2007) Oral lipid-based formulations. Adv Drug Deliv Rev 59:667–676

112. Sciascia Q, Daş G, Metges CC (2016) The pig as a model for humans: effects of nutritional factors on intestinal function and health. J Anim Sci 94:441–452
113. Delosière M, Santé-Lhoutellier V, Chantelauze C, Durand D, Thomas A, Joly C, Pujos-Guillot E, Rémond D, Comte B, Gladine C, Guy A, Durand T, Laurentie M, Dufour C (2016) Quantification of 4-hydroxy-2-nonenal-protein adducts in the in vivo gastric digesta of mini-pigs using a GC-MS/MS method with accuracy profile validation. Food Funct 7:3497–3504
114. Goncharova K, Pierzynowski SG, Grujic D, Kirko S, Szwiec K, Wang J, Kovalenko T, Osadchenko I, Ushakova G, Shmigel H, Fedkiv O, Majda B, Prykhodko O (2014) A piglet with surgically induced exocrine pancreatic insufficiency as an animal model of newborns to study fat digestion. Br J Nutr 112:2060–2067
115. Menard O, Cattenoz T, Guillemin H, Souchon I, Deglaire A, Dupont D, Picque D (2014) Validation of a new in vitro dynamic system to simulate infant digestion. Food Chem 145:1039–1045
116. McCue MD, Welch KC (2016) 13 C-breath testing in animals: theory, applications, and future directions. J Comp Physiol B 186:265–285
117. Szarka LA, Camilleri M (2009) Stomach dysfunction in diabetes mellitus: emerging technology and pharmacology. Am J Physiol Gastrointest Liver Physiol 296:G461–G475
118. Liu D, Parker HL, Curcic J, Schwizer W, Fried M, Kozerke S, Steingoetter A. The visualisation and quantification of human gastrointestinal fat distribution with MRI: a randomised study in healthy subjects. Br J Nutr. https://doi.org/10.1017/S0007114515005188
119. Logan K, Wright AJ, Goff HD (2015) Correlating the structure and in vitro digestion viscosities of different pectin fibers to in vivo human satiety. Food Funct 6:63–71
120. Dietrich CF, Braden B (2009) Sonographic assessments of gastrointestinal and biliary functions. Best Pract Res Clin Gastroenterol 23:353–367
121. Ciuti G, Menciassi A, Dario P (2011) Capsule endoscopy: from current achievements to open challenges. IEEE Rev Biomed Eng 4:59–72
122. Saad RJ (2016) The wireless motility capsule: a one-stop shop for the evaluation of GI motility disorders. Curr Gastroenterol Rep 18:14
123. Maqbool S, Parkman HP, Friedenberg FK (2009) Wireless capsule motility: comparison of the SmartPill® GI monitoring system with scintigraphy for measuring whole gut transit. Dig Dis Sci 54:2167–2174
124. Benard C, Cultrone A, Michel C, Rosales C, Segain J-PP, Lahaye M, Galmiche J-PP, Cherbut C, Blottière HM (2010) Degraded carrageenan causing colitis in rats induces TNF secretion and ICAM-1 upregulation in monocytes through NF-κB activation. PLoS One 5:e8666
125. Robertson MD, Mathers JC (2000) Gastric emptying rate of solids is reduced in a group of ileostomy patients. Dig Dis Sci 45:1285–1292
126. Booijink CCGM, El-Aidy S, Rajilić-Stojanović M, Heilig HGHJ, Troost FJ, Smidt H, Kleerebezem M, De Vos WM, Zoetendal EG (2010) High temporal and inter-individual variation detected in the human ileal microbiota. Environ Microbiol 12:3213–3227
127. Edwards CH, Grundy MM, Grassby T, Vasilopoulou D, Frost GS, Butterworth PJ, Berry SE, Sanderson J, Ellis PR (2015) Manipulation of starch bioaccessibility in wheat endosperm to regulate starch digestion, postprandial glycemia, insulinemia, and gut hormone responses: a randomized controlled trial in healthy ileostomy participants. Am J Clin Nutr 102:791–800

Part II
Lipid Metabolism

Chapter 4
Oral Processing of Lipids

Miodrag Glumac, Xinmiao Wang, and Jianshe Chen

4.1 Introduction

Key to understanding how we perceive fat is of great importance in fighting overweightness and more important obesity, as fat represents the most energy dense nutrient. Fat has the lowest satiation value per calorie, behind carbon hydrates and proteins which have the highest [94]. Obesity has been a dominant concern for decades in western countries, and is also on the rise in developing ones and represents a global trend. This is reflected in the average rise in BMI from 1980's to present day [41]. Moreover, projected number of overweight and obese individuals for 2030 is expected to be 2.16 and 1.12 billion, respectively [59]. Obesity and overweightness represent a risk factor for development of cancer [33], they correlate with development of diabetes [72] and kidney complications, [13] amongst many other diseases.

Understanding how we perceive fat is crucial for the future of human food consumption. Bridging the gap between nutritional challenges of food containing fat and its perception aiming towards a healthier eating behavior. Human perception is a highly complex system which utilizes different strategies to obtain a stimulus (or few stimuli) when perceiving fat in food. Description of these processes requires knowledge from several disciplines, just to name key ones: food science, physics, chemistry, biology, medicine and psychology. Discovering the underlying mechanisms of fat perception will enable us to make healthier lower caloric food which will continue to be appreciated during eating by the consumer.

M. Glumac · X. Wang · J. Chen (✉)
School of Food Science and Biotechnology, Laboratory of Food Oral Processing, Zhejiang Gongshang University, Hangzhou, People's Republic of China
e-mail: jschen@zjgsu.edu.cn

M. M.-L. Grundy, P. J. Wilde (eds.), *Bioaccessibility and Digestibility of Lipids from Food*, https://doi.org/10.1007/978-3-030-56909-9_4

4.2 General Aspects of Food Oral Processing

4.2.1 The Mechanical Actions of Food Processing

Food oral processing is one of the most important steps during food consumption [105]. The oral cavity is the place where the food first makes direct contact with the consumer. Two major factors that play a role during food consumption are individuality of consumers and characteristics of food properties. The first refers to variations of oral physiology, while the second refers food compositions and textural properties, both of them play a role in food oral processing and sensory perception [18]. There are four phases of food oral processing, first bite, mastication, transportation and swallowing [63]. Food is broken down in the human mouth during the eating process, it functions as a sophisticated device with several purposes. First is a food processor for reducing size, cutting, mixing, shearing and smearing food. This creates a food bolus which represents a broken down food ready for swallowing with saliva being incorporated in it. Second is the sensing functions that are made possible by taste buds and mechanoreceptors. Third is transportation function that is shown in a proper control of the bolus flow by opening and closing of the "valves". All these actions associated with food oral processing, produce various stimuli which are detected by different sensing channels. Thus a comprehensive stimulus is formed from the sensory perception and appreciation as a direct result of these actions. This occurs during the whole eating process [19]. As mentioned, mechanical destruction of food is the most important and common mechanisms of oral breakdown. During this process food is reduced from initial shape and size to a much smaller size through chewing and biting. Closing of the jaws, teeth shredding function, alongside tongue movement and pressure are all essential components of this mechanism. Size of food particles gets reduced in order of few magnitudes of length from centimeter to sub-millimeter and even micrometer scale at the swallowing point. Oral destabilization also influences sensory perception of beverages where emulsions droplets are used for carrying flavor and texture modification [20]. Texture is sensed and appreciated during whole mastication process. Food deformation, breaking, size distribution and geometry of fractured particles, and the surface wetting will all contribute to texture sensation. The bulk deformation or fracturing of large particles is sensed in the early stages and here the rheology behavior of food destruction and deformation is highly important. After this stage and with continuation of size reduction, rheology is replaced by tribological behavior and this plays a leading role of textural sensation during the later stage of food oral processing [22].

When dealing with dietary lipids it's important to know that they can be present as bulk oil/fat, or in the form of emulsions, which are responsible for the very different oral behaviour and sensory properties they might provide. When perceiving emulsions based liquid food it's important to know that, if the system gets destabilized after oral processing a newly formed microstructure will lead to a textural experience completely different from the previous stable emulsion. Stable emulsion is perceived as creamy, but a flocculated can lead to coalescence of oil droplets and is perceived as oily and greasy [103]. Moreover, during eating different fat containing food in semi-solid and solid states are converted into a swallowable mass that represents a bolus. This enables food to be swallowed as it was broken down and lubricated during chewing. Many factors influence bolus lubrication properties, including saliva incorporation, dry mater of the bolus and oil released from the food matrix. Saliva is incorporated more into bolus as the time of mastication is prolonged and helps lubrication as it lowers friction due to salivary proteins [28]. Fat in food is sensed by different mechanisms, with detection of aroma/smell (olfactory), taste (gustatory), and texture (somatosensory) properties. In a recent review, Michalski et al. laid out a comprehensive picture about the importance of food structures containing dietary lipids [71]. They concluded that dietary lipids have their physical and chemical properties that are manifested from the molecular, supramolecular and macroscopic scope thus displaying a huge diversity (Fig. 4.1). All of these will influence the nature and behavior of food containing lipids [70]. Macroscopic scale is more important in the very initial stages of the eating process.

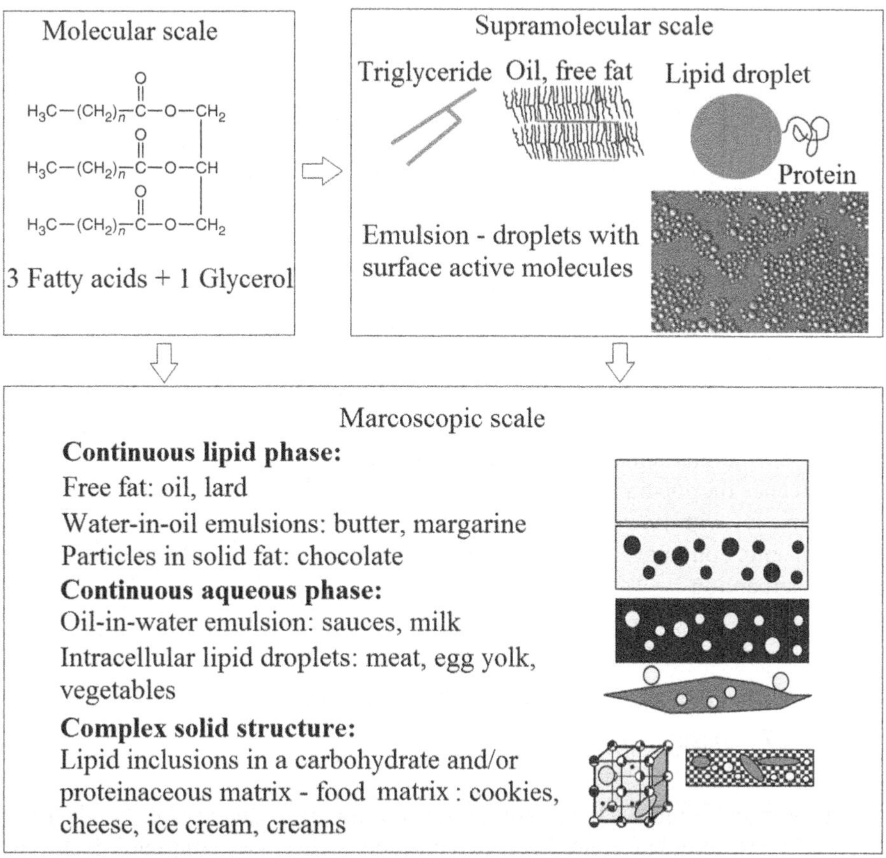

Fig. 4.1 Summary of the various molecular, supramolecular and macroscopic structures of lipids in food products. (Modified from Michalski et al. [71])

Table 4.1 Overview of dietary lipids in food industry. [52]

Type of dietary lipids	Sub-division	Type
Vegetable oils	Produced in high quantity	Palm Soybean Canola Sunflower
	Produced in low quantity	Cottonseed Peanut Olive Corn Coconut
Animal fats		Lard Tallow Fish oil Chicken fat
Food matrix	Dairy	Cream Butter Cheese Milk chocolate Ice cream
	Non-dairy	Margarine Salad oils Salad cream Mayonnaise

However, after food starts to get broken down and mixed with saliva the molecular and supramolecular organization and distribution will also play a role.

Dietary lipids were categorized as oils in liquid state and fats in solids state, and they can be of animal or plant origin. Food matrices (dairy and non-dairy) are more complex structures, in which fat forms emulsions with water [52] (Table 4.1). The importance of fat presence in food is largely due to it providing essential fatty acids, fat soluble vitamins and most importantly having high energy density. Dietary lipids also influence the apperance of a food product, as well as its texture and flavour. Food containing fat with its composition and lipid distribution will be orally processed via mechanic, colloidal and enzymatic mechanisms all playing a role influencing its texture perception.

4.2.2 Saliva Influences on the Oral Processing of Dietary Lipids

Saliva is an important factor during oral processing of fat containing food. It is a natural human biological fluid secreted into the oral cavity continuously by the major and minor salivary glands. Major glands include sublingual, submandibular and parotid, whilst minor glands are scattered inside the oral cavity.

Saliva is a very complex fluid mostly consisting of water (98–99%), and various organic and inorganic substances (1–2%) [2, 36], proteins and enzymes which interact with food components and then affect eating experience [10, 11, 108]. Human saliva is normally distinguished into two types, unstimulated saliva and stimulated saliva, with the later one being more dominated by secreted products from the parotid glands [106]. Moreover, protein and metabolite composition found in these different saliva types varies in both molecular size and quantity [47]. The property and functionality differences between the two types of saliva are reflected in the compositions, viscosity, lubrication, and microstructures found in saliva originating from different glands. Different types of stimulation (e.g. mechanical, gustatory and olfactory) will increase the secretion of SS into the oral cavity [100]. Type of stimulation had an effect of saliva secretion, with citric acid producing the highest amount (2.3 ml/min) [38]. Another study reported 5.5 ml/min on average saliva secretion stimulated by fruit flavored gum [23]. Also, substantial inter-individual differences have been reported for saliva secretion and composition [74]. By one division of salivary proteins there are six major groups, proline-rich proteins, mucins, amylases, cystatins, histatins and statherins with their corresponding roles. Salivary proteins could be classified into several major classes with different biological roles responsible for digestion and lubrication, maintaining calcium ion levels, antimicrobial functions, preventing tissue degradation, taste regulation, tannin neutralization and others [7, 15, 18, 32, 57, 107]. Although, the biological functions and roles during food consumption have been known for some specific salivary proteins [7], for many others, it is still unknown and uncertain [61].

One interesting property of saliva is the high surface activity that can affect its interaction with oil/fat. Food inside the oral cavity gets wetted and coated by a salivary layer. This capability is due to the reduction of the surface energy. Presence of surface active components like phospholipids [60] and proteins statherins [76] influence surface tension by lowering it (50 mN/m for unstimulated saliva) when compared to the surface tension of water (72 mN/m on 25 °C). The characteristic of saliva that could contribute to bolus formation is the viscoelastic property due to the mucin components. Inside the oral cavity saliva helps to prevent food adherence to the oral surfaces due to the present of the mucosa layer [14].

Mosca and Chen [73] in a recent review laid out mechanisms that govern food–saliva interactions on two levels, structural and molecular (Fig. 4.2). They concluded that the bulk properties of food play a smaller role compared to that of food–saliva mixtures in influencing its perception. Saliva will undoubtedly affect all foods consumed and change the perception of what is eaten. Lingual lipase is an enzyme that is present inside the mouth derived from salivary glands close to the taste buds [44]. The enzyme is found in small quantities and probably does not aid in fat digestion, but rather could have a role in fat tasting. Role of lingual lipase for pre-digestion is still debatable [44] because of its low quantity. Lipocalin, a protein from the lipophilic family [49], has a function of aroma detection, in saliva it could have role in acid transportation to the taste buds [44] but this is still uncertain. In one report, it was found that via orthonasal and retronasal way, fatty acid smell was distinguishable for linoleic, oleic and stearic acids. But these three weren't

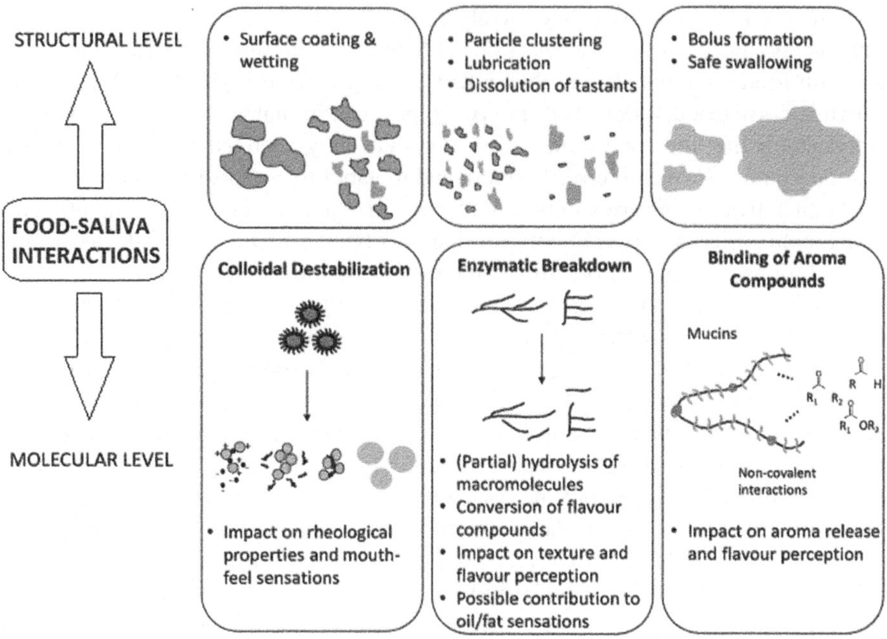

Fig. 4.2 Influence of saliva on food oral processing. (Modified from Mosca and Chen [73])

distinguished just via orally tasting [6]. Role of saliva in fat perception was explored in depth by Feron and Poette [40]. They concluded that saliva had some interesting roles such as: major factor in detection and perception of fat, large inter-individual variability of saliva composition that could influence fat sensory perception, positive and negative role of saliva in fat preference, lingual lipase role and activity. Based on the facts mentioned it is reasonable to conclude that saliva and its proteins are very important component when discussing food oral processing of fat rich foods because of emulsification, detection and wetting capabilities.

4.3 Oral Processing of Dietary Lipids

4.3.1 Oral Behavior of Free Lipids – Oral Emulsification

Facts dealing with ability of human saliva functioning as an effective emulsifier of free lipids and therefore oral emulsification could be an important mechanism for the oral processing of oil/fat, are addressed. Oral emulsification process is still largely unexplored, and literature reports dealing with it are rather scarce and few. One report concluded that bulk oil was emulsified to droplets of 20–30 μm after

30 seconds, with the movement of oil through dentures and between the tongue and soft palate being primary contributors of oral emulsification. Estimated shear forces applied to the droplets reached 50 Pa [24]. Bulk corn and castor oil were orally emulsified and droplets recorded with fluorescent video-endoscopy. Droplet size was in function of viscosity, where lower viscosity oil produced smaller droplets [1]. It was theorized that two phenomena can govern behavior of emulsion system deposited on the tongue surface. One is the influence of aggregation and coalescence, and another one being emulsification with saliva and shear forces inside the oral cavity [97].

Ongoing work is carried out to test the hypothesis that saliva is capable of emulsifying free lipids in the oral cavity [46]. The major focus points investigated were saliva's influence on free lipids, comparing and characterizing in vitro as well as in vivo emulsions, emulsification behavior of both plant oil and animal fat, and the roles of salivary proteins. Although there have been extensive debates about the sensory mechanisms of oil/fat, this work takes a step back by examining the oral behavior of free lipids. These preliminary findings show that saliva can actually function as emulsifier to oil/fat, which means that ingested oil/fat will be dispersed and converted into an emulsion before being swallowed.

4.3.1.1 Saliva Characteristics That Influence Free Lipid Emulsification

Saliva plays one of the main roles during oral processing of oil/fat, with surface activity being of key importance [58]. Saliva obtained from human participant was analyzed for flow rate, pH, surface tension, and total protein content. Flow rate and pH were comparable to other literature reports [45, 69, 75]. Surface tension was similar between unstimulated and stimulated saliva, however showed inter-individual difference with results observed close to those report by Gittings et al. [45]. Protein contents were also found to be significantly different between subjects, protein content in stimulated saliva is generally lower than that in unstimulated saliva, which is caused by more watery parotid saliva secretion after stimulation [18].

It is reasonable to suppose that the low surface tension of human saliva is the result of the presence of salivary proteins. By plotting the surface tension and protein contents for both unstimulated saliva and stimulated saliva samples, one could see a moderate correlation between the two (Fig. 4.3a). However, it is still unknown which specific molecular compounds influence the surface tension. Brauer et al. [8] isolated four groups of salivary gland proteins (A, B, C and D) which could have an ability to act as surfactants. Denny et al. [27] demonstrated that whole human saliva contains 1166 different proteins. Proteins are amphiphilic as that they can distribute on the interface of two immiscible fluids [61]. They can have hydrophobic and hydrophilic part, and they act as emulsifying agents that stabilise lipid droplets against coalescence or flocculation in food systems [68].

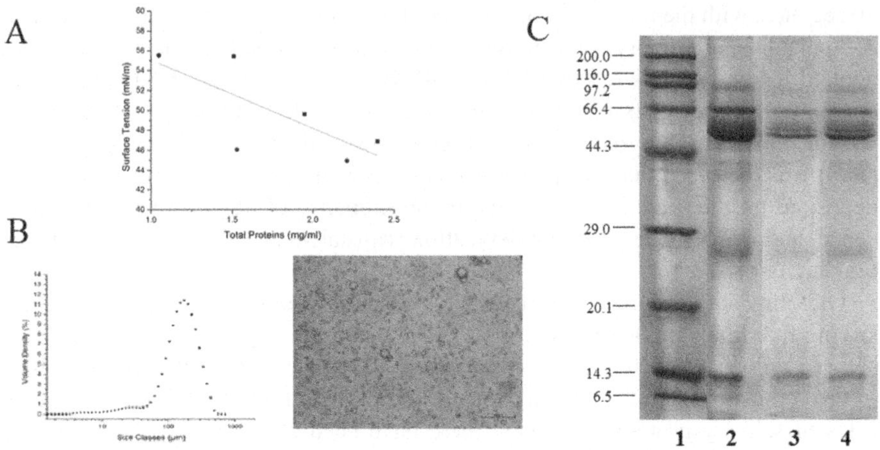

Fig. 4.3 (**a**) The surface tension of saliva shows a moderate correlation ($R^2 = 0.42$) to its protein content for both unstimulated (square) and stimulated (circle) salivas; (**b**) Emulsion made *in situ* after oral processing of pork belly: Particle size distribution (d_{32}); microscope observation (scale bar 100 μm); (**c**) SDS-PAGE lanes: 1. Protein markers (kDa); 2. Unstimulated saliva; 3. Aqueous phase of the *In vitro* sample (10% rapeseed oil; 4000 rpm on 1 minute); 4. Aqueous phase of the *in situ* sample (1 ml rapeseed oil; 1 minute mastication). (From Glumac et al. [46])

4.3.1.2 Saliva Emulsions Made In Vitro and In Vivo

Free lipid–salliva interactions have been investigated by making emulsions with mechanical mixing outside the oral cavity and orally by natural tongue moving against the hard palate. Also, emulsions stabilised by sodium caseinate were made and used as reference, as sodium caseinate is well researched colloidal stabilizer [30]. General trend found was that the mean droplet size was found to decrease with the increase of mixing speed (4000 vs. 1000 and 3000 rpm) when oil was used in lower percentage (5% vs. 10%), showing that enhanced shear condition and lower amount of oil is beneficial to produce a finer saliva emulsion. When comparing saliva to sodium caseinate, it is noticed that saliva is a weaker emulsifying agent judged by droplet size. Emulsions made with oral movements in different quantities of oil (0.25, 0.5 and 1.0 ml) had mean droplet size in order of few tens microns (d_{32}) and close to 100 microns (d_{43}). Zeta potential (ζ) a measure of density of surface charge which gives electric repulsion force once droplets approach each other is useful for emulsion stability characterisation. The higher the zeta potential in absolute values (+ or –), the stronger the repulsive forces and the more stable the emulsion [29]. In vitro emulsions had zeta potentials reaching −18 mV (5% oil) and −12 mV (10% oil), orally made emulsions had zeta potentials of −20 mV for all three cases (0.25, 0.5 and 1.0 ml oil), but sodium caseinate produces the most stable emulsions with zeta potentials reaching −55 mV (5% oil) and −44 mV (10% oil). Zeta potential of saliva emulsion weren't in absolute value as large as that of sodium caseinate emulsions, the electric charge at droplet surface was sufficiently large to give droplets protection against flocculation and coalescence [78].

4.3.2 *Orally Made Emulsions with Animal Fat*

Cooked animal fat (pork belly) was cut into a small piece and masticated continuously. Fat-saliva mixture was then spitted out after oral processing (chewing). Liquid phase was carefully separated out from the solid residues and was examined for droplet size distribution and microstructure observation. Results from samples of fat–saliva mixture is shown in Fig. 4.3b. A stable fat-in-water emulsion was clearly observable, with a mean droplet size of $d_{32} = 24$ μm and $d_{43} = 170$ μm, significantly smaller than that of rapeseed oil emulsions of made mechanically and orally. A much smaller emulsion droplet size could be due to the high shear conditions during thorough chewing of pork belly.

4.3.3 *Emulsification Role of Salivary Proteins*

Salivary proteins contribution to oral emulsification was analyzed by SDS-PAGE. Figure 4.3c shows the electrophoresis results, in which lane 1 is a lane of protein markers, lane 2 represents saliva sample, lane 3 is for proteins in the aqueous phase of the in vitro emulsion sample, and lane 4 is for protein in the aqueous phase of the orally made emulsion sample. One can see that various protein groups exist in human saliva, with molecular weights ranging from as low as few thousands to as high as close to 100 kDa, consistent with those reported by Esser et al. [39]. When comparing the protein bands of the two emulsions (lines 3 and 4) with that of saliva (line 2), one can notice some differences. Firstly the protein bands of the emulsion systems are much weaker than that of the corresponding bands of the saliva, indicating a much lower protein concentration in the continuous phase of the two emulsions. This is an indication that proteins are depleted from the saliva in order to form a monolayer to stabilise oil droplets. Secondly, the biggest depletion of protein bands are those at around 27.0 kDa and 55.0 kDa. It was also noticeable that proteins with molecular weight below 14.3 kDa were missing to a greater extent in lane 4 than in lane 3 sample (Fig. 4.3c). From these results, we can conclude that oral emulsification creates a mixed monolayer consisting of various salivary proteins. The biggest contribution to the monolayer formation could be those with medium molecular weights. These newly discovered facts surrounding oral emulsification lead to a conclusion that human saliva contains functionally appropriate amounts of salivary proteins leading to significant surface activity of the saliva. These findings reveal possibly a very different oral mechanism of oral processing of free lipids inside oral cavity, which can improve our understanding on the controlling mechanisms of fat perception.

4.3.4 Oral Behavior of Dispersed Lipids – Emulsion Destabilisation

Oral processing and behaviour of food containing dispersed lipids in the emulsified form has been extensively researched. Most important parameters that are addressed are droplet composition and structure, in mouth behaviour and their interaction with the oral cavity mainly focusing on destabilisation. When food starts being processed inside the mouth, its structure and behavior is going to be different from the state being ingested [97]. Sensory perception of these foods can be altered by influencing the emulsion characteristics mainly the interface region. Characteristics of these changes to structure are crucial for understanding the sensory properties and to understand mechanism that will govern food design. Perception will be influenced by structure, composition and behavior of emulsion food and droplet layer surface and emulsifier used. As mentioned before, fat containing food can be in liquid, semi-solid or solid state. Some liquid emulsion based foods such as milk, yogurt, shakes, soy-based beverages are actually oil-in-water emulsions. Their texture perception is affected by the status of protein–stabilized droplets, in such way that they influence viscosity and interact with oral cavity. By using different emulsifiers in food formulas it will directly impact behavior and when interacting with saliva with shear forces produced by the tongue also effecting the process. One categorization of the mechanisms influencing emulsions behavior inside mouth, can be on droplet aggregation, droplet adhesion to mucosal surfaces, droplet coalescence, and droplet spreading [99]. All these mechanisms can affect the texture of perceived food emulsions, with the later three tightly connected in their action.

Droplet aggregation occurs with all food grade emulsifiers, it's considered as a general phenomenon. Emulsion–saliva mixtures have an increased viscosity because of saliva induced aggregation of emulsion particles. The viscous layer at the tongue surface consists of aggregated droplets and saliva components [91, 103]. Emulsion destabilization could occur via mechanisms of Oswald ripening, creaming, flocculation and coalescence [29], however it is believed that in the oral environment, the flocculation and coalescence are the most likely mechanisms. Interaction forces that cause this aggregations of droplets are van der Waals forces, electrostatic forces, depletion, and molecular bridging [101]. These authors researched the electrostatic forces influence on the emulsion charge by analyzing emulsion–saliva mixtures. They found that strongly negatively charged ones didn't flocculate because repulsion forces overcame the depletion and van der Waals interactions. Neutral and weakly negatively charged emulsions flocculated but the effect was reversible with addition of water. Depletion interactions induced by mucins with van der Waals interaction and repulsion forces is the most likely reasons of this behavior. Emulsions with a positive charge will irreversibly flocculate as mucins building aggregates with droplets [16, 89, 90]. This aggregation behavior of lippid droplets increases viscosity and can also influence lubrication properties, whereas some thickeners just increase viscosity. Sarkar and Singh concluded that oral destabilization of food emulsions as a result of saliva mixing and oral shear, where the salt-induced aggregation, depletion flocculation, bridging flocculation, and coalescence were the four most important mechanisms [88]. Adhesion of droplets to mucosal surfaces is one of the possible outcomes when dealing with emulsion based food. Dresselhuis et al. [34] demonstrated that emulsions with less protein were more likely to adhere to tongue surface, with saliva playing a role in oil/fat surface retention. This gives rise to after–feel sensation associated with fatty attributes that happened after swallowing. The same researchers concluded that oral coalescence of a food emulsion was also a likely phenomenon affected by emulsion characteristics, oral shear force (applied by the tongue), and oral mucosa characteristics [35]. This retention behavior of emulsion droplets is probably due to a lower amount of proteins at the interface as it was higher when the stability of emulsions decreased. It is recognized that droplet size and distribution are also important factors of emulsion based food [104]. Droplet sizes of low viscosity emulsions had minor effect to fatty sensory characteristics like thickness, creaminess, fattiness, and taste after–feel [102]. It is reasonable to say that sensory perception of emulsions will be effected by droplet size as an indirectly as a consequence before mentioned processes like aggregation, adhesion to oral mucosa, and coalescence stability. Texture perception of liquid emulsion will be directly resulted because of the involvement of the mechanisms of droplet aggregation and change in viscosity, retention and deposition of lipids on tongue, coalescence [4] with droplet size saliva composition also playing a role.

Oral behavior of semi-solid and solid emulsion based food could be far more complicated than their liquid counterpart. In one study, it was discovered that texture perception of a semi-solid custard dessert was influenced by the presence of added small silica particles. Authors concluded based upon the results that the mechanism by which particles affect texture sensation was due to the lubricative properties of food when interacting with oral tissues [37]. Review that dealt with in-mouth behavior of semi-solid food concluded that different time of food processing will give a specific food attribute. Thickness relates to bulk properties and is sensed fast as the food is relatively intact. Fattiness is sensed in a degraded bolus on its surface and is sensed slower, with fatty after-feel being sensed only after swallowing [26]. Solid based food namely cheese products has also been a focus of investigation. Research was done for two types of cheeses, one with full fat and one with reduced fat content. The fat reduced cheese was described as waxy, fracturable, chewy, hard and springy while the full fat one was more sticky, cohesive, meltable and smooth. Fracture strain was not distinguished based on fat content, while the fracture stress values were lower in full fat cheese compared to those of equivalent fat reduced cheese. It was concluded that texture improvement of fat reduced cheese texture requires further understanding of chemical, structural and mechanical properties [53]. Four types of cheese were investigated each containing different fat content. Fracture stress and firmness increased as the fat level decreased, with viscosity and melt time increasing as well [51]. Review that dealt with sensory and mechanical aspect of cheese texture in which they concluded that sensory attributes that describe parameters such as breakdown pattern, adhesiveness and cohesiveness are weakly correlated with rheological properties of cheeses [42]. It was found that in cheeses the percentage of saliva incorporated into the soft cheese product type was 23% while the firmer cheese product had 52%. The increase of fat content per dry mater from 25% in low fat to 50% in high fat cheeses, actually decreased the percentage of saliva incorporation from 41% to 22% [77]. Solid, semi-solid and liquid foods display different behavior when comparing aggregation, lubrication and wetting properties before consuming, during and after mastication and when swallowing. More knowledge is needed from emulsion science to explain these complex relations between food properties and human perception.

4.4 Sensory Perspectives of Dietary Lipid Oral Processing – Fat Perception

Sensation of fat is believed to be achieved through chemical and mechanical perception, with detection of aroma, flavor and texture. Alongside these, visual [95] and audio [92] inputs contribute to form an integrated stimulus processed in the brain [98]. Nose and mouth are organs that play the main role in food perception as it's the place where digestion first starts and where sensory sensors are located. Tongue is important as it possesses the majority of papillae from which fungiform, foliate and circumvallate (contain taste buds) detect taste, while filiform papillae make tongue surface rough and help detect food texture. Five tastes humans can perceive are sweet, salt, sour, biter and umami. All the basic tastes can be viewed as adaptation for perceiving necessary food ingredients (carbon hydrates as sweet; glutamate as umami; sodium salts as salty) or harmful ones (toxins perceived as bitter; acids as sour). They transduce signals via VII, XI and X cranial nerves [9].

Possibility of a sixth fat sense named oleogustus was recently proposed by Running et al. [87]. Large numbers of different types of food contain fat in them. They are categorized as fat–containing food products, in which fat makes an important ingredient and its perception stimulates a pleasant neural response. Sensation of dietary fats is refereed by consumers as "creamy" or "smooth" in product containing milk fat [21]. Although creaminess perception is of great interest to food scientist, the topic is not the main focus of this chapter. For more information surrounding creaminess perception, one should refer to a recent comprehensive review deals with current understanding and facts surrounding it [31]. There are three main aspects which are important when investigating human fat perception. Volatile molecules detected by olfactory receptors in the nasal cavity giving insight to caloric value. Fat is perceived when free fatty acids released from triglycerides are detected by chemoreceptors on the tongue surface in taste buds. Somatosensory receptors detect fat texture and neurons perceive it as texture-specific signal. In brief, we shall discuss these three main aspect alongside giving insight into multimodal perception of fat in food with all cues forming a comprehensive stimulus.

4.4.1 Aroma Aspect of Fat Perception

Smell plays a huge role in the food perception process, and is important when perceiving fat. Aroma usually is the first cue important for determining food type and it gives indication of its caloric content as fat is the aroma compound carrier in most food systems. Brain interprets aroma when molecules are detected with olfactory tissue in the nose. Two routes for aroma detection are before consumption while smelling ortho-nasally and while eating retro-nasally. Interpersonal differences in saliva production, anatomy and eating habits could affect retro-nasal aroma perception [12]. Although there can be some differences in aroma profiles between individuals, each individual can be tuned for their own level of sensory stimuli that induces satiation and end of food intake [86]. Aroma intensity, quality and release rate are needed to be taken into consideration when dealing with fat rich food [25]. In order to make low-fat food the aroma delivery profile has to be very similar or to match the sensory experience of high-fat food products. Aroma plays a role in detecting dietary lipids in food, and will also attribute to sensory perception when interacting with other modalities such as taste and texture.

4.4.2 Taste Aspect of Fat Perception

Mounting evidence points to a chemical based aspect when detecting fat in the mouth [66]. Free fatty acids act as signaling molecules for receptors on the taste bud cell, triggering a molecular signaling cascade transduced to neural endings. This signal travels via VII (chorda tympani) and IX (glosso-pharyngeal) nerves to the brain. Fat has a crucial role in maintaining life functions, because fatty acids play a role in detection, ingestion, and digestion. Large number of facts point to a gustatory and olfactory component beside the dominant textural for detection of fat [64]. Chemo-sensitivity to different 18 C fatty acids was reported with detection of multiple senses. Fat seem to be detected with texture attributes masked, in mixtures with emulsifiers and thickeners. Linoleic, oleic and stearic acids were used for detection threshold tests and oxidized linoleic sensitivity acid was also investigated. Routes through which detection was achieved were ortho-nasal olfactory, retro-nasal olflactory, nasal irritancy, oral irritancy, gustatory, and multimodal route. Obtained results showed that 18 C fatty acids were detected with many different senses in humans. Retro-nasal route showed significantly higher threshold thus having lower sensitivity compared to other routes. Sensitivity was highest for oxidized linoleic acid which is due to them being unpleasant and possible evolutionary context as they can be unsafe nutrients [17]. In another report, different chained fatty acids in an array of concentrations were detected by humans. Four free fatty acids were tested, and threshold of caproic acid had lower compared to stearic acid and lauric acid.

Moreover, the linoleic acid threshold was lower to that of lauric. It was concluded that there is a possibility that humans have multiple mechanisms through which they detect free fatty acids of varying chain length [65].

Criteria by which to classify a taste is still elusive to determine, although there can be certain parameters taste sensation (oil/fat) could fulfill. Provide an adaptive value, fatty acids are essential to nutrition and fats are most nutrition dense food. Also, humans can sense rancid (oxidized) oil/fat that are harmful (cytotoxic and genotoxic). They must have a stimulus on the cell receptors, as free fatty acids could possibly be signaling molecules for fat taste in humans. Moreover, they must have a unique transduction pathways with correlating receptors. Receptors CD36 and GPR120 are capable of detecting free fatty acids of varying size and are present in the oral cavity thus making detection possible during eating. Initiation of peripheral signals detected by nerves that stimulate gustatory centers. Chorda tympani nerve conveys signals from fungiform and foliate papillae, while glossopharyngeal carry signals form foliate and circumvallate papillae in the taste buds. Perceptibility and uniqueness of oil/fat, even when non gustatory cues are minimized or eliminated, is present in the oral cavity. Although inter-individual variability for fat detection such as the possibility of fat and non-fat tasters, the existence of several orders of magnitude for fat thresholds detection is proving to be a key factor of uncertainty. This is partly because of using arbitrary units during testing. Evoking a physiological and behavioral response by oil/fat exposure in mouth has been shown in some cases. Alimentary lipemia is effected by oral exposure to oil/fats, even with stimulus's duration is short and in low volume [67]. It was concluded after many years of research and accumulated scientific evidence points to a unique taste of oil/fat. Though still in dispute oleogustus proposed by Running et al. it might be the sixth basic sense.

There may be a correlation between fat intake and sensitivity to free fatty acids. Hypersensitive persons usually have lower fat intake and BMI values while hyposensitive ones have higher. They were sorted in these two groups (hyper and hypo) by their ability to distinguish a varying range of C 18:1 oleic acid in custard. Hypersensitive persons were able to recognize the sample three times over the control sample while hyposensitive couldn't [93]. Detection of long chain fatty acids seems to be possible only if the chain length is adequate and existence of free carboxylic group. Medium chain fatty acids and short chain fatty acids are also detectable by humans in the fungiform and circumvallate papillae. Most notable proteins involving reception are GPR40 and GPR120 from the G protein-coupled receptor superfamily. The CD36 protein from the B scavenger receptor family is also of particular interest as it is able to transfer signals from attached long chain fatty acids to the taste bud cell. Consumption and detection of oil/fats can be related to the expression of gene for CD36, as it may play an important role in lipid-sensing system. Indirect cooperation hypothesis for long chain fatty acids detection was proposed by Besnard et al. [5]. It was concluded that free fatty acids are poorly soluble in saliva, but they can trigger responses in both the taste buds and somatosensory nerve endings [5]. In report Tucker et al. and based upon review of 7 studies of fat taste to non-esterified fatty acids, it was concluded that there was no significant association between fat taste (oleogustus) and increased BMI (obesity). The number of studies reported in literature is still limited with them having a small sample size [96]. In a population study that tried to estimate heritability of fat taste (oleogustus), experimental data from showed that there is growth in intensity rating for the higher concentrations of linoleic acid. They found evidence that 19% of the variability of taste response to linoleic acid relative to baseline (0% LA), is heritable at the highest concentration tested (0.38% LA) [43]. Taste plays an integral role in fat perception although its exact role is still not fully understood.

4.4.3 Texture Aspect of Fat Perception

Texture properties of oil/fat have been focus of investigation in neurological, perception and food science studies [80, 81]. Neurological studies showed existence of primary and secondary taste cortex in primate brain [83]. Specific taste and intensity of food substances are represented separately in the primary while the pleasantness of food is detected in secondary cortex [82]. Primate neurons respond to fat texture in the orbitofrontal cortex of the brain. Fat texture is detected by somatosensory nerves, neurons respond uni-modally to fat texture while others respond with convergence of taste and some with convergence olfactory [84]. Group of 112 neurons found close to each other and are located in the orbitofrontal taste area display unimodal activity for taste, olfactory and visual inputs by 34%, 13%, and 21%, respectively. Moreover, some showed convergence stimulated with taste and visual 13%, taste and olfactory 13%, and olfactory and visual 5% inputs. Texture is important indicator of oil present in the food, some neurons show activation by presence of cream and milk containing fat. They also showed response when being presented with paraffin and silicone oils. Feeding to satiety points decreases to response to null but when presented with different stimuli by glucose it shows activity [79]. Some neurons can respond to fat without responding to viscosity, others can respond to viscosity and fat based on viscosity of fat and they represent a channel that responds to viscosity from fat and non-fat stimulus. Finally, there are neurons that sense viscosity derived from non-oily sources [80]. Findings show that combination of high fat and pleasant flavor enhance responses in the brain area that responds to fat texture. However, when fat and a less pleasant taste are present in the mouth or when pleasant flavor and low fat values are present they are not active [48]. Most recent data suggest that neurons who respond to fat actually encode the coefficient of sliding friction of food or model compounds present in the mouth. Discovered alongside them, there are neurons who have reduced firing rates when compared to coefficient of sliding friction, so they respond less to oils and emulsions. These two opposing sensing mechanisms of coefficient of sliding friction detection are precise encoders who show whether there is fat texture in the mouth [85]. The evidence strongly suggest that fat is represented in the mouth by fat texture detected by the somatosensory system.

4.4.4 Multimodal Fat Perception

Multimodal perception of fat includes all the possible mechanisms that can be involved when consuming fat containing food. Aroma, taste, texture with viscosity, lubrication, temperature all playing a role. In one report, Heinze et al. discussed how various factors influence human fat perception, they divided them in four categories on sensory, genetic, personal and neuronal [56] (Fig. 4.4). They concluded that more research is needed to understand the connection between sensory detection, genetic and neuronal mechanisms of fat perception. In another, it was concluded that fat perception is a multimodal sensation that includes olfactory, gustatory and tactile modalities. The same authors also summarized processes that govern how spreads are broken down inside the mouth to three phases: melting, inversion and bolus formation [50]. Some evidence points to the fact that sensory and hedonic properties of sweet and fat can vary with BMI. Obese people have reduced sweetness perception that intensifies fat sensation, with also a higher preference for sweet and fat compared to the non-obese [3]. Similar study involving sugar/fat mixtures

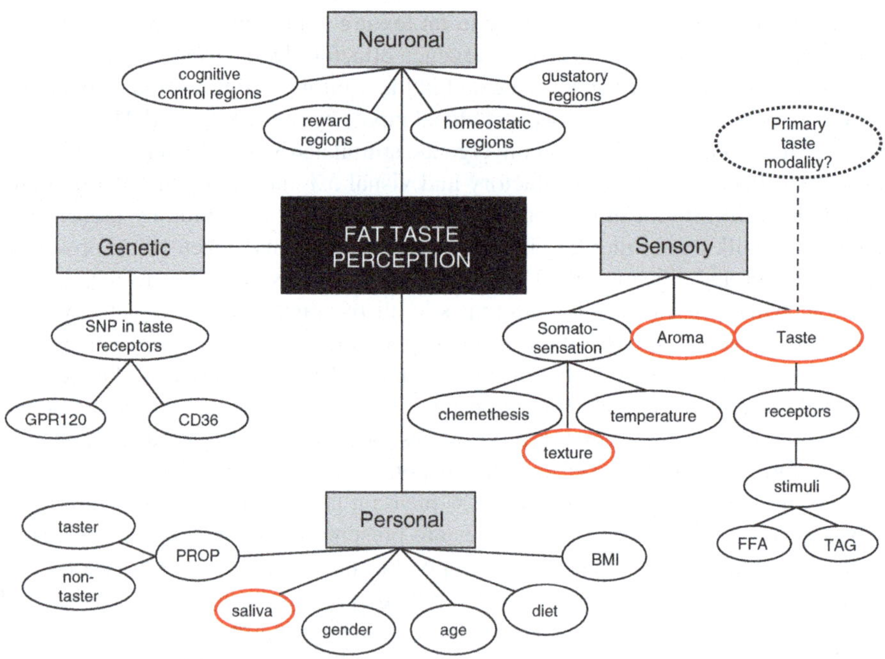

Fig. 4.4 Summary of the most prominent sensory, genetic, personal, and neuronal factors influencing fat taste perception. Factors highlighted in red were the focus of this book chapter. (Modified from Heinze et al. [56])

concluded that the taste phenotype affects the perception of sweetness and creaminess by suppressing or enhancing it [54]. Also, taste phenotype is affected by concentration of sugar and fat required for maximum hedonic liking, with most individual exhibit a better response with higher concentrations [55]. Perception of creaminess in fat-containing food was investigated with examining four food samples single cream, evaporated milk, corn starch and corn starch with added fatty acids. Test involved assessing sample's aroma, taste, taste and tactile, multimodal (all senses). Based on the results obtained, it was concluded that the more senses were involved the perception of creaminess increased. Food creaminess is perceived with several senses and is not based just on chemical or texture properties [21]. Research has been done by Le Calvé et al. [62] where they used model emulsions, milk, yoghurt and white sauce for fat discrimination tests. Also, important factors that influence fat perception were investigated. Rheological methods were used for all products and tribology for yoghurt in light of their physicochemical properties. Model emulsion reference sample had 10%, white sauce 15% and 30%, milk samples 0%, 2.7% and 3.8%, while the yoghurts had 0%, 0.5%, 1.75% and 3.5% fat content. It was noted that interactions between the aroma, taste and texture were crucial for distinguishing between samples. Moreover, relation between viscosity and perception of fattiness proved to be complex [62]. Fat perception mechanisms are subjected to ongoing investigations and are activley researched. Human fat perception is a highly complex multimodal sensory sensation that happens in the oral cavity, which is still not fully understood [109].

4.5 Conclusion

Food oral processing is a dynamic complex process. It involves complex mechanical movements inside the oral cavity where food gets broken down and mixed with saliva. Free lipids and fat-containing food (liquid, semi-solid and solid emulsions) will interact with saliva eliciting specific sensory perception. Fat perception has multiple aspects to it, aroma, taste, texture and multimodal one [109]. Aroma one deals with compounds in fat which are detected and that elicit a specific sensory response. Large amount of evidence points to the fact that free fatty acids activate taste channels and provide a sensory stimulus. Moreover, there is also strong evidence that fat is represented as texture in the mouth and detected and interpreted by neurons in the primary taste cortex in the brain. Multimodal perception of fat with all possible senses that can impact consumer perception while consuming these types of foods. Fat rich food can have complex structure that during oral processing will be detected in a multimodal way. More fundamental knowledge regarding human fat perception, is still needed for understanding perception of food products containing dietary lipids. This will enable food scientist to develop fat-reduced food formulas while maintaining preferable sensory atributes and consumer satisfaction.

References

1. Adams S, Singleton S, Juskaitis R, Wilson T (2007) In-vivo visualisation of mouth-material interactions by video rate endoscopy. Food Hydrocoll 21(5–6):986–995
2. Aps JKM, Martens LC (2005) Review: the physiology of saliva and transfer of drugs into saliva. Forensic Sci Int 150:119–131
3. Bartoshuk LM, Duffy VB, Hayes JE, Moskowitz HR, Snyder DJ (2006) Psychophysics of sweet and fat perception in obesity: problems, solutions and new perspectives. Philos Trans R Soc B 361(1471):1137–1148
4. Benjamins J, Vingerhoeds MH, Zoet FD, de Hoog EHA, van Aken GA (2009) Partial coalescence as a tool to control sensory perception of emulsions. Food Hydrocoll 23:102–115
5. Besnard P, Passilly-Degrace P, Khan NA (2016) Taste of fat: a sixth taste modality? Physiol Rev 96(1):151–176
6. Bolton B, Halpern BP (2010) Orthonasal and retronasal but not oral-cavity-only discrimination of vapor-phase fatty acids. Chem Senses 35(3):229–238
7. Brandão E, Soares S, Mateus N, Freitas V (2014) Human saliva protein profile: influence of food ingestion. Food Res Int 64:508–513
8. Brauer L, Meschter S, Beileke S, Jeger K, Garreis F, Paulsen FP (2009) Human parotid and submandibular glands express and secrete surfactant proteins A, B, C and D. Histochem Cell Biol. 132(3):331–338
9. Breslin PAS (2013) An evolutionary perspective on food and human taste. Curr Biol 23(9):409–418
10. Bridges J, Smythe J, Reddrick R (2017) Impact of salivary enzyme activity on the oral perception of starch containing foods. J Texture Stud 48:288–293
11. Brossard N, Cai H, Osorio F, Bordeu E, Chen J (2016) “Oral” tribological study on the astringency sensation of red wines. J Texture Stud 47:392–402
12. Buettner A, Beer A, Hannig C, Settles M, Schieberle P (2002) Physiological and analytical studies on flavor perception dynamics as induced by the eating and swallowing process. Food Qual Prefer 13(7–8):497–504
13. Calder CL, Ortega G, Vij A, Chawla K, Nnamdi-Emetarom C, Stephanie S, Callender CO, Fullum TM (2016) Morbid obesity is an independent risk factor for postoperative renal dysfunction in young adults: a review of the American College of Surgeons National Surgical Quality Improvement Program Database. Am J Surg 211(4):772–777
14. Carpenter GH (2012) Role of saliva in the oral processing of food. In: Food oral processing. Wiley-Blackwell, UK, pp 45–60 (Chapter 3)
15. Carpenter GH (2013) The secretion, components, and properties of saliva. Annu Rev Food Sci Technol 4:267–276
16. Çelebioğlu HY, Lee S, Chronakis IS (2019) Interactions of salivary mucins and saliva with food proteins: a review. Crit Rev Food Sci Nutr 11:1–20
17. Chalé-Rush A, Burgess JR, Mattes RD (2007) Multiple routes of chemosensitivity to free fatty acids in humans. Am J Physiol Gastrointest Liver Physiol 292(5):1206–1212
18. Chen J (2009) Food oral processing – a review. Food Hydrocoll 23(1):1–25
19. Chen J (2014) Food oral processing: some important underpinning principles of eating and sensory perception. Food Struct 1(2):91–105
20. Chen J (2015) Food oral processing: mechanisms and implications of food oral destruction. Trends Food Sci Technol 45(2):222–228
21. Chen J, Eaton L (2012) Multimodal mechanisms of food creaminess sensation. Food Funct 3:1265–1270
22. Chen J, Stokes JR (2012) Rheology and tribology: two distinctive regimes of food texture sensation. Trends Food Sci Technol 25(1):4–12
23. Dawes C, Kubieniec K (2004) The effects of prolonged gum chewing on salivary flow rate and composition. Arch Oral Biol 49(8):665–669

24. de Bruijne DW, Hendrickx HACM, Alderliesten L, de Looff J (1993) Mouthfeel of foods. In: Food colloids and polymers: stability and mechanical properties. Royal Society of Chemistry, Cambridge, UK, pp 204–213 (Chapter 3)
25. de Hoog EHA, Ruijschop RMAJ, Pyett SP, Kok PMT (2011) The functional attributes fat brings to food. In: Reducing saturated fats in foods. Wiley-Blackwell, UK, pp 29–46 (Chapter 2)
26. de Wijk RA, Janssen AM, Prinz JF (2011) Oral movements and the perception of semi-solid foods. Physiol Behav 104(3):423–428
27. Denny P, Hagen FK, Hardt M, Liao L, Yan W, Arellannom M et al (2007) The proteomes of human parotid and submandibular/sublingual gland salivas collected as the ductal secretions. J Proteome Res 7(5):1994–2006
28. Devezeaux de Lavergne M, Van de Velde F, Stieger M (2017) Bolus matters: the influence of food oral breakdown on dynamic texture perception. Food Funct 8(2):464–480
29. Dickinson E (1992) An introduction to food colloids. Oxford University Press, UK. (Chapter 4)
30. Dickinson E (2009) Hydrocolloids as emulsifiers and emulsion stabilizers. Food Hydrocoll 23(6):1473–1482
31. Dickinson E (2018) On the road to understanding and control of creaminess perception in food colloids. Food Hydrocoll 77:372–385
32. Dodds MWJ, Johnson DA, Yeh CK (2005) Health benefits of saliva: a review. J Dent 33(3):223–233
33. Donohoe CL, Lysaght J, O'Sullivan J, Reynolds JV (2017) Emerging concepts linking obesity with the hallmarks of cancer. Trends Endocrinol Metab 28(1):46–62
34. Dresselhuis DM, Stuart MAC, van Aken GA, Schipper RG, de Hoog EHA (2008) Fat retention at the tongue and the role of saliva: adhesion and spreading of 'protein-poor' versus 'protein-rich' emulsions. J Colloid Interface Sci 321:21–29
35. Dresselhuis DM, de Hoog EHA, Stuart MAC, Vingerhoeds MH, van Aken GA (2008) The occurrence of in-mouth coalescence of emulsion droplets in relation to perception of fat. Food Hydrocoll 22:1170–1183
36. Edgar WM (1992) Saliva: its secretion, composition and functions. Br Dent J 172:305–312
37. Engelen L, de Wijk RA, van Der Bilt A, Prinz JF, Janssen AM, Bosman F (2005) Relating particles and texture perception. Physiol Behav 86(1–2):111–117
38. Engelen L, van den Keybus PAM, de Wijk RA, Veerman ECI, Nieuw Amerongen AV, Bosman F, Prinz JF, van der Bilt A (2007) The effect of saliva composition on texture perception of semi-solids. Arch Oral Biol 52:518–525
39. Esser D, Alvarez-Llamas G, de Vries MP, Weening D, Vonk RJ, Roelofsen H (2008) Sample stability and protein composition of saliva: implications for its use as a diagnostic fluid. Biomark Insights 3:25–27
40. Feron G, Poette J (2013) In-mouth mechanism leading to the perception of fat in humans: from detection to preferences. The particular role of saliva. OCL - Oleagineux Corps Gras Lipides 20(2):102–107
41. Finucane MM, Stevens GA, Cowan MJ, Danaei G, Lin JK, Paciorek CJ, Singh GM, Gutierrez HR, Lu Y, Bahalim AN, Farzadfar F, Riley LM, Ezzati M (2011) National, regional, and global trends in body-mass index since 1980: systematic analysis of health examination surveys and epidemiological studies with 960 country-years and 9·1 million participants. Lancet 377(9765):557–567
42. Foegeding EA, Brown J, Drake MA, Daubert CR (2003) Sensory and mechanical aspects of cheese texture. Int Dairy J 13(8):585–591
43. Garneau NL, Nuessle TM, Tucker RM, Yao M, Santorico SA, Mattes RD, on behalf of The Genetics of Taste Lab Citizen Scientists (2017) Taste responses to linoleic acid: a crowdsourced population study. Chem Senses 42(9):769–775
44. Gilbertson TA (1998) Gustatory mechanisms for the detection of fat. Curr Opin Neurobiol 8(4):447–452

45. Gittings S, Turnbull N, Henry B, Roberts CJ, Gershkovich P (2015) Characterization of human saliva as a platform for oral dissolution medium development. Eur J Pharm Biopharm 91:16–24
46. Glumac M, Qin L, Chen J, Ritzoulis C (2018) Saliva could act as an emulsifier during oral processing of oil/fat. J Texture Stud 50(1):83–89
47. Gouveia-Figueira S, Nording ML, Öhman C, Öhman A, Figueira J, Lif Holgerson P (2017) Metabolite quantification by NMR and LC-MS/MS reveals differences between unstimulated, stimulated, and pure parotid saliva. J Pharm Biomed Anal 140:295–300
48. Grabenhorst F, Rolls ET (2014) The representation of oral fat texture in the human somatosensory cortex. Hum Brain Mapp 35(6):2521–2530
49. Grzyb J, Latowski D, Strzalka K (2006) Lipocalins – a family portrait. J Plant Physiol 163(9):895–915
50. Guichard E, Galindo-Cuspinera V, Feron G (2018) Physiological mechanisms explaining human differences in fat perception and liking in food spreads-a review. Trends Food Sci Technol 74:46–55
51. Guinee TP, Auty MAE, Fenelon MA (2000) The effect of fat content on the rheology, microstructure and heat-induced functional characteristics of Cheddar cheese. Int Dairy J 10(4):277–288
52. Gunstone FD (2008) Major sources of oils and fats; pp 11–25 (Chapter 2); Major edible uses of oils and fats; pp 119–136 (Chapter 8): In: Oils and fats in the food industry. Wiley-Blackwell, UK.
53. Gwartney EA, Foegeding EA, Larick DK (2002) The texture of commercial full-fat and reduced-fat cheese. J Food Sci 67(2):812–816
54. Hayes JE, Duffy VB (2007) Revisiting sugar-fat mixtures: sweetness and creaminess vary with phenotypic markers of oral sensation. Chem Senses 32(3):225–236
55. Hayes JE, Duffy VB (2008) Oral sensory phenotype identifies level of sugar and fat required for maximal liking. Physiol Behav 95(1–2):77–87
56. Heinze JM, Preissl H, Fritsche A, Frank S (2015) Controversies in fat perception. Physiol Behav 152:479–493
57. Hu S, Xie Y, Ramachandran P, Loo RRO, Li Y, Loo JA, Wong DT (2005) Large-scale identification of proteins in human salivary proteome by liquid chromatography/mass spectrometry and two-dimensional gel electrophoresis-mass spectrometry. Proteomics 5:1714–1728
58. Kazakov VN, Udod AA, Zinkovych II, Fainerman VB, Miller R (2009) Dynamic surface tension of saliva: general relationships and application in medical diagnostics. Colloids Surf B: Biointerfaces 74(2):457–461
59. Kelly T, Yang W, Chen C, Reynolds K, He J (2008) Global burden of obesity in 2005 and projections to 2030. Int J Obes 32(9):1431–1437
60. Kirkness JP, Christenson HK, Wheatley JR, Amis TC (2005) Application of the 'pull-off' force method for measurement of surface tension of upper airway mucosal lining liquid. Physiol Meas 26(5):677–688
61. Kupirovič UP, Elmadfa I, Juillerat MA, Raspor P (2017) Effect of saliva on physical food properties in fat texture perception. Crit Rev Food Sci Nutr 57(6):1061–1077
62. Le Calvé B, Saint-Léger C, Babas R, Gelin JL, Parker A, Erni P, Cayeux I (2015) Fat perception: how sensitive are we? J Texture Stud 46(3):200–211
63. Liu D, Deng Y, Sha L, Abul Hashem M, Gai S (2017) Impact of oral processing on texture attributes and taste perception. J Food Sci Technol 54(8):2585–2593
64. Mattes RD (2005) Fat taste and lipid metabolism in humans. Physiol Behav 86(5):691–697
65. Mattes RD (2009) Oral detection of short-, medium-, and long-chain free fatty acids in humans. Chem Senses 34(2):145–150
66. Mattes RD (2010) Fat taste in humans: is it a primary? In: Fat detection taste, texture, and post ingestive effects. CRC Press, Boca Raton, USA pp 167–193 (Chapter 7)
67. Mattes RD (2011) Accumulating evidence supports a taste component for free fatty acids in humans. Physiol Behav 104(4):624–631

68. McClements DJ (2016) Emulsion stability. In: 3rd Edition Food emulsions: principles, practice and techniques. CRC Press, Boca Raton, USA pp 289–382 (Chapter 7)
69. Méjean C, Morzel M, Neyraud E, Issanchou S, Martin C, Bozonnet S et al (2015) Salivary composition is associated with liking and usual nutrient intake. PLoS One 10:e0137473
70. Meynier A, Genot C (2017) Molecular and structural organization of lipids in foods: their fate during digestion and impact in nutrition. Oilseeds and Fats, Crops and Lipids 24(2):D202
71. Michalski MC, Genot C, Gayet C, Lopez C, Fine F, Joffre F, Vendeuvre JL, Bouvier J, Chardigny JM, Raynal-Ljutovac K (2013) Multiscale structures of lipids in foods as parameters affecting fatty acid bioavailability and lipid metabolism. Prog Lipid Res 52(4):354–373
72. Minges KE, Whittemore R, Weinzimer SA, Irwin ML, Redeker NS, Grey M (2017) Correlates of overweight and obesity in 5529 adolescents with type 1 diabetes: the T1D exchange clinic registry. Diabetes Res Clin Pract 126:68–78
73. Mosca AC, Chen J (2017) Food-saliva interactions: mechanisms and implications. Trends Food Sci Technol 66:125–134
74. Neyraud E, Prinz J, Dransfield E (2003) NaCl and sugar release, salivation and taste during mastication of salted chewing gum. Physiol Behav 79(4–5):731–737
75. Neyraud E, Palicki O, Schwartz C, Nicklaus S, Feron G (2012) Variability of human saliva composition: possible relationships with fat perception and liking. Arch Oral Biol 57(5):556–566
76. Proctor GB, Hamdan S, Carpenter GH, Wilde P (2005) A statherin and calcium enriched layer at the air interface of human parotid saliva. Biochem J 389(1):111–116
77. Repoux M, Labouré H, Courcoux P, Andriot I, Sémon É, Yven C, Feron G, Guichard E (2012) Combined effect of cheese characteristics and food oral processing on *in vivo* aroma release. Flavour Fragr J 27(6):414–423
78. Roland I, Piel G, Delattre L, Evrard B (2003) Systematic characterization of oil-in-water emulsions for formulation design. Int J Pharm 263(1–2):85–94
79. Rolls ET (2008) The representation of flavor in the brain. The Senses - A Comprehensive Reference. In: Olfaction and taste. Academic Press, San Diego, USA, 26(4):469–478 (Chapter 4)
80. Rolls ET (2010) Neural representation of fat texture in the mouth. In: Fat detection taste, texture, and post ingestive effects. CRC Press, Boca Raton, USA, pp 197–223 (Chapter 8)
81. Rolls ET (2011) The neural representation of oral texture including fat texture. J Texture Stud 42(2):137–156
82. Rolls ET (2013) Emotion and decision-making explained. Oxford University Press, United Kingdom, pp 224–275. (Chapter 5)
83. Rolls ET, Yaxley S, Sienkiewicz ZJ (1990) Gustatory responses of single neurons in the caudolateral orbitofrontal cortex of the macaque monkey. J Neurophysiol 64(4):1055–1066
84. Rolls ET, Critchley HD, Browning AS, Hernadi I, Lenard L (1999) Responses to the sensory properties of fat of neurons in the primate orbitofrontal cortex. J Neurosci 19(4):1532–1540
85. Rolls ET, Mills T, Norton AB, Lazidis A, Norton IT (2018) The neuronal encoding of oral fat by the coefficient of sliding friction in the cerebral cortex and amygdala. Cereb Cortex 28(11):4080–4089
86. Ruijschop RMAJ, Burgering MJM, Jacobs MA, Boelrijk AEM (2009) Retro-nasal aroma release depends on both subject and product differences: a link to food intake regulation? Chem Senses 34(5):395–403
87. Running CA, Craig BA, Mattes RD (2015) Oleogustus: the unique taste of fat. Chem Senses 40(7):507–516
88. Sarkar A, Singh H (2012) Oral behaviour of food emulsions. In: Food oral processing: fundamentals of eating and sensory perception. Wiley-Blackwell, Oxford, UK, pp 111–138 (Chapter 6)
89. Silletti E, Vingerhoeds MH, Norde W, van Aken GA (2007) The role of electrostatics in saliva-induced emulsion flocculation. Food Hydrocoll 21:596–606

90. Silletti E, Vingerhoeds MH, Norde W, van Aken GA (2007) Complex formation in mixtures of lysozyme-stabilized emulsions and human saliva. J Colloid Interface Sci 313(2):485–493
91. Silletti E, Vingerhoeds MH, van Aken GA, Norde W (2008) Rheological behavior of food emulsions mixed with saliva: effect of oil content, salivary protein content, and saliva type. Food Biophys 3(3):318–328
92. Spence C (2015) Eating with our ears: assessing the importance of the sounds of consumption on our perception and enjoyment of multisensory flavour experiences. Flavour 4:1–14
93. Stewart JE, Feinle-Bisset C, Golding M, Delahunty C, Clifton PM, Keast RSJ (2010) Oral sensitivity to fatty acids, food consumption and BMI in human subjects. Br J Nutr 104(1):145–152
94. Stubbs J, Ferres S, Horgan G (2000) Energy density of foods: effects on energy intake. Crit Rev Food Sci Nutr 40(6):481–515
95. Toepel U, Knebel JF, Hudry J, le Coutre J, Murray MM (2009) The brain tracks the energetic value in food images. NeuroImage 44(3):967–974
96. Tucker RM, Kaiser KA, Parman MA, George BJ, Allison DB, Mattes RD (2017) Comparisons of fatty acid taste detection thresholds in people who are lean vs. overweight or obese: a systematic review and meta-analysis. PLoS One 12(1):1–15
97. van Aken GA, de Hoog EHA, Vingerhoeds MH (2009) Oral processing and perception of food emulsions: the relevance for fat reduction in food. In: Designing functional foods. Measuring and controlling food structure breakdown and nutrient absorption. CRC Press, USA, pp 481–501 (Chapter 19)
98. van den Oever GJ, Busch J, van der Linden E, Smit G, Zuidam NJ (2009) Design of foods for the optimal delivery of basic tastes. In: Designing functional foods: measuring and controlling food structure breakdown and nutrient absorption. CRC Press, USA, pp 453–480 (Chapter 18)
99. van Vliet T, van Aken GA, de Jongh HHJ, Hamer RJ (2009) Colloidal aspects of texture perception. Adv Colloid Interf Sci 150(1):27–40
100. Veerman ECI, van den Keybus PAM, Vissink A, van Nieuw Amerongen A (1996) Human glandular salivas: their separate collection and analysis. Eur J Oral Sci 104:346–352
101. Vingerhoeds MH, Blijdenstein TBJ, Zoet FD, van Aken GA (2005) Emulsion flocculation induced by saliva and mucin. Food Hydrocoll 19(5):915–922
102. Vingerhoeds MH, van Aken GA, Nixdorf RR, Zoet FD, de Wijk RA (2007) How emulsion composition and structure affect sensory perception of low-viscosity model emulsions. Food Hydrocoll 22(4):631–646
103. Vingerhoeds MH, Silletti E, de Groot J, Schipper RG, van Aken GA (2009) Relating the effect of saliva-induced emulsion flocculation on rheological properties and retention on the tongue surface with sensory perception. Food Hydrocoll 23(3):773–785
104. Walstra P (2005) Fundamentals of interface and colloid science. Soft solids. Elsevier Academic Press, the Netherlands, pp 643–736 (Chapter 8)
105. Wang X, Chen J (2017) Food oral processing : recent developments and challenges. Curr Opin Colloid Interface Sci 28:22–30
106. Weber F, Barrantes A (2017) Real-time formation of salivary films onto polymeric materials for dental applications: differences between unstimulated and stimulated saliva. Colloids Surf B: Biointerfaces 154:203–209
107. Yakubov GE, Macakova L, Wilson S, Windust JHC, Stokes JR (2015) Aqueous lubrication by fractionated salivary proteins: synergistic interaction of mucin polymer brush with low molecular weight macromolecules. Tribol Int 89:34–45
108. Young AK, Cheong JN, Foster KD, Hedderley DI, Morgernstern MP, James BJ (2016) Exploring the links between texture perception and bolus properties throughout oral processing. Part 1: breakdown paths. J Texture Stud 47:461–473
109. Glumac M, Chen J (2020) Contribution analysis of sensory cues to oil/fat perception. J Sens Stud 35(3):e12566

Chapter 5
Physiological Aspects of Lipid Digestion: Luminal and Mucosal Events

Alan Mackie

5.1 Introduction

Lipids are important nutrients in their own right but can also act as carriers of other bioactive dietary compounds such as vitamins. They differ from other macro and some micronutrients in that they are at best only sparingly soluble in the aqueous environment of the gastrointestinal (GI) tract. As a result, the products of hydrolysis self-assemble in a rather uncontrolled way in the gut lumen; are disassembled in order to be absorbed and are then reassembled into chylomicrons in a more controlled way in the absorptive cell of the intestine. In this chapter, the behaviour of lipids in the gastrointestinal lumen as well interactions with the intestinal mucosa will be discussed. The association of cholesterol and saturated fats with obesity and cardiovascular disease has led to fat as being seen as a potentially problematic constituent of the diet despite the fact that there is little evidence for that link [1]. However, the replacement of lipids by carbohydrates, especially sugar sweetened beverages has simply changed the emphasis to a different set of metabolic disorders such as type 2 diabetes [2]. It should be remembered that food is not simply a collection of macronutrients but has structure that may be altered through digestion. This structure can then affect the extent and rate of nutrient release. Studies have shown that food structure can modify the nutritional properties of a food [3, 4]. For example, plant-based foods contain cellular structures that need to be broken down before the encapsulated nutrients and bioactive compounds can be released and absorbed. This breakdown may be achieved by processing (e.g., industrial processing and cooking) and by oral mastication. Despite this, the bioavailability of nutrients may be limited in some foods [5]. An example is almonds, for which the food matrix attenuates postprandial lipemia after consumption [3]. A slower release rate

A. Mackie (✉)
School of Food Science and Nutrition, University of Leeds, Woodhouse Lane, Leeds, UK
e-mail: A.R.Mackie@leeds.ac.uk

M. M.-L. Grundy, P. J. Wilde (eds.), *Bioaccessibility and Digestibility of Lipids from Food*, https://doi.org/10.1007/978-3-030-56909-9_5

of nutrients from naturally encapsulated systems was shown to increase satiety after consumption [6]. This paradigm may also apply to other foods such as dairy products [7].

5.2 The Gastrointestinal Digestion of Lipids

Digestion is normally considered to start in the mouth with the comminution of solid foods and mixing with saliva. Humans, unlike rodents, do not secrete lingual lipase and oral residence time is very limited [8]. Consequently, there is little lipid hydrolysis in the oral phase of digestion. Oral processing will effectively mix the components of the food eaten in a single bite into a bolus prior to swallowing. Once swallowed, the bolus passes down the oesophagus into the stomach where it is further digested. At this point, the behaviour of the lipid component of the food becomes less predictable. The gastric behaviour of the lipids in a food may vary significantly depending on the nature of the food and indeed the lipid. The primary form of lipid ingested is as triglycerides but significant amounts of other forms of lipids are also consumed such as phospholipids and galactolipids. Where lipids are insoluble, as in the case of most triglycerides, hydrolysis can only take place at the interface between the water where the enzyme resides and the oil that is the substrate. Thus, the available interfacial area is a strong determinant of the rate of hydrolysis. For triglycerides, a key factor for digestion is the stability of the lipid to a range of biophysical challenges such as flocculation, coalescence and phase separation (creaming). For example, the limited mixing in the main body of the stomach may allow lipid to phase separate [7]. Clearly, the conditions for this to happen depend on the size of the lipid droplets and the viscosity of the surrounding chyme. The acidic environment in combination with lipase and protease may lead to the depletion of the original interfacial layer leading to coalescence and then subsequently to phase separation [9]. In general, it is thought that some 10–20% of lipid may be hydrolysed in the gastric phase of digestion [10]. The limiting factor tends to be the solubility of the hydrolysis products as the typical gastric pH is below the pKa of fatty acids and they are not dissociated. This may also be true of long chain fatty acids in the small intestine, as even polyunsaturated C18 fatty acids have a pKa above pH 8 [11].

Regardless of gastric behaviour, lipid is not absorbed in the stomach and so passes through the pylorus into the duodenum where the pH is raised through the secretion of bicarbonate from Brunner's glands and the chyme mixes with bile and pancreatic enzymes. The lipid content of the chyme at any given time will be dependent on the lipid content of the food but also the extent of phase separation that occurred in the gastric phase. Free lipid will tend to float to the top of the stomach and be emptied last while lipid entrapped in denser material may be dragged to the antral region of the stomach and be emptied first [7]. This will be discussed in more detail later. In the small intestine, gastric lipase may still be active but it is joined by pancreatic enzymes and bile. Bile is primarily comprised of bile acids, phospholipids, cholesterol and bilirubin. The bile acids are either recycled from the distal ileum or formed from cholesterol in the liver and stored in the gall bladder. The combination of higher pH and the presence of bile salts helps to solubilise the lipid digestion products in the form of self-assembled mixed micelles. These range widely in size but are typically tens of nanometres in diameter. Mixing in the intestinal lumen is vital for transporting these micelles to the intestinal mucus layer where they are free to diffuse through to the underlying enterocytes for absorption of most of their constituents [12]. Bile acids are not absorbed in the proximal small intestine giving a consistent background concentration of bile acids that gradually decreases more distally.

Lipids are at best poorly soluble and are transported into the absorptive cells using a number of different mechanisms that are still rather poorly understood. Certainly, short and some medium chain fatty acids are passively absorbed by so called flip-flop diffusion through the cell membrane. However, a number of fatty acid transport proteins (FATPs) are also thought to be involved. Once in the cell the digestion process is reversed and the Golgi reform the fatty acids into triglycerides that are then incorporated into chylomicrons along with other oil soluble nutrients. These are subsequently exported from the basolateral side of the cell into the lymphatic system where they are made available more systemically. In addition to absorptive enterocytes, the intestinal mucosa also contains a range of other cell types. A small proportion of these (~1%) will be endocrine cells that secrete gastrointestinal hormones in response to binding of nutrients onto G-protein coupled receptors. These hormones have a range of physiological roles that are important in controlling digestion and appetite.

In the following sections we will look at a range of different factors that affect the processes described above. In particular, it is important to consider what food composition, structure and processing might be used to alter rates and patterns of digestion and absorption.

5.3 Factors Affecting Absorption

5.3.1 The Role of the Gastric Phase of Digestion

The gastric compartment acts as a storage vessel, allowing food to be gradually released into the small intestine in a well-controlled way. That release is controlled by a number of different factors such as the rheological / structural properties of the gastric chyme and the nutrient content of the chyme passing through the pylorus into the duodenum. During the gastric phase of digestion, lipids are exposed to gastric lipase that can hydrolyse up to 25% of the lipid in a meal. Human gastric lipase (HGL) is a 379-amino acid serine hydrolase (EC 3.1.1.3) with a molecular weight of 43 kDa [13]. Native HGL purified from human gastric secretions is highly N-glycosylated (15% w/w), which results in an increase in molecular weight to ~50 kDa and the existence of four major isoforms with isoelectric points ranging from 6.8 to 7.4 [14]. As you might expect, HGL is highly stable under acidic conditions including gastric secretions at pH values ranging from 7 down to 2. It is particularly stable between pH 3 and 5 with half-inactivation times >24 hours [15] but this stability decreases below pH 2 and above pH 7. The maximum activity of HGL is at pH 5.0–5.4 on long chain triacylglycerol (TAG). Thus, HGL is primarily active in the early phase of meal digestion while gastric pH is likely to be relatively high. This is typically during the first 50% of gastric residence [15].

The activity of gastric lipase is a function of concentration in the chyme and the local pH as outlined above but it is also a function of the available lipid surface area. Thus, hydrolysis rates will significantly decrease coalescence or phase separation of lipids. The creaming of lipid to the top of the stomach can also have an impact the timing of emptying into the small intestine. Gastric emptying is a function of caloric load and can be significantly altered by phase separation in the stomach [16, 17]. In a pioneering study of the gastric behaviour of acid stable versus acid unstable emulsions, the authors used MRI to show the effect of phase separation (creaming of lipid to the top of the stomach) on gastric emptying and gall bladder volume. The acid unstable emulsion was emptied much more quickly from the stomach than the acid stable meal. Although the main difference between the two meals was the phase separation of the acid unstable emulsion, this not only altered the spatial distribution of the emulsion droplets but also their size distribution. In this case, the droplets of the acid unstable emulsion stabilised by Span 80 would have coalesced, which would have lowered the interfacial area and increased the creaming rate as was observed. Under these circumstances, the lipid phase at the top of the gastric contents was emptied last but this also means that any hydrolysis by gastric lipase would necessarily have been very limited in comparison to the acid stable emulsion.

A number of studies have used imaging to determine gastric structuring of lipid emulsions and its effect on gastrointestinal function [18]. Using the idea of lipid partitioning, some studies have attempted to use different methods to control the gastric stability of emulsion systems [19, 20]. In these studies, both interfacial composition and oil composition were investigated. The key feature is the ability of a system to retain structure into the intestinal phase where most of the lipolysis and lipid absorption take place. For example, the use of different interfacial composition to control flocculation and coalescence may lead to significant phase separation and emulsion coarsening. However, if the gastric lipid is re-emulsified as it is sheared as a result of passing through the pylorus then the impact on lipolysis rate due to the reduction in surface area is likely to be lost. The use of solid fat seems to maintain the benefits and be consistently more slowly hydrolysed [19].

In a study by Armand et al. [21], emulsions with different droplet size (0.7 μm vs 10 μm) were fed to healthy participants. Although the droplet size of the fine emulsion increased in the stomach it retained smaller size droplets and its lipolysis was greater than that of the coarse emulsion. There were differences in gastric emptying rates with the coarse emulsion emptied faster. In the duodenum, lipolysis of the fine emulsion was higher. The overall 0–7-h plasma and chylomicron responses given by the areas under the curve were not significantly different between the emulsions, but the triacylglycerol peak was delayed with the fine emulsion (3 h 56 min

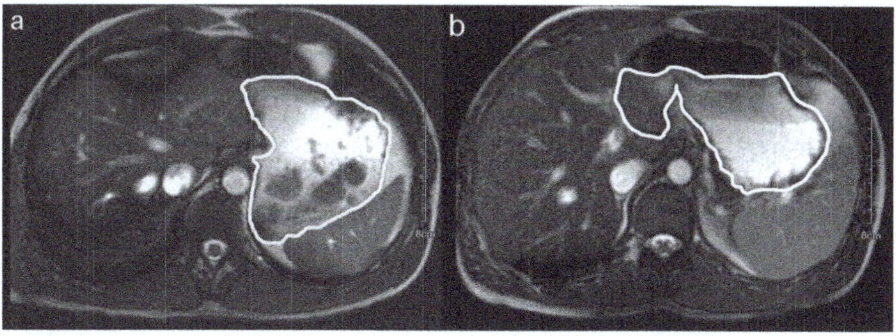

Fig. 5.1 Examples of sedimentation (**a**) and creaming (**b**) in the stomach (outlined in white) of participants lying on their back in the scanner a few minutes after consuming either a semi-solid (**a**) or liquid (**b**) meal. (The images are taken from Mackie et al. [7])

compared with 2 h 50 min). Although the authors did not make much of the differences, they highlight some essential issues around the caloric control of gastric emptying. Almost certainly, the coarse emulsion would have exhibited creaming in the stomach to a much greater extent than the fine emulsion. Thus, as in other studies where the phase separation was observed by MRI [7], the differences in caloric density of the chyme are likely to have had an impact on the gastric emptying rate and leading to slower emptying, digestion and absorption of the fine emulsion. Consequently, it is clear that factors affecting phase behaviour or local caloric density can potentially affect lipid digestion as well as appetite. Two MRI images illustrating this point are shown in Fig. 5.1 using MRI sequences in which high water is represented as bright areas and low water content by darker areas.

Mechanical mixing in the main body of the stomach is rather limited as is apparent from Fig. 5.1. However, shear forces in the antrum and through the pylorus can be much higher. Computational fluid dynamics (CFD) has shown the effect of chyme rheological properties (viscosity) on the patterns of shear and flow and in particular, on the extent of retropulson [22]. This latter flow pattern is a result of peristaltic waves moving along the antrum when the pyloric sphincter is closed. Thus chyme is forced back into the main body of the stomach under high shear and breaking up particles and lipid droplets in the flow. As a result of these actions lipid may be emulsified as it passes from the stomach into the duodenum.

5.3.2 *Intestinal Behaviour*

The small intestine is the main site of digestion. It is about 3–5 m long in an adult and comprises three regions, the duodenum, the jejunum and the ileum. The proximal small intestine (duodenum) is the main site of lipid hydrolysis and absorption and unlike the gastric compartment of the GI tract, the contents are normally well mixed [23]. Mixing within the lumen is not turbulent but results from localised folding and kneading of the contents. The architecture of the small intestine changes along its length, with long villi maximising absorptive area in the duodenum and much shorter villi in the ilium. Although it is likely that gastric lipase retains its activity in the duodenum, the suboptimal pH environment will limit its effectiveness. The primary lipase in the small intestine is human pancreatic lipase (HPL) [24], which has a co-factor (co-lipase), which helps anchor lipase to the triacylglyceride (TAG) substrate. The biosurfactant, bile is also present as a result of secretion from the liver and storage in the gall bladder. Bile is important in aiding the transport of the products of hydrolysis from the TAG interface into the lumen [25]. It is clear that different bile salts play slightly different roles in this process but the end result is the spontaneous self-assembly of a range of structures that are considered mixed micelles. Lipid in the diet takes many forms but the largest amount is in the form of TAGs. These are hydrolysed by HPL into fatty acids and monoglycerides. The typical pH in the small intestine is around neutral and so the majority of short and medium chain fatty acids will be fully dissociated. Additionally, there will normally be small amounts of diglyceride present as an intermediate product of hydrolysis. As all the product of lipid hydrolysis are essentially insoluble, they self-assemble into mixed micelles and start their journey to becoming chylomicrons as shown in Fig. 5.2. Human pancreatic lipase preferentially cleaves at the Sn-1 and Sn-3 positions and has a pH optimum of 6.5–7.5 [26] and will generate two fatty acids and one monoglyceride for every triglyceride molecule. Other lipases present are pancreatic lipase related protein 1 and 2 as well as Carboxyl ester lipase and phospholipase A2 [27]. These can hydrolyse a wide range of dietary lipids including phospholipids and galactolipids.

Rates of hydrolysis are governed by many factors but all are fundamentally related to the accessibility of the cleavage site on the substrate for the active site on the enzyme. This in turn is affected by acyl chain length, degree of saturation, available surface area, composition of the interfacial layer, diffusion of the enzymes, formation of mixed micelles, transport of micellar phase away from the site of hydrolysis, etc. As has already been shown above, the available triglyceride surface area has a significant effect on rates of hydrolysis and thus the stability and size distribution of the lipid droplets may be key factors [28] or may have little impact on the final absorption [21]. The effect of stability and interfacial layer composition on lipid hydrolysis has been explored extensively by others [29, 30] and so it will not be repeated here. However, there are a few key features that it is worth highlighting and the following have been shown to decrease the rate of lipid hydrolysis *in vitro*. Increased emulsion droplet size, solid lipid as opposed to liquid fat in the emulsion, thicker interfacial layers (especially when involving polysaccharides) and finally encapsulation. However, it is worth bearing in mind that many of these effects have not been seen *in vivo* in humans.

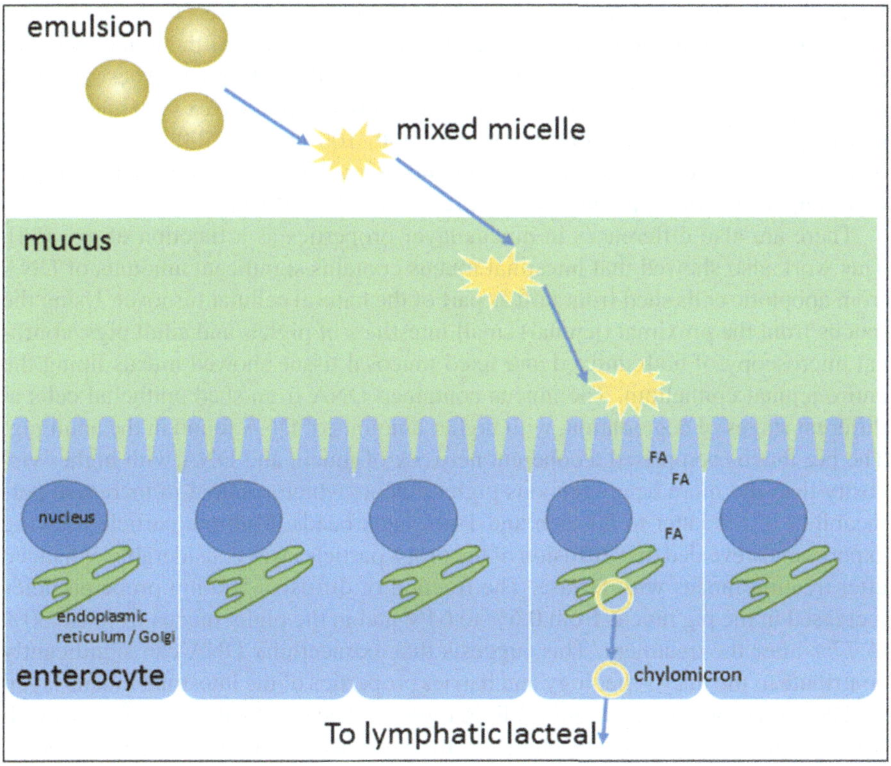

Fig. 5.2 This schematic diagram shows the journey taken by lipids from the intestinal lumen to the lymphatic system. The triglycerides are hydrolysed into fatty acids (FA) and monoglycerides, which together with bile form mixed micelles, which must diffuse through the mucus layer to the enterocytes. The absorbed material is then converted into chylomicrons and secreted into the lymphatic system

5.3.3 Mucosal Behaviour

The intestinal mucosa is not only the site of nutrient absorption but represents a barrier to potentially harmful dietary components such as bacteria. The mucus layer presents an unstirred region of the gut, through which all nutrients must pass. This mucus layer varies in thickness and composition along the GI tract [31]. In is thickest in the colon, where the bacterial load is highest and the layer is thinnest in the duodenum, which is the main site of nutrient absorption. The intestinal architecture plays a role in changing the heterogeneity of the mucus layer and thus its permeability. As the bacterial load in the small intestine is estimated to be between 10^4 and 10^8 colony forming units (CFU) per mL [32], a continuous mucus layer would provide a plausible barrier for preventing contact with the epithelium. The mucosal structure and muscular contractions, such as peristalsis, result in a highly dynamic environment at the villus tips [33]. The resulting flow and mixing aids nutrient availability but could disrupt mucus structure such that only the layers closest to the epithelium remain intact (tightly-adherent layer) and causing disruption of the mucin network resulting in areas of agglomeration in the loosely-adherent mucus. These areas of mucin agglomeration have been observed previously in porcine ex vivo mucus [34], and may explain the immobile fraction of both 100 nm and 500 nm latex beads observed in the particle tracking experiments as well as the mobile fraction of particles larger than the apparent mucus pore size of 100–200 nm [35].

There are also differences in mucus layer properties as a function of age [34]. This work also showed that intestinal mucus contains significant amounts of DNA from apoptotic cells shed from villi as part of the natural cellular turnover. Using the mucus from the proximal (jejunal) small intestines of piglets and adult pigs, confocal microscopy of both unfixed and fixed mucosal tissue showed mucus lining the entire jejunal epithelium. The mucus contained DNA from shed epithelial cells at different stages of degradation, with higher amounts of DNA found in the adult pig. The pig mucus comprised a coherent network of mucin and DNA with higher viscosity than the more heterogeneous piglet mucus, which resulted in increased permeability of the latter to 500-nm and 1-μm latex beads. Multiple-particle tracking experiments revealed that diffusion of the probe particles was considerably enhanced after treating mucus with DNase. The fraction of diffusive 500-nm probe particles increased in the pig mucus from 0.6% to 64% and in the piglet mucus from ca. 30% to 77% after the treatment. This suggests that extracellular DNA can significantly contribute to the micro-rheology and barrier properties of the intestinal mucus layer.

These findings are important for lipid transport because by their nature, the products of lipid digestion are insoluble in the aqueous environment of the GI tract. Thus, they self-assemble into a range of different structures generally known as mixed micelles that need to diffuse through the mucus layer to the site of absorption. As the pore size mentioned above is of the order of 100 nm, one might expect that mucus permeability to self-assembled particles small than that would be unaffected. However, this would appear not to be so clear cut as there is some evidence that mucus permeability to lipid digesta can be affected by dietary fibre consumption [36]. In this study, intestinal mucus from pigs fed on a diet containing 10% oatbran was less permeable to lipid from that diet than the mucus from pigs fed the control diet. The dietary fibre, oat β-glucan, has been shown to lower cholesterol by reducing bile recycling. The data shown in this study suggested that reducing the diffusion of bile and lipid through the intestinal mucus layer may be one way in which this decrease in bile reabsorption is enabled and that postprandial lipid absorption is prolonged.

The mixed micelles comprise fatty acids, monoglycerides, bile acids and phospholipids as well as any other amphipathic or hydrophobic small molecules in the intestinal environment. Once the mixed micelles have diffused through the mucus layer to the underlying epithelium, the fatty acids and monoglycerides will be absorbed [37]. Intestinal fatty acid absorption is a multistep process that is regulated by multiple genes at the enterocyte level. The molecular mechanisms for the transport of water-insoluble fatty acids across membranes have been debated for many years and is thought to be either by a simple passive diffusion mechanism, a protein-mediated mechanism, or both. A protein-mediated fatty acid uptake system is now believed to be the dominant means by which fatty acids are taken up by membrane-associated fatty acid-binding proteins on the apical membrane of enterocytes [38]. A large number of fatty acid transporters have been identified, which can facilitate the cellular uptake of fatty acids and these include cluster determinant 36 (CD36), plasma membrane-associated fatty acid–binding protein (FABPpm), and a family of fatty acid transport proteins 1–6 (FATP1-6). It is still unclear whether there is a difference in the intestinal uptake among saturated, monounsaturated and polyunsaturated fatty acids and among short-chain, middle-chain and long-chain fatty acids.

The bile will primarily be absorbed only in the distal ileum by apical sodium dependant bile acid transporters (ASBT) [39]. As a result there will be a rather consistent concentration of bile from the lumen to the intestinal epithelium but potentially gradients of fatty acids and other lipid soluble compounds. This is likely to result in a highly dynamic environment for the composition of mixed micelles as they diffuse to the site of absorption.

5.3.4 Nutrient Sensing and Feedback

The passage of food along the GI tract is controlled by a number different feedback systems that have developed to optimise digestion efficiency. Nutrients are sensed by G-protein coupled receptors on endocrine cells that represent about 1% of enterocytes (the cells lining the gut). There are a number of different types of receptors responsible for the secretion of different gastrointestinal hormones [40]. These hormones in turn have a range of effects associated with gastrointestinal motility, enzyme secretion and appetite. There are more than 30 GI hormones [41] that have a range of functions and it is not the purpose of this chapter to describe them all. However, it is worth describing the effects of some of the more relevant ones that are secreted from the small intestine. Cholecystokinin (CCK) is secreted almost exclusively from I-cells in the proximal small intestine in response to protein or fat and to a much lesser extent glucose [40]. The density of I-cells decreases from the proximal to the distal small intestine. This hormone has been shown to have a short term suppressive effect on appetite and to increase the secretion of pancreatic enzymes and contractions of the gall bladder leading to more bile entering the small intestine. Lipid has been shown to stimulate CCK release and longer chain fatty acids are more effective. There is also evidence that the number of cells immune-labelled with CCK is inversely correlated with BMI, such that duodenal CCK (I-cell) and GLP-1 (L-cell) cell density decreased with increasing BMI [42]. GLP-1 is perhaps the main appetite supressing hormone but the density of L-cells increases more distally along the intestine. However, GLP-1 secretion is not stimulated in response to lipid as much as protein or carbohydrate. Both of these GI hormones have the ability to alter appetite in the short term (satiation), and longer term (satiety).

5.3.5 Conclusions and Perspectives

In this chapter I have described the primary physiological events in lipid digestion. These run from oral processing to intestinal absorption and include passage through a range of biochemical and biophysical environments that can alter the rates of lipid digestion, transport and absorption. As fat is the most calorie dense macronutrient, it offers a range of interesting possibilities for modulating appetite and thus energy intake in response to the increasing incidence of overweight and obesity.

References

1. Siri-Tarino PW et al (2010) Meta-analysis of prospective cohort studies evaluating the association of saturated fat with cardiovascular disease. Am J Clin Nutr 91(3):535–546
2. Malik VS et al (2010) Sugar-sweetened beverages, obesity, type 2 diabetes mellitus, and cardiovascular disease risk. Circulation 121(11):1356–1364
3. Grundy MML et al (2016) The role of plant cell wall encapsulation and porosity in regulating lipolysis during the digestion of almond seeds. Food Funct 7(1):69–78
4. Mackie AR et al (2017) Oatmeal particle size alters glycemic index but not as a function of gastric emptying rate. Am J Physiol Gastrointest Liver Physiol 313(3):G239–G246
5. Tydeman EA et al (2010) Effect of carrot (Daucus carota) microstructure on carotene bioaccessibilty in the upper gastrointestinal tract. 1. In vitro simulations of carrot digestion. J Agric Food Chem 58(17):9847–9854
6. Hull S et al (2015) A mid-morning snack of almonds generates satiety and appropriate adjustment of subsequent food intake in healthy women. Eur J Nutr 54(5):803–810
7. Mackie AR et al (2013) Specific food structures supress appetite through reduced gastric emptying rate. Am J Physiol Gastrointest Liver Physiol 304(11):G1038–G1043
8. Lai WYW et al (2019) Analysis of the lipolytic activity of whole-saliva and site-specific secretions from the oral cavity of healthy adults. Nutrients 11(1):191
9. Mulet-Cabero AI et al (2017) Dairy food structures influence the rates of nutrient digestion through different in vitro gastric behaviour. Food Hydrocoll 67:63–73
10. Bakala N'Goma JC et al (2012) Understanding the lipid-digestion processes in the GI tract before designing lipid-based drug-delivery systems. Ther Deliv 3(1):105–124
11. Kanicky JR, Shah DO (2002) Effect of degree, type, and position of unsaturation on the pK(a) of long-chain fatty acids. J Colloid Interface Sci 256(1):201–207
12. Bakalis S et al (2010) Modelling of physical processes in the small intestine. In: Williams PA, Phillips GO (eds) Gums and stabilisers for the food Industry 15. Royal Society of Chemistry, Cambridge, pp 377–383
13. Bodmer MW et al (1987) Molecular-cloning of a human gastric lipase and expression of the enzyme in yeast. Biochim Biophys Acta 909(3):237–244
14. Moreau H et al (1992) Isoform purification of gastric lipases - towards crystallization. J Mol Biol 225(1):147–153
15. Sams L et al (2016) Relevant pH and lipase for in vitro models of gastric digestion. Food Funct 7(1):30–45
16. Marciani L et al (2007) Enhancement of intragastric acid stability of a fat emulsion meal delays gastric emptying and increases cholecystokinin release and gallbladder contraction. Am J Physiol Gastrointest Liver Physiol 292(6):G1607–G1613
17. Marciani L et al (2006) Magnetic resonance imaging of the behaviour of oil-in-water emulsions in the gastric lumen of man. Br J Nutr 95(2):331–339

18. Steingoetter A et al (2015) Imaging gastric structuring of lipid emulsions and its effect on gastrointestinal function: a randomized trial in healthy subjects. Am J Clin Nutr 101(4):714–724
19. Day L et al (2014) Tailoring the digestion of structured emulsions using mixed monoglyceride-caseinate interfaces. Food Hydrocoll 36:151–161
20. Wooster TJ et al (2014) Impact of different biopolymer networks on the digestion of gastric structured emulsions. Food Hydrocoll 36:102–114
21. Armand M et al (1999) Digestion and absorption of 2 fat emulsions with different droplet sizes in the human digestive tract. Am J Clin Nutr 70(6):1096–1106
22. Ferrua MJ, Kong FB, Singh RP (2011) Computational modeling of gastric digestion and the role of food material properties. Trends Food Sci Technol 22(9):480–491
23. Lentle RG, de Loubens C (2015) A review of mixing and propulsion of chyme in the small intestine: fresh insights from new methods. J Comp Physiol B 185(4):369–387
24. Carriere F et al (2000) The specific activities of human digestive lipases measured from the in vivo and in vitro lipolysis of test meals. Gastroenterology 119(4):949–960
25. Maldonado-Valderrama J et al (2011) The role of bile salts in digestion. Adv Colloid Interf Sci 165(1):36–46
26. Carriere F et al (1998) Structural basis for the substrate selectivity of pancreatic lipases and some related proteins. Biochim Biophys Acta 1376(3):417–432
27. Whitcomb DC, Lowe ME (2007) Human pancreatic digestive enzymes. Dig Dis Sci 52(1):1–17
28. Marciani L et al (2006) Delaying gastric emptying and enhancing cholecystokinin release and satiety by using acid stable fat emulsions. Gastroenterology 130(4):A227–A227
29. McClements DJ, Decker EA, Park Y (2009) Controlling lipid bioavailability through physicochemical and structural approaches. Crit Rev Food Sci Nutr 49(1):48–67
30. McClements DJ, Li Y (2010) Structured emulsion-based delivery systems: controlling the digestion and release of lipophilic food components. Adv Colloid Interf Sci 159(2):213–228
31. Atuma C et al (2001) The adherent gastrointestinal mucus gel layer: thickness and physical state in vivo. Am J Physiol Gastrointest Liver Physiol 280(5):G922–G929
32. Gibson PR, Barrett JS (2010) The concept of small intestinal bacterial overgrowth in relation to functional gastrointestinal disorders. Nutrition 26(11–12):1038–1043
33. Lim YF et al (2015) Flow and mixing by small intestine villi. Food Funct 6(6):1787–1795
34. Macierzanka A et al (2014) Transport of particles in intestinal mucus under simulated infant and adult physiological conditions: impact of mucus structure and extracellular DNA. PLoS One 9(4):11
35. Bajka BH et al (2015) The influence of small intestinal mucus structure on particle transport ex vivo. Colloids Surf B Biointerfaces 135:73–80
36. Mackie A et al (2016) Increasing dietary oat fibre decreases the permeability of intestinal mucus. J Funct Foods 26:418–427
37. Wang TY et al (2013) New insights into the molecular mechanism of intestinal fatty acid absorption. Eur J Clin Investig 43(11):1203–1223
38. Cifarelli V, Abumrad NA (2018) Intestinal CD36 and other key proteins of lipid utilization: role in absorption and gut homeostasis. Compr Physiol 8(2):493–507
39. Zwicker BL, Agellon LB (2013) Transport and biological activities of bile acids. Int J Biochem Cell Biol 45(7):1389–1398
40. Karhunen LJ et al (2008) Effect of protein, fat, carbohydrate and fibre on gastrointestinal peptide release in humans. Regul Pept 149(1–3):70–78
41. Rehfeld JF (1998) The new biology of gastrointestinal hormones. Physiol Rev 78(4):1087–1108
42. Little TJ et al (2014) Characterization of duodenal expression and localization of fatty acid-sensing receptors in humans: relationships with body mass index. Am J Physiol Gastrointest Liver Physiol 307(10):G958–G967

Chapter 6
Lipids and Cardiovascular Disease Risks with a Focus on Dairy Foods

D. I. Givens

6.1 Introduction

Despite improved prevention and treatment programmes, cardiovascular diseases (coronary heart disease and stroke, CVD) remain a major cause of death and morbidity in the European Union (EU) [38] and was responsible for 72% of the 54.7 million deaths in 2016 worldwide [9]. Although CVD mortality is now decreasing in most European countries, there are some 49 million people living with CVD in the EU with a cost of some €210 billion/year [38]. Moreover, it is concerning that recent data from the US indicate a levelling off of the decline in CVD mortality, thought probably a result of increasing obesity and related type 2 diabetes [39]. This highlights the risk of complacency and the continued need for modifiable risk factors for CVD such as diet are used to their full advantage.

A high intake of saturated fatty acids (SFA) has for many years been regarded as an important risk factor for CVD leading to public health targets recommending a dietary SFA intake that should not be exceeded. In the UK, this recommendation, recently reviewed and confirmed [26] is that SFA intake should not exceed 11% of food energy intake or 10% of total energy intake but, as shown in Table 6.1, intake for all age groups in the *National Diet and Nutrition Survey* [25] exceeds these targets. Milk and dairy products (including butter) are often the major sources of dietary SFA (UK contribution in Table 6.1) and this has led to proposals to reduce dairy product consumption in order to reduce SFA intake. Recent evidence suggests that this approach is likely to be too simplistic [31] and dairy foods are also key sources of other nutrients making it important to understand these complex foods in detail.

D. I. Givens (✉)
Institute of Food, Nutrition and Health, University of Reading, Reading, UK
e-mail: d.i.givens@reading.ac.uk

M. M.-L. Grundy, P. J. Wilde (eds.), *Bioaccessibility and Digestibility of Lipids from Food*, https://doi.org/10.1007/978-3-030-56909-9_6

Table 6.1 Mean daily intake of saturated fatty acids (SFA) and as a percentage of food energy (FE) and total energy (TE) by gender and age together with the contribution of dairy foods to SFA intake in the UK from NDNS years 7–8

	Males (years of age)				Females (years of age)			
	4–10	11–18	19–64	65+	4–10	11–18	19–64	65+
SFA (g/day)	21.5	26.7	27.5	27.3	20.3	21.4	22.8	21.8
SFA % FE	12.7	12.6	12.3	13.4	13.2	12.3	12.7	14.1
SFA % TE	12.7	12.5	11.6	12.8	13.2	12.2	12.2	13.7
Contribution of dairy foods[a] to total SFA intake, mean of males and females (%)	34	26	27	32				

From Roberts et al. [25]
[a]Including butter

6.2 Effects of Saturated Fat Consumption

A high intake of SFA has for many years been regarded as an important risk factor for CVD although recent meta-analyses of prospective studies [2, 5, 28] have raised doubts about the validity of this. The specific effect of SFA on CVD risk is well-established and is mainly mediated via increases in blood lipids, particularly low density lipoprotein cholesterol (LDL-C). There is strong evidence implicating LDL-C as a risk factor in CVD events [24], with randomised controlled trials (RCT) with LDL-C lowering interventions confirming a causal relationship between blood LDL-C and CVD, particularly coronary heart disease (CHD; [18]). However, as highlighted by Mensink et al., [19], SFA can also increase high density lipoprotein cholesterol (HDL-C), generally regarded as being associated with reduced CVD risk. It should also be noted that blood LDL-C contains the cholesterol present in all LDL particles and the small, dense LDL particles (sdLDL) whilst containing less cholesterol per particle than other LDL particles, are more strongly associated with increased CHD risk than the larger less dense LDL particles [12]. One cross-sectional study in healthy Swedish men indicated that dietary fatty acids typically found in dairy foods gave rise to significantly less sdLDL particles [29]. In particular, SFA 4:0 to 10:0 and 14:0 ($P < 0.05$) in the diet and 15:0 and 17:0 (markers of dairy fat intake) in serum phospholipids ($P < 0.05$) were associated with fewer sdLDL particles. This study suggests that LDL particle size distribution may be favourably modified by dairy products in the diet but this needs further investigation.

There is now increasing evidence that the effect of SFA depends on the source. For example the prospective studies of de Oliveira Otto et al. [4] and Vissers et al. [34] both showed that whilst a positive association with CVD risk was seen with SFA from meat, this was not the case with SFA from dairy. This effect may be related to food-specific fatty acids or other aspects of the foods, but clearly this requires RCTs to be carried out. Similarly, within dairy foods there are now a number of RCTs that show a positive TC and LDL-C response to dairy SFA in butter but a moderated, neutral or negative effect when the same mass of SFA is consumed in hard cheese (e.g. [6]). This effect is now regarded as being a function of the dairy food matrix [31] and Feeney et al. [6] explored the matrix effect in a stepwise-response fashion. The study was a 6-week randomised parallel design where subjects consumed 40 g of dairy fat/d in macronutrient-matched food matrices as either (A) 120 g full-fat cheddar cheese, (B) 120 g reduced-fat cheddar cheese plus 31 g of butter, or (C) 49 g butter plus 30 g calcium caseinate and 500 mg of calcium as $CaCO_3$. The blood lipid responses are summarised in Table 6.2 but the results did indeed show a stepwise-matrix effect with significantly lower post-intervention TC and LDL-C concentrations seen when all of the dairy fat was provided by the cheese.

Various mechanisms have been suggested as being responsible the potentially health-related benefits of this so-called matrix effect. The study of Lorenzen and Astrup [17] identified mechanisms which could contribute to the differential effects of cheese and butter on blood cholesterol. They randomised study participants on to four isoenergetic diets comprising low (25% energy intake) and high fat diets (40% energy intake), each of which contained either low (700 mg/d) or high (2800 mg/d) calcium. The key results are summarised in Table 6.3. Increasing calcium intake increased faecal fat and bile acid excretion in both low and high fat diets, although the high fat diets also increased faecal fat loss. Lorenzen and Astrup [17] proposed that the key effect of calcium on faecal fat and bile acid excretion is likely to be the result of formation of insoluble calcium-fatty acid soaps and/or of hydrophobic aggregations with phosphorus and bile acids. Increased faecal fat excretion is indicative of reduced fat absorption and hence a moderated blood cholesterol response, whilst increased bile acid excretion would reduce the amount that was recycled via the normal enterohepatic circulation. Reduced bile acid recycling would lead to increased metabolism of cholesterol for bile acid synthesis thus reducing that in the circulation.

These studies indicate that the impact of SFA in dairy foods on CVD risk factors may be dependent on the food matrix, even for dairy foods that supply the same amount of dairy fat and SFA. However, as concluded by Thorning et al. [31], further work is needed to fully understand food matrix effects on health and the nutritional value of dairy products should therefore be considered as 'the biofunctionality of the sum of nutrients within the dairy matrix structures'.

Table 6.2 Changes in blood lipids from baseline to end of 6 weeks intervention when 40 g/d of dairy fat was provided as 120 g Cheddar cheese (T1), 120 g low fat Cheddar cheese with 21 g butter (T2) or 49 g butter with 30 g calcium caseinate and 500 mg calcium (as $CaCO_3$) (T3)

	Treatment			
Blood lipid (mmol/L)	T1	T2	T3	P (overall effect[a])
Total cholesterol	−0.52	−0.37	−0.15	0.033
HDL-cholesterol	0	−0.07	+0.05	0.284
LDL-cholesterol	−0.45	−0.27	−0.14	0.016
Triaclyglycerols	−0.15	−0.05	−0.12	0.386

From Feeney et al. [6]

[a]For controlling factors included in model see original paper

Table 6.3 The effect of diets high or low in fat and calcium (Ca) on faecal losses of fat, calcium and bile acids

	High fat		Low fat		P for	
Faecal losses	Low Ca	High Ca	Low Ca	High Ca	Ca	Fat
Fat (g/day)	6.6	11.3	5.5	8.0	***	**
Ca (mg/day)	549	2477	576	2478	***	NS
Bile acid (μmol/day)	274	393	178	346	**	NS

From Lorenzen and Astrup [17]
***P < 0.001, **P < 0.01, NS; non-significant P > 0.05

The evidence regarding the differential effect of SFA from different foods was highlighted by Astrup et al. [1] in their response to the World Health Organisation's draft report of dietary SFA [37].

6.3 Substitution of Dietary SFA with Unsaturated Fatty Acids

As noted above, data from prospective studies [2, 5] has cast doubts on the traditional view of SFA as a risk factor for CVD. Wang and Hu [35] confirmed that whilst the current evidence supports the idea that different types of dietary fatty acids have different effects on CVD risk, it is vital to realise that the effects are also dependent on the replacement macronutrient. Nettleton et al. [22] concluded that the greatest reduction in CHD risk was observed when dietary SFA are replaced with *cis*-polyunsaturated fatty acids (*cis*-PUFA) although with little effect on risk of stroke. They cited RCTs which showed that replacement of 10% and 5% of energy from SFA by *cis*-PUFA reduced CVD events by 27% and 10% respectively. It is also noteworthy that replacement of SFA with refined carbohydrate is associated with no improvement or even an increase in CVD risk [20].

Nettleton et al. [22] reported that there are insufficient data to draw conclusions on the effect of replacing SFA with *cis*-MUFA, although there are a number of important studies. A cross-sectional population study by Miura et al. [21] reported a linear inverse association between *cis*-MUFA (especially oleic acid) intake and diastolic blood pressure, whilst in the recruited population study of Hassan et al. [11] dietary oleic acid intake was negatively associated with systolic and diastolic blood pressure. More recently the RCT of Liu et al. [15] compared diets high in *cis*-PUFA with those high in *cis*-MUFA and showed that the latter treatment reduced abdominal obesity with related reductions in blood pressure and blood triacylglycerols (TAG). The *Reading, Imperial, Surrey, Cambridge, and Kings* (*RISCK*) study [13] showed that compared with replacing SFA with carbohydrate, replacing with *cis*-MUFA reduced the TC:HDL ratio and more recently, the *Dietary Intervention and VAScular function* (*DIVAS*) study [32] showed that replacement of about 9.6% of energy from SFA by MUFA lowered serum TC by 8.4%, LDL-C by 11.3% and TC:HDL-C by 5.6% (P < 0.001). In addition, Williams et al. [36] reported on an

RCT involving manufactured foods high in MUFA and showed that diets where SFA have been partially replaced by MUFA led to significant reductions in TC and LDL-C even when total fat and energy intakes were equal. Thus, overall, there is increasing evidence that replacing dietary SFA with *cis*-MUFA can be beneficial for reducing the risk of CVD.

6.4 Does Partial Replacement of SFA by MUFA in Dairy Products Provide Health Benefits?

There is now considerable evidence from meta-analyses of prospective studies ([3, 10, 30]; Table 6.4) which confirm that increased consumption of milk and dairy foods is not associated with increased risk of CVD and indeed for some outcomes high intake is significantly associated with a reduced risk. Because of the long

Table 6.4 A selection of dose-response meta-analyses examining the relative risk (RR) of cardiometabolic disease in relation to consumption of dairy foods

Dairy food	Outcome	RR (95% CI[a])	References
Milk (per 244 g/day)	All-cause mortality	1.00 (0.93–1.07)	Guo et al. [10]
Milk (per 244 g/day)	CVD	1.01 (0.93–1.10)	Guo et al. [10]
Cheese (per 10 g/day)	CVD	0.98 (0.95–1.00)	Guo et al. [10]
Yoghurt (per 50 g/day)	CVD	1.03 (0.97–1.09)	Guo et al. [10]
Milk (per 200 g/day)	Stroke	0.93 (0.88–0.98)	de Goede et al. [3]
Milk (per 200 g/d)	Stroke	0.92 (0.88–0.97)	Soedamah-Muthu and de Goede [30]
Cheese (per 40 g/day)	Stroke	0.97 (0.94–1.01)	de Goede et al. [3]
Yoghurt (per 80 g/day)	Type 2 diabetes	0.86 (0.83–0.90)	Gijsbers et al. [7]
Yoghurt (per 100 g/day)	Type 2 diabetes	0.94 (0.91–0.97)	Soedamah-Muthu and de Goede [30]
Butter (per 14 g/day)	All-cause mortality	1.01 (1.00–1.03)	Pimpin et al. [23]
Butter (per 14 g/day)	CVD	1.00 (0.98–1.02)	Pimpin et al. [23]
Butter (per 14 g/day)	CHD	0.99 (0.96–1.03)	Pimpin et al. [23]
Butter (per 14 g/day)	Stroke	1.01 (0.93–0.99)	Pimpin et al. [23]
Butter (per 14 g/day)	Type 2 diabetes	0.96 (0.93–0.99)	Pimpin et al. [23]

[a]Confidence interval

standing association between SFA intake and plasma cholesterol these results are counterintuitive to many. However based on other data reviewed earlier, it is often asked whether changing the FA composition of dairy foods by replacing some SFA with *cis*-PUFA and/or *cis*-MUFA could provide a useful way to reduce SFA intake whilst retaining the nutritional and cardiometabolic benefits of milk. Indeed using published models of cholesterol responses to SFA intake it was predicted that, in the EU, reductions of some 10,500 and 3900 deaths from CHD and stroke per annum, respectively, could result from such a strategy [8], although it is now clear that such simple models are not likely to apply to all dairy foods for some of the reasons discussed above.

Livingstone et al. [16] reviewed the RCT-based evidence on the effect on CVD risk markers of consuming dairy products with modified fatty acid composition obtained by changing the nutrition of the dairy cow. Of the nine RCT studies identified, most used modified butter as the main test food and used changes in blood cholesterol as the primary risk marker. Of the six comparable studies that measured blood cholesterol, five and four reported significant reductions in TC and LDL-C respectively, one showed a rise in TC and two showed no effect on LDL-C following chronic consumption of modified milk and dairy products. One study [27] was of particular interest as it was the only study to compare the effects of modified dairy products in normocholesterolaemic (NC) and hypercholesterolaemic (HC) subjects. Dairy cows were fed a diet containing *cis*-MUFA-rich rapeseed cake to achieve the changes in milk FA concentrations and the milk was used to produce butter and yoghurt. In a 13-week, three-phase, randomised, cross-over, controlled study, subjects were fed modified milk, butter and yogurt, as well as non-modified milk, margarine and yogurt. The results showed that, overall, LDL-C was reduced (−0.38 mmol/L; $P < 0.05$) and HDL-C was increased (+0.33 mmol/L; $P < 0.05$) when the modified products were consumed, resulting in a reduced LDL: HDL ratio (−0.43 mmol/L; $P < 0.05$). These changes were seen in both the HC and NC subjects; with interestingly, a greater change in the NC subjects. Seidel et al. [27] proposed that this may be due to lipid metabolism in HC subjects being more complex and/or dietary intervention being less effective.

It is now clear that the use of a single marker (e.g. blood cholesterol) for predicting the cardiometabolic risk/benefit associated with consumption of fatty acid-modified dairy foods may not be adequate and that a wider range and more holistic markers are needed. A study of this type was recently undertaken at the University of Reading in the *REplacement of SaturatEd fat in dairy on Total cholesterol* (*RESET*) study (ClinicalTrials.gov NCT02089035), a 3.5-year project funded by the UK Medical Research Council. It investigated the impact of reducing SFA in milk, cheese and butter on vascular function and a wide range of cardiometabolic risk biomarkers, after both acute and chronic consumption. The findings on the fatty acid composition of the dairy foods used were presented by Kliem et al. [14] and the results of the chronic study by Vasilopoulou et al. [33]. In the chronic study the diet containing the modified dairy foods attenuated the rise in fasting LDL-C seen with the control diet (0.03 ± 0.06 mmol/L and 0.19 ± 0.05 mmol/L, respectively; $P = 0.03$). Relative to baseline, the flow-mediated dilation (FMD) response measured in the brachial artery increased after the consumption of the diet containing the modified foods (0.35% ± 0.15%), whereas a decrease was observed after the control diet (−0.51% ± 0.15%; $P < 0.0001$). In agreement with the FMD response, fasting plasma nitrite concentrations increased after the modified diet, yet decreased after the control diet (0.02 ± 0.01 μmol/L and − 0.03 ± 0.02 μmol/L, respectively; $P = 0.01$). Overall therefore, the dairy foods with a proportion of SFA replaced by cis-MUFA led to improvements in blood lipids and in endothelial health. Given the nature of the study it was not possible to assess whether there were differential effects of the three dairy foods but this clearly needs study.

6.5 Conclusions

As discussed, it is now clear that the effects of SFA consumption are complex and will depend on a number of factors including their source and a moderating effect of food matrices. It is also the case that the effects of reducing SFA intake are best interpreted by a knowledge of the macronutrients that replace them, with reduced CVD risk being associated with the replacement of SFA with *cis*-PUFA and *cis*-MUFA. Most judgements have, however, been made using TC and LDL-C and/or TC:HDL-C as the primary markers of CVD risk and because of potentially counter-balancing effects in dairy food (e.g. effect of protein on blood lipids and blood pressure; matrix effects on fat bioavailability) a wider range of markers are needed. Nevertheless, whilst prospective evidence shows no increase in CVD risk associated with high dairy consumption, it is still uncertain whether replacing a proportion of SFA in dairy fat with *cis*-MUFA will consistently lead to reduced CVD risk although the relatively few RCTs that have examined this using TC and LDL-C as risk markers, provide an indication of reduced CVD risk from consumption of milk and dairy products with modified fatty acid composition, compared with those of normal milk fat composition. In addition a very recent RCT involving diets containing milk, cheese and butter has shown benefits in blood lipids and other vascular CVD risk markers although futher work is needed to identify any differential effects of the three foods.

Conflict of Interest The author declares no conflicts of interest.

References

1. Astrup A, Bertram HCS, Bonjour J-P, de Groot LCP, de Oliveira Otto MC, Feeney EL, Garg ML, Givens DI, Kok FJ, Krauss RM, Lamarche B, Lecerf J-M, Legrand P, McKinley M, Micha R, Michalski M-C, Mozaffarian D, Soedamah-Muthu SS (2019) WHO draft guidelines on dietary saturated and *trans* fatty acids: time for a new approach? Br Med J 365:l4137
2. Chowdhury R, Warnakula S, Kunutsor S, Crowe F, Ward HA, Johnson L, Franco OH, Butterworth AS, Forouhi NG, Thompson SG, Khaw KT, Mozaffarian D, Danesh J, Di

Angelantonio E (2014) Association of dietary, circulating, and supplement fatty acids with coronary risk: a systematic review and meta-analysis. Ann Intern Med 160:398e406

3. de Goede J, Soedamah-Muthu SS, Pan A, Gijsbers L, Geleijnse JM (2016) Dairy consumption and risk of stroke: a systematic review and updated dose-response meta-analysis of prospective cohort studies. J Am Heart Assoc 5:e002787
4. de Oliveira Otto MC, Mozaffarian D, Kromhout D, Bertoni AG, Sibley ST, Jacobs DR Jr, Nettleton JA (2012) Dietary intake of saturated fat by food source and incident cardiovascular disease: the Multi-Ethnic Study of Atherosclerosis. Am J Clin Nutr 96:397–404
5. de Souza RJ, Mente A, Maroleanu A, Cozma AI, Ha V, Kishibe T, Uleryk E, Budylowski P, Schünemann H, Beyene J, Anand SS (2015) Intake of saturated and *trans* unsaturated fatty acids and risk of all-cause mortality, cardiovascular disease, and type 2 diabetes: systematic review and meta-analysis of observational studies. Br Med J 351:h3978
6. Feeney EL, Barron R, Dible V, Hamilton Z, Power Y, Tanner L, Flynn C, Bouchier P, Beresford T, Noronha N, Gibney ER (2018) Dairy matrix effects: response to consumption of dairy fat differs when eaten within the cheese matrix-a randomized controlled trial. Am J Clin Nutr 108:1–8
7. Gijsbers L, Ding EL, Malik VS, de Goede J, Geleijnse JM, Soedamah-Muthu SS (2016) Consumption of dairy foods and diabetes incidence: a dose-response meta-analysis of observational studies. Am J Clin Nutr 103:1111–1124
8. Givens DI (2008) Impact on CVD risk of modifying milk fat to decrease intake of saturated fatty acids and increase intake of *cis*-monounsaturates. Proc Nutr Soc 67:419–427
9. GBD (2017) Global Burden of Disease, Causes of Death Collaborators. Global, regional, and national age-sex specific mortality for 264 causes of death, 1980–2016: a systematic analysis for the Global Burden of Disease Study 2016. Lancet 390:1151–1210
10. Guo J, Astrup A, Lovegrove JA, Gijsbers L, Givens DI, Soedamah-Muthu SS (2017) Milk and dairy consumption and risk of cardiovascular diseases and all-cause mortality: dose-response meta-analysis of prospective cohort studies. Eur J Epidemiol 32:269–287
11. Hassan NE, El Shebini SM, Ahmed NH, Selim Mostafa M (2015) Association between macronutrients intake, visceral obesity and blood pressure in a sample of obese Egyptian women. Maced J Med Sci 3:184–188
12. Hoogeveen RC, Gaubatz JW, Sun W, Dodge RC, Crosby JR, Jiang J, Couper D, Virani SS, Kathiresan S, Boerwinkle E, Ballantyne CM (2014) Small dense low-density lipoprotein-cholesterol concentrations predict risk for coronary heart disease: the Atherosclerosis Risk in Communities (ARIC) study. Arterioscler Thromb Vasc Biol 34:1069–1077
13. Jebb SA, Lovegrove JA, Griffin BA, Frost GS, Moore CS, Chatfield MD, Bluck LJ, Williams CM, Sanders TA, RISCK Study Group (2010) Effect of changing the amount and type of fat and carbohydrate on insulin sensitivity and cardiovascular risk: the RISCK trial. Am J Clin Nutr 92:748–758
14. Kliem KE, Humphries D, Markey O, Vasilopoulou D, Fagan C, Grandison A, Jackson K, Todd S, Givens DI, Lovegrove JA (2019) Food chain approach to lowering the saturated fat of milk and dairy products. Int J Dairy Technol 72:100–109
15. Liu X, Kris-Etherton PM, West SG, Lamarche B, Jenkins DJ, Fleming JA, McCrea CE, Pu S, Couture P, Connelly PW, Jones PJ (2016) Effects of Canola and high-oleic-acid Canola oils on abdominal fat mass in individuals with central obesity. Obesity 24:2261–2268
16. Livingstone KM, Lovegrove JA, Givens DI (2012) The impact of substituting SFA in dairy products with MUFA or PUFA on CVD risk: evidence from human intervention studies. Nutr Res Rev 25:193–206
17. Lorenzen JK, Astrup A (2011) Dairy calcium intake modifies responsiveness of fat metabolism and blood lipids to a high-fat diet. Br J Nutr 105:1823–1831
18. Marz W, Scharnagl H, Gouni-Berthold I, Silbernagel G, Dressel A, Grammer TB, Landmesser U, Dieplinger H, Windler E, Laufs U (2016) LDL-cholesterol: standards of treatment: a German perspective. Am J Cardiovasc Drugs 16:323–336

19. Mensink RP, Zock PL, Kester AD, Katan MB (2003) Effects of dietary fatty acids and carbohydrates on the ratio of serum total to HDL cholesterol and on serum lipids and apolipoproteins: a meta-analysis of 60 controlled trials. Am J Clin Nutr 77:1146–1155
20. Micha R, Mozaffarian D (2010) Saturated fat and cardiometabolic risk factors, coronary heart disease, stroke, and diabetes: a fresh look at the evidence. Lipids 45:893–905
21. Miura K, Stamler J, Brown IJ, Ueshima H, Nakagawa H, Sakurai M, Chan Q, Appel LJ, Okayama A, Okuda N, Curb JD, Rodriguez BL, Robertson C, Zhao L, Elliott P, INTERMAP Research Group (2013) Relationship of dietary monounsaturated fatty acids to blood pressure: the International Study of Macro/Micronutrients and Blood Pressure. J Hypertens 31:1144–1150
22. Nettleton JA, Brouwer IA, Geleijnse JA, Hornstra G (2017) Saturated fat consumption and risk of coronary heart disease and ischemic stroke: a science update. Ann Nutr Metab 70:26–33
23. Pimpin L, Wu JHY, Haskelberg H, Del Gobbo L, Mozaffarian D (2016) Is butter back? A systematic review and meta-analysis of butter consumption and risk of cardiovascular disease, diabetes, and total mortality. PLoS One 11:e0158118
24. Ridker PM (2014) LDL cholesterol: controversies and future therapeutic directions. Lancet 384:607–617
25. Roberts C, Steer T, Maplethorpe N, Cox L, Meadows S, Nicholson S, Page P, Swan G (2018) National diet and nutrition survey. results from years 7–8 (combined) of the rolling programme (2014/15 to 2015/16). PHE Publication gateway number: 2017851 Public Health England, London
26. SACN (2019) Saturated fats and health. Scientific Advisory Committee on Nutrition. Public Health England. Available at: https://www.gov.uk/government/publications/saturated-fats-and-health-sacn-report
27. Seidel C, Deufel T, Jahreis G (2005) Effects of fat-modified dairy products on blood lipids in humans in comparison with other fats. Ann Nutr Metab 49:42–48
28. Siri-Tarino PW, Sun Q, Hu FB, Krauss RM (2010) Meta-analysis of prospective cohort studies evaluating the association of saturated fat with cardiovascular disease. Am J Clin Nutr 91:535–546
29. Sjogren P, Rosell M, Skoglund-Andersson C, Zdravkovic S, Vessby B, de Faire U, Hamsten A, Hellenius ML, Fisher RM (2004) Milk-derived fatty acids are associated with a more favorable LDL particle size distribution in healthy men. J Nutr 134:1729–1735
30. Soedamah-Muthu SS, de Goede J (2018) Dairy consumption and cardiometabolic diseases: systematic review and updated meta-analyses of prospective cohort studies. Curr Nutr Rep 7:171–172
31. Thorning TK, Bertram HC, Bonjour J-P, de Groot L, Dupont D, Feeney E, Ipsen R, Lecerf J-M, Mackie A, McKinley MC, Michalski M-C, Rémond D, Risérus U, Soedamah-Muthu SS, Tholstrup T, Weaver C, Astrup A, Givens DI (2017) Whole dairy matrix or single nutrients in assessment of health effects: current evidence and knowledge gaps. Am J Clin Nutr 105:1033–1045
32. Vafeiadou K, Weech M, Altowaijri H, Todd S, Yaqoob P, Jackson KG, Lovegrove JA (2015) Replacement of saturated with unsaturated fats had no impact on vascular function but beneficial effects on lipid biomarkers, E-selectin, and blood pressure: results from the randomized, controlled Dietary Intervention and VAScular function (DIVAS) study. Am J Clin Nutr 102:40–48
33. Vasilopoulou D, Markey O, Kliem KE, Fagan CC, Grandison AS, Humphries DJ, Todd S, Jackson KG, Givens DI, Lovegrove JA (2020) Reformulation initiative for partial replacement of saturated with unsaturated fats in dairy foods attenuates the increase in low-density lipoprotein cholesterol and improves flow-mediated dilatation compared with conventional dairy: the randomized, controlled RESET study. Am J Clin Nutr 111:739–748
34. Vissers LET, Rijksen J, Boer JMA, Verschuren WMM, van der Schouw YT, Sluijs I (2019) Fatty acids from dairy and meat and their association with risk of coronary heart disease. Eur J Nutr 58:2639–2647

35. Wang DD, Hu FB (2017) Dietary fat and risk of cardiovascular disease: recent controversies and advances. Annu Rev Nutr 37:423–446
36. Williams CM, Francis-Knapper JA, Webb D, Brookes CA, Zampelas A, Tredger JA, Wright J, Meijer G, Calder PC, Yaqoob P, Roche H, Gibney MJ (1999) Cholesterol reduction using manufactured foods high in monounsaturated fatty acids: a randomized crossover study. Br J Nutr 81:439–446
37. WHO (2018) World Health Organisation. Draft guidelines on saturated fatty acid and trans-fatty acid intake for adults and children. Public Consultation May to June 2018. https://extranet.who.int/dataform/upload/surveys/666752/files/Draft%20WHO%20SFA-TFA%20guidelines_04052018%20Public%20Consultation(1).pdf
38. Wilkins E, Wilson L, Wickramasinghe K, Bhatnagar P, Rayner M, Townsend N, Leal J, Luengo-Fernandez R, Burns R (2017) European cardiovascular disease statistics 2017. European Heart Network, Brussels
39. Zia A, Siddiqui HU, Mohiuddin H, Gul S (2018) Levelling-off of declining trend of cardiovascular disease-related mortality in the USA: the challenge to rein in obesity and diabetes epidemic. Cardiovasc Endocrinol Metab 7:54–55

Part III
Food Structure

Chapter 7
Plant Food Structure and Lipid Digestibility

Myriam M.-L. Grundy and Peter J. Wilde

Lipid intakes, especially saturated fats, are, in many developed countries such as the UK, above the current recommendations (total fat intake should be <35% of the total energy intake). An overconsumption of fats can lead to hyperlipidaemia, elevated levels of lipids and cholesterol in the blood [53], which is an important risk factor for atherosclerosis and therefore cardiovascular disease (CVD) [72, 124, 126]. Elevated plasma fatty acids has also been associated with insulin resistance [9, 119] and obesity [11]. Despite their rather negative reputation, lipids have incontestable nutritional values. Lipids play various roles in the human body, in addition to being a source of energy they are structural components of cell and organelles (e.g. mitochondria, chloroplast and oil bodies) membranes, mainly as phospholipids arranged into a bilayer. Omega-3 fatty acids (e.g. eicosapentaenoic acid, EPA; docosahexaenoic acid, DHA; and α-linolenic acid, ALA) that can be found in marine sources (i.e. EPA and DHA in oily fish and algae) and nuts and seeds (ALA), are involved in the synthesis of eicosanoids that are important metabolites in inflammatory processes [15, 120] and neuronal functions [10, 75, 132]. The current evidence suggests that the consumption of essential fatty acids (EFA), in particular the

M. M.-L. Grundy (✉)
PEGASE, INRAE, Institut Agro, Saint Gilles, France
e-mail: myriam.grundy@inrae.fr

P. J. Wilde
Quadram Institute Bioscience, Norwich Research Park, Colney, Norwich, UK
e-mail: peter.wilde@quadram.ac.uk

M. M.-L. Grundy, P. J. Wilde (eds.), *Bioaccessibility and Digestibility of Lipids from Food*, https://doi.org/10.1007/978-3-030-56909-9_7

omega-3 polyunsaturated fatty acids (PUFA), benefits in cardiovascular diseases, certain cancers, age-related cognitive decline, and depression [117]. Dietary lipids are also required for the transport and absorption of lipid soluble vitamins (A, D, E and K) and they are the precursors of steroid hormones (oestrogens, testosterone, adrenal hormones) [62]. Finally, from an organoleptic point of view, lipids provide palatability to the food by contributing to its texture (mouthfeel) and also have the capacity to carry aromatic compounds [82].

When it comes to plant-based foods, in particular the dietary fibres they contain and their positive effect on health, a lot of attention has been devoted to the events that occur in the colonic phase [1, 14, 29, 50, 73]. This is mainly because dietary fibres by definition are carbohydrate that are not hydrolysed in the upper gut of humans but degraded (fermented) by the microorganisms present in the large intestine. A striking example is how dietary fibre can modulate the gut microbiota, the products of their activity (short-chain fatty acids, SCFA) and consequently the risk factors of diet-related diseases such as CVD [26]. However, what is happening in the upper gut can also significantly affects lipid metabolism and CVD risk profile. Indeed, the structure of plant foods has an impact on postprandial nutrient and hormone responses including postprandial lipaemia [20, 49, 105]. The major sources of dietary fibre are plant cell walls. The composition, structure and porosity of the cell walls and the way there are degraded before (as a result of food processing) and during digestion in the upper gut has been demonstrated to be of critical importance for lipid release and digestibility, and thereby their health effects [18, 33, 56, 80]. Furthermore, the type and form of lipids contained within a plant food influence lipid metabolism.

Lipids are an extremely diverse class of compounds that are classified based on their physical properties: extracted/solubilised in nonpolar organic solvents [104]. Therefore, they are mostly hydrophobic and associate by noncovalent interactions when in contact with water (e.g. saliva and gastrointestinal fluids). However, some lipids, such as phospholipids and galactolipids, possess a hydrophobic and a hydrophilic region, which enables them to be in direct contact with both lipid and aqueous environments. Because of this particular chemical property, amphipathic molecules can organise themselves so that they form mono- or bilayers, micelles and vesicles. Lipids can adopt different behaviour in the gastrointestinal tract (GIT) based on their physicochemical properties, polarity as described above but also physical state (liquid vs solid), level of saturation, and chain length. During digestion, they can therefore change physical state, particle size, aggregation state, composition and interfacial properties [91].

In this chapter, we will first present the different sources of lipids from plant origin and their organisation within the plant tissues. We will then provide an overview of the current literature that shows evidence of the mechanisms by which the plant structure and its constituents influence lipid bioaccessibility and digestibility.

7.1 Lipids in Plant

7.1.1 Forms of Lipids

7.1.1.1 Composition

Triacylglycerol (TAG) is the most abundant form of lipid present in plants (up to 97%) but other glycerolipids (diacylglycerols, monoacylglycerols) and free fatty acids are also found [93, 104]. Phospholipids, glycolipids such as galactolipids, and isoprenoid compounds (e.g. sterols, carotenoids, tocols and saponins) are present in smaller amount mainly in cell membranes but are of great nutritional importance due to their distinctive physicochemical properties (amphiphilic character and precursors of steroids hormones). Waxes and the cell wall components, cutin and suberin, form a protective coat for the plant that controls water flow and transport [113].

Fatty acids can be classified according to the number of double bonds they contain: saturated (none, SFA), monounsaturated (one, MUFA) or polyunsaturated (two or more, PUFA) fatty acids. EFAs belong to the PUFA category; they are fatty acids that cannot be synthesised by the human body, EFA should therefore be provided from the diet. They include two families: the omega-3 such as ALA and its derivatives EPA and DHA, and the omega-6 such as linoleic acid (LA) a precursor of arachidonic acid (AA). Lipids from plants contain a wider variety of fatty acids compared with animal sources in terms of both chain length and degrees of unsaturation. Plant lipids tend to be liquid at room temperature because of the high levels of unsaturated fatty acids that they generally contain. Exception to this generalisation are coconut and palm oils that are high in saturated fats. Plants are also a source of omega 3 fatty acids, ALA but not the long chain EPA and DHA. Overall, lipid and fatty acid composition fluctuates greatly between cell types and plant species due to their diverse roles and functions [123].

7.1.1.2 Organisation

Dietary relevant plant lipids, like animal lipids, tend to be organised in two main forms, as either storage or functional lipids.

Storage Lipids

Lipids as discussed earlier are an excellent source of energy, therefore some plants utilise lipids instead of starch as an energy store. Oil bodies are ubiquitous in plants, they are not only used as an energy store, but also to provide the substrate for synthesising functional lipids and hormones and molecules associated with stress response and pathogen resistance [27]. In plants that do utilise lipids for energy storage, the lipids are mainly in the form of TAGs, located in oil bodies primarily in seeds, although oil bodies can be found throughout the plant, but in much lower concentrations. Oil bodies are essentially droplets of primarily TAG, stabilised by a surface layer of phospholipids and proteins [45]. The proteins, such as the oleosins, steroleosins and caleosins, help to both physically stabilise the droplets as well as regulate their biogenesis and utilisation [118]. Most seed oil bodies range in diameter from 0.5 to 2 μm [129]. Oil bodies are composed primarily (94–98%) of neutral TAG that forms the core of the lipid droplet, and about 0.5–2% polar lipids, largely forming the outer stabilising layer [129]. Oil bodies may also contain small quantities of lipid soluble nutrients such as tocopherols (vitamin E) and polyphenols [43, 44].

Functional Lipids

Lipids perform a wide variety of additional roles in plants from cell membranes structure, signalling, hormone precursors, moisture retention (waxes), antioxidants and photosynthesis [113]. From a dietary perspective, most functional lipids consumed in the diet are polar lipids which are mainly in the form of membrane structures [93]. Cell membranes are largely composed of a bilayer containing a mixture of polar lipids including phospholipids, galactolipids and sphingomyelin. Phospholipid membranes are found in a range of organelles, including the outer cell membrane, Golgi apparatus, endoplasmic reticulum and chloroplasts. These structures contain most of the polar lipids in plants and are vital for the majority of essential physiological functions. These structures also contain bioactive lipophilic molecules such as tocopherols, carotenoids, phytosterols and polyphenols.

7.1.2 Plant Sources

Lipids are a major form of carbon storage in many plant species, and they are predominantly found in the grains and seeds of plants in the form of lipid or oil bodies [113]. However, as discussed above, some lipids can also be present in other part of the plant, for instance leaves in the form of oil bodies, but also in membrane structures such as chloroplasts that contain high levels of polar lipids.

7.1.2.1 Seeds and Nuts

Seeds and nuts are usually rich in oils, with oil contents often around 50–75% by dry weight (USDA data tables https://fdc.nal.usda.gov/index.html). As described earlier, seeds contain significant stores of lipids to aid germination, growth and other physiological functions. Around 95% of the lipid in most seeds and nuts is in the form of TAG, with the remainder mainly comprising polar lipids. Some seeds and nuts are widely consumed in their native form or as ingredients in other foodstuffs [93]. However, due to their high oil content, the oil is extracted, particularly from seeds such as soybean, rapeseed, sunflower, peanut, camelina and sesame, and used as a food ingredient [74, 110]. The extracted oil is normally at least 95% TAG, but depending on the method of processing and extraction, the presence of polar lipids and antioxidants may vary.

7.1.2.2 Cereals and Pseudo-Cereals

Cereals and pseudo-cereals have in general a rather low in lipid content, however oat and amaranth can contain 10% of fat or more depending on the variety [5, 130]. This is because the grains generally use starch as their primary energy source, but there are low levels of oil bodies in the grain as a lipid source during germination [112]. Oil is often extracted from maize, even though the lipid content is only 3–5% of the grain. However, 80–85% of the lipid is in the germ, which can be separated from the rest of the grain [135].

7.1.2.3 Fruit and Vegetables

Botanically, the word "fruit" is a generic term that refers to the part of a flowering plant that contains the seeds (mature ovary). Whereas, in culinary terms fruits are categorised as fruit and vegetables mainly depending on their taste, sweet for fruit and savoury for vegetables [122]. Fruit and vegetables contain a wide variety and amounts of lipids due to the numerous physiological forms and functions of the different kinds of fruits and vegetable consumed. The section below describes the main lipids found in different fruit and vegetable structures.

Some fruits store significant quantities of lipid within their structure, the most notable being olive and avocado [110]. Their lipids are located in the pericarp and their content is around 10–15%, but this is due to the high water content of these fruits, which corresponds to around 50% oil by dry weight [101]. Contrary to seeds these fruits do not possess oil bodies, instead the TAG fuse to form large droplets (~30 μm). For highly pigmented fruits and vegetables such as tomatoes, peppers and carrots, although their oil content is low, it acts as a carrier and a rich source of carotenoids and other lipophilic phytochemicals thought to benefit health [96].

Waxes present in the cuticle of plant fruits, stems and leaves offer a protective barrier to the environment, thereby preventing excessive water and nutrients loss [107].

Green leafy vegetables such as spinach and kale are rich in polar lipids, especially galactolipids due to the high concentration of chloroplasts in this part of the plant [69]. These structures are also a source of lipophilic carotenoids and flavonoids.

There are a wide variety of species and morphologies of algae as potential food sources [136] and their nutritional quality and lipid content will depend on many factors. Nevertheless, many algae species, given the right growing conditions can be a rich and sustainable source of long chain PUFAs.

7.2 Mechanisms Affecting Lipid Bioaccessibility and Digestibility

Beyond the nutritional composition of a plant-based food, its macro- and micro-structure influences the release of nutrients and subsequently their digestion in the upper GIT (Fig. 7.1). To enable and/or speed up their digestion by human beings, many plants, especially cereals and legumes, undergo some form of processing before being consumed [4, 42]. Food processing can be defined as any mechanical or chemical transformation applied to a raw food material or ingredient (post-harvest) before its ingestion. This includes cleaning and removal of outer layers (e.g. dehulling), particle size reduction (e.g. milling/grinding), heat treatment (e.g. drying, boiling, baking and roasting), soaking, canning, fermentation or enzymatic treatment. In addition to render the plant-based food edible, these processing

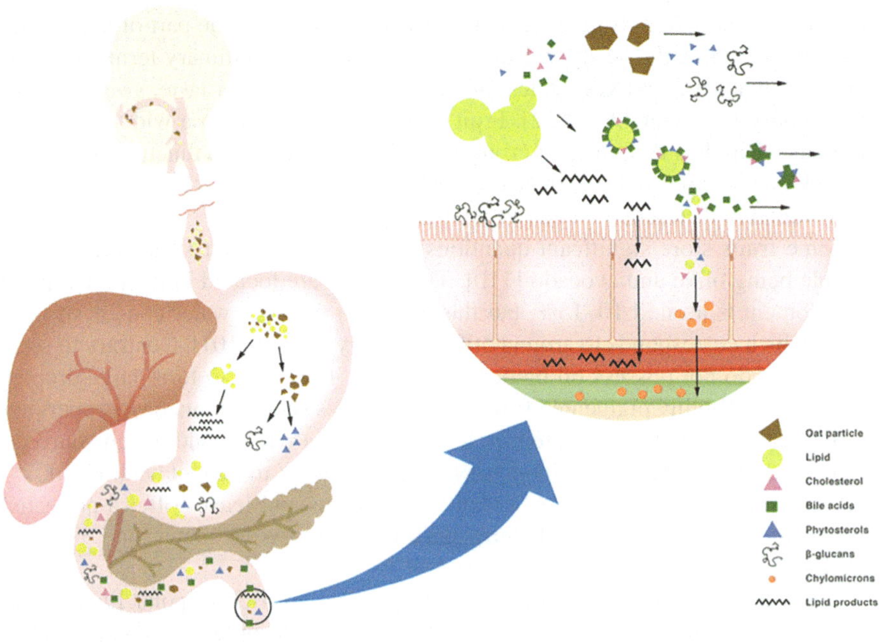

Fig. 7.1 Schematic of potential set of events (macro- and microscopic scales) that can take place during lipid digestion in oats

methods aim at improving its palatability, safety, shelf life, and in certain cases, its nutritional value (enhanced bioavailability) [4].

Then, during digestion itself the structure of the food will further be altered due to mechanical (chewing, gastric contractions, peristalsis), chemical (pH, ionic strength) and biochemical (enzymatic activity) processes. As discussed earlier, lipids are generally insoluble in water, and dietary lipids from plants are in two main structural forms, membrane lipids in structures such as thylakoids in chloroplasts, and oil bodies. Because these are large, super-molecular structures the effect of processing on the properties of these structures as well as the plant tissue matrix itself will be presented in the sections below.

7.2.1 Macroscale: Degradation of the Food Matrix and Nutrients/Compounds Release

Processing techniques are numerous and, as mentioned above, aimed at transforming raw food plants to make them palatable and safe to consume [4, 42]. Starch granules, for instance, gelatinise during cooking, and thus become susceptible to amylase hydrolysis. Milling and grinding, by reducing the size of the plant tissue (increase surface area to volume ratio), enhance the bioaccessibility of macronutrients and the chance of contact between the enzymes and their substrates [34, 54, 64]. Fermentation could also increase the bioavailability of nutrients by degrading the cell wall [102, 140]. However, processing will have diverse effect based on the original plant tissue and the nutrient or food compound.

Another important aspect to consider is the maturation of fruit and vegetables. During maturation and ripening, the tissue and cellular structure of the plant is transformed notably via softening due to enzymes activity and degradation of certain plant compounds such as cell wall polysaccharides [108, 142].

7.2.1.1 Cell Wall Barrier

Depending on the part of the plant organ or tissue (parenchyma, sclerenchyma and collenchyma) where it comes from, a food will have cell walls with specific physicochemical properties, and therefore behaviour in the GIT [92]. These properties are determined by the types of polysaccharides that compose the cell wall and how they are organised. Cell walls are made of a complex combination of polysaccharides (cellulose, hemicellulose, and pectin), proteins (e.g. extensin) and phenolic compounds (e.g ferulic acid) arranged into two or three layers. The first, outer layer is the middle lamella that is shared between adjacent cells. This layer is composed of pectic compounds and proteins, and it is relatively thin apart from the cell corner regions. The second layer is the primary cell wall. This is a dynamic structure, the spatial arrangement of the polymer network and molecular composition is modified during growth and by attack from pathogens. It has various functions including maintenance of cell shape and ultimately plant physiology, through control of cell expansion and cell signalling notably via plasmodesmata [12]. The third layer, the secondary cell wall, that is the closest to the plasma membrane, is synthesised when the cell has ceased to enlarge [144]. Contrary to the primary cell wall, the secondary cell wall is predominantly present in specific cells, sclerenchyma cells, which are found in tracheary elements (in vascular plant) as well as fibres of wood. The secondary cell wall is a rigid layer that provides mechanical support and defence against pathogens to the plant; the sclerenchyma cells also permit the transport of water.

The structure and layout of the primary cell wall components can be classified as type I and type II [23]. Dicotyledons (i.e. most fruit and vegetables), non-commelinid monocotyledons (e.g. asparagus and onion), and conifers have a type I cell wall, whereas type II cell walls are found in cereals and grasses.

Given the complexity of its organisation (tight packing of cellulose microfibrils that obstruct the access of enzymes to the linkages) and the fact that human do not secrete the enzymes that can hydrolyse the cell wall polysaccharides into their constituent monosaccharides, the cell wall has to be degraded by processing or mastication for the lipase to access its substrate. Indeed, the cell wall of plant-based foods behaves as a filter that regulates the exchange of molecules between the intra and extracellular compartments.

Based on the cross-links existing between the cell wall components, the cells within a plant tissue can behave in two different ways during processing and mastication: they can either rupture or separate [12]. A food with cells that rupture will have its content released during processing and/or digestion, as it is the case for almonds [37] or raw carrots [128], whereas in soft plant-based food cells separate (e.g. cooked pulses) and their content is not available as the cell walls remain intact [133]. Encapsulation of nutrients by the cell wall therefore delay or prevent the access of the lipases to their substrate, as demonstrated in nuts [19, 55]. Processing will increase the bioaccessibility of the lipids and lipid soluble molecules, in particular particle size reduction and thermal treatment can damage the cell wall and allow access of the cell content to the enzymes [17, 54, 77, 128].

However, the cell wall is not an impermeable barrier and communication between the cell and the outside environment can be assured in certain cases, notably through pores such as plasmodesmata, creating an opening at the cell surface. Only molecules with a diameter smaller than the pores, the size of which is specific to the plant-based food studied, would have the ability to permeate the plasma membrane [6, 24, 35]. The diffusion of a molecule through the cell wall not only depends on the size of the pore but also on its conformation and flexibility. Recent studies performed on separated cells demonstrated that enzymes were unable to diffuse through the cell walls of raw, native plant cells [8, 55, 80] and thermal processing was necessary for nutrient digestion to occur [98].

7.2.1.2 Viscosity

Water-soluble polysaccharides that made up cell walls can be released during processing and digestion [57], which has consequences on lipid hydrolysis [31, 58, 89]. The released polysaccharides will increase the viscosity of the content of the stomach and small intestine, the extent of which will rely on the molecular weight and concentration of these polysaccharides [76, 138]. Increase viscosity of the digesta will reduce mixing and the polysaccharides will form a network that will decrease the probability of enzymes and substrates to come in contact [78]. The network formed can also affect the motility of the lipid droplets that have been released from the food matrix by a mechanism known as depletion flocculation [3, 40, 59, 139]. The end result of this is a delay in lipid digestion.

Components of the food matrix, other than non-starch polysaccharides, such as proteins and starch can also generate viscous solutions, particularly when cooked in excess water [94, 114]. Therefore, particles with rheological properties, such as starch granules, can increase the viscosity of the gastrointestinal content [78, 125]. However, depending on the nature of those particles the impact on the overall rheology can fluctuate very rapidly once exposed to digestive enzymes and mixing occurring in the GIT [63]. For instance, hydrolysis of starch by α-amylase will reduce the bolus and digesta viscosity in just a few seconds [41].

7.2.1.3 Overall Food Matrix Effect

Beside encapsulation by cell wall and viscosity the overall structure and organisation of the plant-food matrix will govern lipid bioaccessibility and digestibility. Gastric emptying depends on the size (it should be under ~3 mm to pass to the intestine) and amount of particulates present in the chyme [127]. The physicochemical properties of that chyme will also determine the rate of gastric emptying: calorie content, density, lipid or water solubility, and viscosity [86, 87]. In order for the lipids to be digested, the food consumed has to be transformed into particles that are diluted into the gastrointestinal fluids and of small enough size to allow the penetration of digestive agents (i.e. lipases, colipase, and bile salts). Starch and aggregated proteins may prevent the release and digestion of lipids [16, 79]. Similarly to cell wall they can create a physical barrier.

The phenomena described above will dictate the rate and extent of lipid released within the food matrix and subsequently their digestibility. If these are too slow, they will reach the colonic compartment where they will then be degraded by the microbiota. Part of that material/digesta may not be fully degraded and will then be excreted in the faeces [36].

7.2.2 Microscale: Digestion of Lipid-Rich Organelles and Structures Formed During Lipolysis

As discussed in Chap. 2, the 'quality' of the interface (i.e. presence of surface active agents) affects the enzymic lipolysis; the organisation and composition of the lipid system (e.g. micelles, emulsions and oil bodies) have direct implication on the surface area/volume ratio of the oil droplet [22, 99, 131]. Emulsification, the dispersion of oil in water by creating a suspension of oil droplets, is an essential step in lipid digestion that produces a drastic rise in the ratio of surface area to droplet volume. The ultrastructure of the emulsion and the size of its droplets regulate lipase activity [85, 121]. According to the type of fatty acids present on the glycerol molecule backbone, in particular positions sn-1 and sn-3, the lipolysis rate will be different [2, 68]. In other words, lipolysis is governed by environmental conditions and compounds that can alter the structure and/or the integrity of the interface (e.g. surface-active molecules) [25, 38]. Some of those compounds are originally present in the food consumed (e.g. proteins, galactolipids, phospholipids and phytosterols) while others are produced during the digestion process (diacylglycerols, monoacylglycerols and free fatty acids) or secreted in the GIT (e.g. bile salts, colipase and calcium).

7.2.2.1 Oil Bodies and Chloroplasts Digestion

In principle, oil bodies are readily digestible as they are already emulsified (Fig. 7.2). However when transiting throughout the GIT they may lose their integrity, in particular in the gastric environment [47], and combine with other food components [137]. Upon processing, the integrity of the oil bodies can also be compromised, as

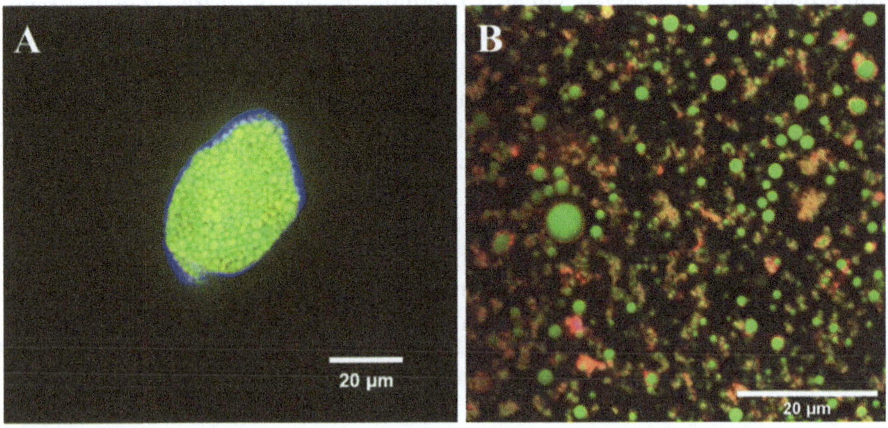

Fig. 7.2 Almond oil bodies enclosed within the cell (**a**), Amaranth oil bodies surrounded by proteins (**b**). Lipids are stained with Nile red (green), cell wall are stained with Calcofluor White (blue), and proteins are with Fast Green (red)

observed in roasted almond, hazelnut and soy [19, 32, 54]. Those events will delay or reduce the digestibility of the lipids.

Moreover, oil bodies are stabilised by a layer of phospholipids and amphiphilic proteins (e.g. oleosins, caleosins, and steroleosin) [67, 103]. The proteins are well embedded into the lipid phase and difficult to displace, therefore proteolysis is required to allow lipase access to the interface [47, 55]. Some of the proteins can remain undigested due to interaction with other food components, for example phenolic compounds, nevertheless lipolysis occurs [7, 46].

Chloroplasts are an often-ignored source of dietary lipids, in particular galactolipids [48, 116]. Their digestion will contribute to free fatty acids, predominantly α-linolenic acid, production [134].

7.2.2.2 Interactions Between Plant Lipids and Other Compounds During Lipid Digestion

Lipid molecules self-assemble into different complex architectures – oil bodies, membranes, vesicles and micelles of different shapes, in order to avoid contact of lipophilic regions with water [100, 109]. During digestion, amphiphilic molecules are produced (di- and monoacylglycerols) and the lipid droplets go through transition leading to new self-assembled structures [111]. Also, bile salts present in the GIT interact with these self-assembled lipids, resulting in colloidal structures with specific physicochemical properties (solubility and interfacial quality) [83]. These dynamic processes tailor the rate and extent of the digestion of lipids and their absorption.

Therefore, the initial properties and the organisation of the lipids will determine their fate in the GIT. While being digested, the lipid droplets can become unstable as the composition of the interface (hydrolysis of proteins, phospholipids and galactolipids) and their size change. These instabilities can cause coalescence, flocculation, creaming, and sedimentation of the droplets [90]. Lipids coalescence and flocculation will reduce the interfacial surface available for the lipases and bile salts to adsorb [52]. On the other hand, the formation of lipolytic products results in a decline in interfacial tension thereby promoting lipase anchoring onto the interface [106].

Surface active molecules present in plants, such as phytosterols and saponins, have been shown to delay lipolysis [137]. The precise mechanisms involved are complex and not fully elucidated. Nonetheless, it was revealed that phytosterols affected the interfacial properties of lipid droplets and interacted with bile salts and micelles formation. Saponins, for their part, are thought to act as lipase inhibitor [97, 143].

Soluble polysaccharides are thought to directly interact with lipases and impact their activity. Some studies have found that some dietary fibers were able to decrease lipase activity [39, 115]. However, the mechanisms behind this potential inhibitory effect warrant further work.

Finally, throughout digestion, lipids can change state and form crystals that either remain at the water-oil interface or precipitate [22]. This phenomena is triggered notably by the presence of calcium [46, 65].

7.2.2.3 Micelles Formation, Transport and Bile Salt Recycling

To help solubilise lipolysis products and lipophilic compounds, and transport them to the enterocytes where they are absorbed, bile salts create amphiphilic structures in the form of mixed micelles [83]. Mixed micelles cover a range of organisations from simple bile salt micelles to vesicle like structures [22, 100]. Pure bile salts can form simple micelles, however, due to their unusual structure, compared to other surfactants, their aggregation behaviour is not conventional [84]. Bile salts are derived from cholesterol, so they have opposing hydrophilic and hydrophobic faces. At low concentrations they form primary micelles with aggregation numbers between 3 and 30 molecules. At higher concentrations they form larger structures, for which many models have been proposed including ellipsoid, rod-like, disc-like and helical [21, 84]. Bile also contains phospholipids and cholesterol. The ratio between phospholipids and bile salts, will influence the proportion of smaller bile salt micelles, and larger vesicle structures [81, 88]. Like detergents, bile salts will facilitate the solubilisation of lipophilic compounds from the lipid phase and their partitioning into micelles or vesicles. Different lipophilic molecules will have affinities for the different structures depending on their polarity, charge and size [30]. For example, it has been shown that vitamin E is more soluble in mixed micelles and vesicles with a higher phospholipid content [141]. Similarly, other studies have shown the many poorly soluble drugs are associated with mixed micelles, but the more hydrophobic molecules tend to associate more with vesicle structures [71]. In addition, the formation of these structures is dynamic during lipid digestion, as free fatty acids and monoglycerides are produced during lipolysis, they will be incorporated into the mixed micelles and vesicles, changing the overall composition and structure as digestion progresses. Kleberg et al. showed that the incorporation of some poorly soluble drugs with mixed micelles was influenced by the proportion of fatty acids to monoglycerides [70].

The micelles then diffuse to the gut wall to facilitate absorption of their payload [22, 83]. The precise mechanisms are not known, but hypotheses exist regarding the bile increasing cell membrane permeability, and there is also evidence that bile salts enhance paracellular transport via tight junctions [95]. During this process bile salts are reabsorbed by the epithelium via bile acid binding protein transport in the ileum, where around 95% of the bile acids are reabsorbed and recirculated to the liver for secretion back into the duodenum (enterohepatic circulation) [66]. Disruption of this recirculation pathway impacts on whole body lipid metabolism [28]. For example, fibre binding to bile salts can disrupt this by depleting the amount of recirculating bile triggering the synthesis of bile from cholesterol, thus reducing serum cholesterol levels [60, 61]. Also, the presence of plant sterols and stanols can disrupt this process by interfering with cholesterol uptake [13, 51]. For more details about these mechanisms, the reader is referred to a recent review [83].

7.3 Conclusion

Plants are a source of wide variety of lipids, organised primarily in oil bodies and to a lesser extent in chloroplasts (found in leaves). The matrix of plant-based food often limits lipid bioaccessibility and digestibility. Indeed, its components (e.g. dietary fibre, protein, starch and phytosterols) can interact with the digestive agents, such as bile salts and lipases. The impact of cell walls has been particularly studied in recent years; they can encapsulate the lipids and prevent their release from the food matrix and thus their digestibility. Progress has been made in our understanding on how lipid from plant get digested. However, more research is needed to fully comprehend some of the mechanisms involved.

References

1. Anderson JW, Baird P, Davis RH Jr, Ferreri S, Knudtson M, Koraym A, Waters V, Williams CL (2009) Health benefits of dietary fiber. Nutr Rev 67(4):188–205
2. Armand M, Borel P, Ythier P, Dutot G, Melin C, Senft M, Lafont H, Lairon D (1992) Effects of droplet size, triacylglycerol composition, and calcium on the hydrolysis of complex emulsions by pancreatic lipase: an in *vitro* study. J Nutr Biochem 3(7):333–341
3. Asakura S, Oosawa F (1958) Interaction between particles suspended in solutions of macromolecules. J Polym Sci 33(126):183–192
4. Augustin MA, Riley M, Stockmann R, Bennett L, Kahl A, Lockett T, Osmond M, Sanguansri P, Stonehouse W, Zajac I, Cobiac L (2016) Role of food processing in food and nutrition security. Trends Food Sci Technol 56:115–125
5. Banas A, Debski H, Banas W, Heneen WK, Dahlqvist A, Bafor M, Gummeson PO, Marttila S, Ekman A, Carlsson AS, Stymne S (2007) Lipids in grain tissues of oat (Avena sativa): differences in content, time of deposition, and fatty acid composition. J Exp Bot 58(10):2463–2470
6. Baron-Epel O, Gharyal PK, Schindler M (1988) Pectins as mediators of wall porosity in soybean cells. Planta 175(3):389–395
7. Beisson F, Ferte N, Voultoury R, Arondel V (2001) Large scale purification of an almond oleosin using an organic solvent procedure. Plant Physiol Biochem 39(7–8):623–630
8. Bhattarai RR, Dhital S, Mense A, Gidley MJ, Shi Y-C (2018) Intact cellular structure in cereal endosperm limits starch digestion in vitro. Food Hydrocoll 81:139–148
9. Boden G (1997) Role of fatty acids in the pathogenesis of insulin resistance and NIDDM. Diabetes 46(1):3–10
10. Bourre JM (2006) Effects of nutrients (in food) on the structure and function of the nervous system: update on dietary requirements for brain. Part 2: macronutrients. J Nutr Health Aging 10(5):386–399
11. Bray GA, Paeratakul S, Popkin BM (2004) Dietary fat and obesity: a review of animal, clinical and epidemiological studies. Physiol Behav 83(4):549–555
12. Brett C, Waldron K (1996) Physiology and biochemistry of plant cell walls, 2nd edn. Chapman & Hall, Cambridge

13. Brufau G, Canela MA, Rafecas M (2008) Phytosterols: physiologic and metabolic aspects related to cholesterol-lowering properties. Nutr Res 28(4):217–225
14. Burkitt DP, Walker ARP, Painter NS (1972) Effect of dietary fibre on stools and transit-times, and its role in the causation of disease. Lancet 300(7792):1408–1411
15. Calder PC (2010) Omega-3 fatty acids and inflammatory processes. Nutrients 2(3):355–374
16. Calvo-Lerma J, Fornés-Ferrer V, Heredia A, Andrés A (2018) In vitro digestion of lipids in real foods: influence of lipid organization within the food matrix and interactions with non-lipid components. J Food Sci 83(10):2629–2637
17. Calvo-Lerma J, Paz-Yepez C, Asensio-Grau A, Heredia A, Andres A (2020) Impact of processing and intestinal conditions on *in vitro* digestion of chia (Salvia hispanica) seeds and derivatives. Foods 9(3):290
18. Capuano E (2017) The behavior of dietary fiber in the gastrointestinal tract determines its physiological effect. Crit Rev Food Sci Nutr 57(16):3543–3564
19. Capuano E, Pellegrini N, Ntone E, Nikiforidis CV (2018) In vitro lipid digestion in raw and roasted hazelnut particles and oil bodies. Food Funct 9(4):2508–2516
20. Cara L, Dubois C, Borel P, Armand M, Senft M, Portugal H, Pauli AM, Bernard PM, Lairon D (1992) Effects of oat bran, rice bran, wheat fiber, and wheat germ on postprandial lipemia in healthy adults. Am J Clin Nutr 55(1):81–88
21. Carey MC, Small DM (1970) The characteristics of mixed micellar solutions with particular reference to bile. Am J Med 49:590–608
22. Carey MC, Small DM, Bliss CM (1983) Lipid digestion and absorption. Annu Rev Physiol 45:651–677
23. Carpita NC, Gibeaut DM (1993) Structural models of primary-cell walls in flowering plants - consistency of molecular-structure with the physical-properties of the walls during growth. Plant J 3(1):1–30
24. Carpita NC, Sabularse D, Montezinos D, Delmer DP (1979) Determination of the pore size of cell walls of living plant cells. Science 205(4411):1144–1147
25. Carriere F, Moreau H, Raphel V, Laugier R, Benicourt C, Junien J-L, Verger R (1991) Purification and biochemical characterization of dog gastric lipase. Eur J Biochem 202(1):75–83
26. Chambers ES, Preston T, Frost G, Morrison DJ (2018) Role of gut microbiota-generated short-chain fatty acids in metabolic and cardiovascular health. Curr Nutr Rep 7(4):198–206
27. Chapman KD, Dyer JM, Mullen RT (2012) Biogenesis and functions of lipid droplets in plants. J Lipid Res 53(2):215–226
28. Chiang JYL (2009) Bile acids: regulation of synthesis. J Lipid Res 50(10):1955–1966
29. Cummings JH, Stephen AM (1980) The role of dietary fibre in the human colon. Can Med Assoc J 123(11):1109–1114
30. Desmarchelier C, Rosilio V, Chapron D, Makky A, Preveraud DP, Devillard E, Legrand-Defretin V, Borel P (2018) Molecular interactions governing the incorporation of cholecalciferol and retinyl-palmitate in mixed taurocholate-lipid micelles. Food Chem 250:221–229
31. Dikeman CL, Fahey GC (2006) Viscosity as related to dietary fiber: a review. Crit Rev Food Sci Nutr 46(8):649–663
32. Ding J, Wen J, Wang J, Tian R, Yu L, Jiang L, Zhang Y, Sui X (2020) The physicochemical properties and gastrointestinal fate of oleosomes from non-heated and heated soymilk. Food Hydrocoll 100:105418
33. Do DT, Singh J, Oey I, Singh H (2018) Biomimetic plant foods: structural design and functionality. Trends Food Sci Technol 82:46–59
34. Edwards CH, Grundy MM-L, Grassby T, Vasilopoulou D, Frost GS, Butterworth PJ, Berry SEE, Sanderson J, Ellis PR (2015) Manipulation of starch bioaccessibility in wheat endosperm to regulate starch digestion, postprandial glycemia, insulinemia, and gut hormone responses: a randomized controlled trial in healthy ileostomy participants. Am J Clin Nutr 102(4):791–800

35. Ehwald R, Woehlecke H, Titel C (1992) Cell-wall microcapsules with different porosity from suspension cultured *Chenopodium album*. Phytochemistry 31(9):3033–3038
36. Elia M, Cummings JH (2007) Physiological aspects of energy metabolism and gastrointestinal effects of carbohydrates. Eur J Clin Nutr 61(Suppl 1):S40–S74
37. Ellis PR, Kendall CW, Ren Y, Parker C, Pacy JF, Waldron KW, Jenkins DJ (2004) Role of cell walls in the bioaccessibility of lipids in almond seeds. Am J Clin Nutr 80(3):604–613
38. Embleton JK, Pouton CW (1997) Structure and function of gastro-intestinal lipases. Adv Drug Deliv Rev 25(1):15–32
39. Espinal-Ruiz M, Parada-Alfonso F, Restrepo-Sánchez L-P, Narváez-Cuenca C-E (2014) Inhibition of digestive enzyme activities by pectic polysaccharides in model solutions. Bioact Carbohydr Diet Fibre 4(1):27–38
40. Espinal-Ruiz M, Restrepo-Sánchez L-P, Narváez-Cuenca C-E, McClements DJ (2016) Impact of pectin properties on lipid digestion under simulated gastrointestinal conditions: comparison of citrus and banana passion fruit (*Passiflora tripartita* var. *mollissima*) pectins. Food Hydrocoll 52:329–342
41. Evans ID, Haisman DR, Elson EL, Pasternak C, McConnaughey WB (1986) The effect of salivary amylase on the viscosity behaviour of gelatinised starch suspensions and the mechanical properties of gelatinised starch granules. J Sci Food Agric 37(6):573–590
42. Fellows PJ (2016) Food processing technology: principles and practice. Elsevier Science & Technology, New York
43. Fisk ID, Gray DA (2011) Soybean (Glycine max) oil bodies and their associated phytochemicals. J Food Sci 76(9):C1349–C1354
44. Fisk ID, White DA, Carvalho A, Gray DA (2006) Tocopherol - an intrinsic component of sunflower seed oil bodies. J Am Oil Chem Soc 83(4):341–344
45. Frandsen GI, Mundy J, Tzen JTC (2001) Oil bodies and their associated proteins, oleosin and caleosin. Physiol Plant 112(3):301–307
46. Gallier S, Tate H, Singh H (2013) *In vitro* gastric and intestinal digestion of a walnut oil body dispersion. J Agric Food Chem 61(2):410–417
47. Gallier S, Acton D, Garg M, Singh H (2017) Natural and processed milk and oil body emulsions: bioavailability, bioaccessibility and functionality. Food Struct 13:13–23
48. Gedi MA, Briars R, Yuseli F, Zainol N, Darwish R, Salter AM, Gray DA (2017) Component analysis of nutritionally rich chloroplasts: recovery from conventional and unconventional green plant species. J Food Sci Technol 54(9):2746–2757
49. Gemen R, de Vries JF, Slavin JL (2011) Relationship between molecular structure of cereal dietary fiber and health effects: focus on glucose/insulin response and gut health. Nutr Rev 69(1):22–33
50. Gill S, Chater PI, Wilcox MD, Pearson JP, Brownlee IA (2018) The impact of dietary fibres on the physiological processes of the large intestine. Bioact Carbohydr Diet Fibre 16:62–74
51. Gleize B, Nowicki M, Daval C, Koutnikova H, Borel P (2016) Form of phytosterols and food matrix in which they are incorporated modulate their incorporation into mixed micelles and impact cholesterol micellarization. Mol Nutr Food Res 60(4):749–759
52. Golding M, Wooster TJ (2010) The influence of emulsion structure and stability on lipid digestion. Curr Opin Colloid Interface Sci 15(1–2):90–101
53. Grundy SM, Denke MA (1990) Dietary influences on serum lipids and lipoproteins. J Lipid Res 31(7):1149–1172
54. Grundy MM-L, Wilde PJ, Butterworth PJ, Gray R, Ellis PR (2015) Impact of cell wall encapsulation of almonds on *in vitro* duodenal lipolysis. Food Chem 185:405–412
55. Grundy MM-L, Carriere F, Mackie AR, Gray DA, Butterworth PJ, Ellis PR (2016a) The role of plant cell wall encapsulation and porosity in regulating lipolysis during the digestion of almond seeds. Food Funct 7:69–78
56. Grundy MM-L, Edwards CH, Mackie AR, Gidley MJ, Butterworth PJ, Ellis PR (2016b) Re-evaluation of the mechanisms of dietary fibre and implications for macronutrient bioaccessibility, digestion and postprandial metabolism. Br J Nutr 116(5):816–833

57. Grundy MM-L, Quint J, Rieder A, Ballance S, Dreiss CA, Butterworth PJ, Ellis PR (2017a) Impact of hydrothermal and mechanical processing on dissolution kinetics and rheology of oat β-glucan. Carbohydr Polym 166:387–397
58. Grundy MM-L, Quint J, Rieder A, Ballance S, Dreiss CA, Cross KL, Gray R, Bajka BH, Butterworth PJ, Ellis PR, Wilde PJ (2017b) The impact of oat structure and β-glucan on in vitro lipid digestion. J Funct Foods 38(Part A):378–388
59. Grundy MML, McClements DJ, Ballance S, Wilde PJ (2018) Influence of oat components on lipid digestion using an in vitro model: impact of viscosity and depletion flocculation mechanism. Food Hydrocoll 83:253–264
60. Gunness P, Gidley MJ (2010) Mechanisms underlying the cholesterol-lowering properties of soluble dietary fibre polysaccharides. Food Funct 1(2):149–155
61. Gunness P, Michiels J, Vanhaecke L, De Smet S, Kravchuk O, Van de Meene A, Gidley MJ (2016) Reduction in circulating bile acid and restricted diffusion across the intestinal epithelium are associated with a decrease in blood cholesterol in the presence of oat β-glucan. FASEB J 30(12):4227–4238
62. Gurr MI, Harwood JL, Frayn KN (2002) Lipid biochemistry: an introduction, 5th edn. Wiley-Blackwell, Oxford
63. Hardacre AK, Lentle RG, Yap S-Y, Monro JA (2016) Does viscosity or structure govern the rate at which starch granules are digested? Carbohydr Polym 136:667–675
64. Heaton KW, Marcus SN, Emmett PM, Bolton CH (1988) Particle size of wheat, maize, and oat test meals - effects on plasma-glucose and insulin responses and on the rate of starch digestion *in vitro*. Am J Clin Nutr 47(4):675–682
65. Hernell O, Staggers JE, Carey MC (1990) Physical-chemical behavior of dietary and biliary lipids during intestinal digestion and absorption. 2. Phase analysis and aggregation states of luminal lipids during duodenal fat digestion in healthy adult human beings. Biochemistry 29(8):2041–2056
66. Hofmann AF (2009) The enterohepatic circulation of bile acids in mammals: form and functions. Front Biosci (Landmark Ed) 14:2584–2598
67. Huang AHC (1992) Oil bodies and oleosins in seeds. Annu Rev Plant Biol 43:177–200
68. Jandacek RJ, Whiteside JA, Holcombe BN, Volpenhein RA, Taulbee JD (1987) The rapid hydrolysis and efficient absorption of triglycerides with octanoic acid in the 1 and 3 positions and long-chain fatty acid in the 2 position. Am J Clin Nutr 45(5):940–945
69. Juhler RK, Andreasson E, Yu S-G, Albertsson P-Å (1993) Composition of photosynthetic pigments in thylakoid membrane vesicles from spinach. Photosynth Res 35(2):171–178
70. Kleberg K, Jacobsen F, Fatouros DG, Mullertz A (2010) Biorelevant media simulating fed state intestinal fluids: colloid phase characterization and impact on solubilization capacity. J Pharm Sci 99(8):3522–3532
71. Kossena GA, Boyd BJ, Porter CJH, Charman WN (2003) Separation and characterization of the colloidal phases produced on digestion of common formulation lipids and assessment of their impact on the apparent solubility of selected poorly water-soluble drugs. J Pharm Sci 92(3):634–648
72. Kris-Etherton PM, Krummel D, Russell ME, Dreon D, Mackey S, Borchers J, Wood PD (1988) The effect of diet on plasma lipids, lipoproteins, and coronary heart disease. J Am Diet Assoc 88(11):1373–1400
73. Kristek A, Schär MY, Soycan G, Alsharif S, Kuhnle GGC, Walton G, Spencer JPE (2018) The gut microbiota and cardiovascular health benefits: a focus on wholegrain oats. Nutr Bull 43(4):358–373
74. Kumar A, Sharma A, Upadhyaya KC (2016) Vegetable oil: nutritional and industrial perspective. Curr Genomics 17(3):230–240
75. Lavialle M, Laye S (2010) Polyunsaturated fatty acids (omega 3, omega 6) and brain functions. Innov Agronomiques 10:25–42
76. Lazaridou A, Biliaderis CG (2007) Molecular aspects of cereal β-glucan functionality: physical properties, technological applications and physiological effects. J Cereal Sci 46(2):101–118

77. Lemmens L, Van Buggenhout S, Van Loey AM, Hendrickx ME (2010) Particle size reduction leading to cell wall rupture is more important for the beta-carotene bioaccessibility of raw compared to thermally processed carrots. J Agric Food Chem 58(24):12769–12776
78. Lentle RG, Janssen PWM (2008) Physical characteristics of digesta and their influence on flow and mixing in the mammalian intestine: a review. J Comp Physiol B 178(6):673–690
79. Li W, Dobraszczyk BJ, Wilde PJ (2004) Surface properties and locations of gluten proteins and lipids revealed using confocal scanning laser microscopy in bread dough. J Cereal Sci 39(3):403–411
80. Li H, Gidley MJ, Dhital S (2019) Wall porosity in isolated cells from food plants: implications for nutritional functionality. Food Chem 279:416–425
81. Long MA, Kaler EW, Lee SP (1994) Structural characterization of the micelle-vesicle transition in lecithin-bile salt solutions. Biophys J 67(4):1733–1742
82. Lucca PA, Tepper BJ (1994) Fat replacers and the functionality of fat in foods. Trends Food Sci Technol 5(1):12–19
83. Macierzanka A, Torcello-Gómez A, Jungnickel C, Maldonado-Valderrama J (2019) Bile salts in digestion and transport of lipids. Adv Colloid Interf Sci 274:102045
84. Madenci D, Egelhaaf SU (2010) Self-assembly in aqueous bile salt solutions. Curr Opin Colloid Interface Sci 15(1–2):109–115
85. Marangoni AG (1994) Enzyme kinetics of lipolysis revisited: the role of lipase interfacial binding. Biochem Biophys Res Commun 200(3):1321–1328
86. Marciani L, Gowland PA, Spiller RC, Manoj P, Moore RJ, Young P, Al-Sahab S, Bush D, Wright J, Fillery-Travis AJ (2000) Gastric response to increased meal viscosity assessed by echo-planar magnetic resonance imaging in humans. J Nutr 130(1):122–127
87. Marciani L, Gowland PA, Spiller RC, Manoj P, Moore RJ, Young P, Fillery-Travis AJ (2001) Effect of meal viscosity and nutrients on satiety, intragastric dilution, and emptying assessed by MRI. Am J Physiol Gastrointest Liver Physiol 280(6):G1227–G1233
88. Mazer NA, Benedek GB, Carey MC (1980) Quasielastic light-scattering studies of aqueous biliary lipid systems. Mixed micelle formation in bile salt-lecithin solutions. Biochemistry 19(4):601–615
89. McClements DJ (2000) Comments on viscosity enhancement and depletion flocculation by polysaccharides. Food Hydrocoll 14(2):173–177
90. McClements DJ (2007) Critical review of techniques and methodologies for characterization of emulsion stability. Crit Rev Food Sci Nutr 47(7):611–649
91. McClements DJ (2018) The biophysics of digestion: lipids. Curr Opin Food Sci 21:1–6
92. McDougall GJ, Morrison IM, Stewart D, Hillman JR (1996) Plant cell walls as dietary fibre: range, structure, processing and function. J Sci Food Agric 70(2):133–150
93. Meynier A, Genot C (2017) Molecular and structural organization of lipids in foods: their fate during digestion and impact in nutrition. OCL - Oilseeds Fats Crops Lipids 24(2) D202
94. Miles MJ, Morris VJ, Orford PD, Ring SG (1985) The roles of amylose and amylopectin in the gelation and retrogradation of starch. Carbohydr Res 135(2):271–281
95. Moghimipour E, Ameri A, Handali S (2015) Absorption-enhancing effects of bile salts. Molecules 20(8):14451–14473
96. Müller H (1997) Determination of the carotenoid content in selected vegetables and fruit by HPLC and photodiode array detection. Zeitschrift für Lebensmitteluntersuchung und-Forschung A 204(2):88–94
97. Navarro del Hierro JN, Herrera T, Fornari T, Reglero G, Martin D (2018) The gastrointestinal behavior of saponins and its significance for their bioavailability and bioactivities. J Funct Foods 40:484–497
98. Pallares Pallares A, Alvarez Miranda B, Truong NQA, Kyomugasho C, Chigwedere CM, Hendrickx M, Grauwet T (2018) Process-induced cell wall permeability modulates the *in vitro* starch digestion kinetics of common bean cotyledon cells. Food Funct 9(12):6544–6554
99. Panaiotov I, Verger R (2000) Enzymatic reactions at interfaces: interfacial and temporal organization of enzymatic lipolysis. In: Baszkin A, Norde W (eds) Physical chemistry of biological interfaces. CRC Press, New York, pp 359–400

100. Phan S, Salentinig S, Prestidge CA, Boyd BJ (2014) Self-assembled structures formed during lipid digestion: characterization and implications for oral lipid-based drug delivery systems. Drug Deliv Transl Res 4(3):275–294
101. Platt KA, Thomson WW (1992) Idioblast oil cells of avocado: distribution, isolation, ultrastructure, histochemistry, and biochemistry. Int J Plant Sci 153(3, Part 1):301–310
102. Poutanen K, Flander L, Katina K (2009) Sourdough and cereal fermentation in a nutritional perspective. Food Microbiol 26(7):693–699
103. Purkrtova Z, Jolivet P, Miquel M, Chardot T (2008) Structure and function of seed lipid body-associated proteins. C R Biol 331(10):746–754
104. Quinn PJ, Williams WP (1979) Plant lipids and their role in membrane function. Prog Biophys Mol Biol 34:109–173
105. Redard CL, Davis PA, Schneeman BO (1990) Dietary fiber and gender: effect on postprandial lipemia. Am J Clin Nutr 52(5):837–845
106. Reis P, Holmberg K, Watzke H, Leser ME, Miller R (2009) Lipases at interfaces: a review. Adv Colloid Interf Sci 147–148:237–250
107. Reynhardt EC, Riederer M (1994) Structures and molecular dynamics of plant waxes. Eur Biophys J 23(1):59–70
108. Rongkaumpan G, Amsbury S, Andablo-Reyes E, Linford H, Connell S, Knox JP, Sarkar A, Benitez-Alfonso Y, Orfila C (2019) Cell wall polymer composition and spatial distribution in ripe banana and mango fruit: implications for cell adhesion and texture perception. Front Plant Sci 10:858
109. Sagalowicz L, Michel M, Blank I, Schafer O, Leser ME (2017) Self-assembly in food — a concept for structure formation inspired by Nature. Curr Opin Colloid Interface Sci 28:87–95
110. Salas JJ, Sánchez J, Ramli US, Manaf AM, Williams M, Harwood JL (2000) Biochemistry of lipid metabolism in olive and other oil fruits. Prog Lipid Res 39(2):151–180
111. Salentinig S, Sagalowicz L, Leser ME, Tedeschi C, Glatter O (2011) Transitions in the internal structure of lipid droplets during fat digestion. Soft Matter 7(2):650–661
112. Salisbury FB, Ross CW (1992) Growth and development. In: Plant physiology, 4th edn. Wadsworth Publishing Company, Belmont. pp 329–356
113. Salisbury FB, Ross CW (1992) Lipids and other natural products. In: Plant physiology, 4th edn. Wadsworth Publishing Company, Belmont. pp 308–326
114. Sathe SK, Salunkhe DK (1981) Functional properties of the Great Northern Bean (Phaseolus vulgaris L.) proteins: Emulsion, foaming, viscosity, and gelation properties. J Food Sci 46(1):71–81
115. Schneeman BO, Gallaher D (1985) Effects of dietary fiber on digestive enzyme activity and bile acids in the small intestine. Exp Biol Med 180(3):409–414
116. Schwertner HA, Biale JB (1973) Lipid composition of plant mitochondria and of chloroplasts. J Lipid Res 14(2):235–242
117. Shahidi F, Ambigaipalan P (2018) Omega-3 polyunsaturated fatty acids and their health benefits. Annu Rev Food Sci Technol 9:345–381
118. Shimada TL, Hayashi M, Hara-Nishimura I (2018) Membrane dynamics and multiple functions of oil bodies in seeds and leaves. Plant Physiol 176(1):199–207
119. Shulman GI (2000) Cellular mechanisms of insulin resistance. J Clin Investig 106(2):171–176
120. Simopoulos AP (2002) Omega-3 fatty acids in inflammation and autoimmune diseases. J Am Coll Nutr 21(6):495–505
121. Singh H, Ye AQ, Horne D (2009) Structuring food emulsions in the gastrointestinal tract to modify lipid digestion. Prog Lipid Res 48(2):92–100
122. Smith SA, Campbell DR, Elmer PJ, Martini MC, Slavin JL, Potter JD (1995) The University of Minnesota Cancer Prevention Research Unit vegetable and fruit classification scheme (United States). Cancer Causes Control 6(4):292–302
123. Somerville C, Browse J (1991) Plant lipids: metabolism, mutants, and membranes. Science 252(5002):80–87

124. Stone NJ (1990) Diet, lipids, and coronary heart disease. Endocrinol Metab Clin N Am 19(2):321–344
125. Takahashi T (2011) Flow behavior of digesta and the absorption of nutrients in the gastrointestine. J Nutr Sci Vitaminol 57(4):265–273
126. Temple NJ (1994) Diet, blood lipids and coronary heart disease - current controversies. S Afr Med J Suppl:30–31
127. Turgeon SL, Rioux L-E (2011) Food matrix impact on macronutrients nutritional properties. Food Hydrocoll 25(8):1915–1924
128. Tydeman EA, Parker ML, Faulks RM, Cross KL, Fillery-Travis A, Gidley MJ, Rich GT, Waldron KW (2010) Effect of carrot (*Daucus carota*) microstructure on carotene bioaccessibility in the upper gastrointestinal tract. 2. *In vivo* digestions. J Agric Food Chem 58(17):9855–9860
129. Tzen JTC, Cao YZ, Laurent P, Ratnayake C, Huang AHC (1993) Lipids, proteins, and structure of seed oil bodies from diverse species. Plant Physiol 101(1):267–276
130. Venskutonis PR, Kraujalis P (2013) Nutritional components of amaranth seeds and vegetables: a review on composition, properties, and uses. Compr Rev Food Sci Food Saf 12(4):381–412
131. Verger R, de Haas GH (1976) Interfacial enzyme kinetics of lipolysis. Annu Rev Biophys Bioeng 5:77–117
132. Wainwright PE (2002) Dietary essential fatty acids and brain function: a developmental perspective on mechanisms. Proc Nutr Soc 61(01):61–69
133. Waldron KW, Smith AC, Parr AJ, Ng A, Parker ML (1997) New approaches to understanding and controlling cell separation in relation to fruit and vegetable texture. Trends Food Sci Technol 8(7):213–221
134. Wattanakul J, Sahaka M, Amara S, Mansor S, Gontero B, Carrière F, Gray D (2019) In vitro digestion of galactolipids from chloroplast-rich fraction (CRF) of postharvest, pea vine field residue (haulm) and spinach leaves. Food Func 10(12):7806–7817
135. Weinstock BA, Janni J, Hagen L, Wright S (2006) Prediction of oil and oleic acid concentrations in individual corn (Zea mays L.) kernels using near-infrared reflectance hyperspectral imaging and multivariate analysis. Appl Spectrosc 60(1):9–16
136. Wells ML, Potin P, Craigie JS, Raven JA, Merchant SS, Helliwell KE, Smith AG, Camire ME, Brawley SH (2017) Algae as nutritional and functional food sources: revisiting our understanding. J Appl Phycol 29(2):949–982
137. Wilde PJ, Garcia-Llatas G, Lagarda MJ, Haslam RP, Grundy MM-L (2019) Oat and lipolysis: food matrix effect. Food Chem 278:683–691
138. Wood PJ (2002) Relationships between solution properties of cereal β-glucans and physiological effects — a review. Trends Food Sci Technol 15(6):313–320
139. Wooster TJ, Day L, Xu M, Golding M, Oiseth S, Keogh J, Clifton P (2014) Impact of different biopolymer networks on the digestion of gastric structured emulsions. Food Hydrocoll 36:102–114
140. Xiang H, Sun-Waterhouse D, Waterhouse GIN, Cui C, Ruan Z (2019) Fermentation-enabled wellness foods: a fresh perspective. Food Sci Human Wellness 8(3):203–243
141. Yang Y, McClements DJ (2013) Vitamin E and Vitamin E acetate solubilization in mixed micelles: physicochemical basis of bioaccessibility. J Colloid Interface Sci 405:312–321
142. Yang S, Hallett I, Oh HE, Woolf AB, Wong M (2019) The impact of fruit softening on avocado cell microstructure changes monitored by electrical impedance and conductivity for cold-pressed oil extraction. J Food Process Eng 42(4):e13068
143. Yu H, Dong S, Wang L, Liu Y (2017) The effect of triterpenoid saponins on pancreatic lipase *in vitro*: activity, conformation, kinetics, thermodynamics and morphology. Biochem Eng J 125:1–9
144. Zhong R, Ye Z-H (2009) Secondary cell walls. In: Encyclopedia of life sciences. Wiley, Chichester

Chapter 8
Dairy Products

A. -I. Mulet-Cabero and A. Brodkorb

Abbreviations

CVD	Cardiovascular disease
FA	Fatty acid
GI	Gastrointestinal
HDL	High-density lipoprotein
LDL	Low-density lipoprotein
MFGM	Milk fat globule membrane
TAG	Triaglycerol
UHT	Ultra-high temperature

8.1 Introduction

Dairy products have been widely consumed for their richness in nutrients. However, they have received much negative attention for their role in human nutrition because milk fat has a relatively high content of saturated fatty acids (FAs). A high consumption of saturated FA content was linked to increased low-density lipoprotein (LDL) cholesterol, thereby promoting the risk of cardiovascular diseases (CVD) [20]. However, recent meta-analyses of prospective cohort studies and randomised controlled studies have consistently indicated that the consumption of dairy products has neutral association or even a reduction in the risk of CVDs [19]. This discrepancy is probably due to the simplistic approach of evaluating the health benefits of food according to its individual components. The effect of the dairy matrix on digestion and health has long been neglected and needs to be considered as a whole in relation to nutritional and physiological effects [54].

A. -I. Mulet-Cabero (✉)
Quadram Institute Bioscience, Norwich Research Park, Norwich, Nolfolk, UK
e-mail: Anabel.Mulet-Cabero@quadram.ac.uk

A. Brodkorb
Teagasc Food Research Centre, Moorepark, Fermoy, Co. Cork, Ireland

M. M.-L. Grundy, P. J. Wilde (eds.), *Bioaccessibility and Digestibility of Lipids from Food*, https://doi.org/10.1007/978-3-030-56909-9_8

The main constituents of milk are proteins (3–4%), lipids (3–4%), carbohydrates (4–5%), minerals (0.8%) and vitamins (0.1%) [36]. Milk is a naturally occurring emulsion, where the lipids are secreted in the form of milk fat globules. The size distribution of milk fat globules ranges from 0.2 to 10 μm with a mean diameter of around 4 μm [40]. The triacylglycerols (TAGs) in milk fat (97–98%) contain about 400 different FAs [35] and are surrounded by a biological membrane, called the milk fat globule membrane (MFGM). The MFGM is organised as a trilayer of 4–10 nm in thickness and is composed of mainly phospholipids as well as sphingolipids, glycoproteins, cholesterol, enzymes and other minor components [30]. The presence of the MFGM allows the stabilisation of the native milk emulsion. This structural layout of lipids in which there is a TAG core surrounded by an interface or membrane is also present in egg yolk and oilseeds. However, the composition and organisation of the interface is different. For instance, in plant seeds, lipids are stored in oleosomes, in which the lipid core rich in TAG is surrounded by a monolayer of phospholipids and proteins. These oleosomes are enclosed by cell walls, which can limit their bioaccessibility. In meat and muscle foods, TAGs are mainly present as droplets in the adipocytes of the adipose tissue, and phospholipids are found as bilayers in the cell membranes. As with milk, processing can modulate the structure and organisation of lipids within the food matrix and therefore affect lipid digestibility.

There is a vast range of dairy products that originate from milk following processing, the most common products being cheese, butter, fermented milk (e.g. yoghurt) and powders. These complex dairy matrices are obtained by applying physical (e.g. thermal treatment, membrane filtration and drying), mechanical (e.g. homogenisation and churning) and biochemical (e.g. pH, ionic strength, microbes and enzymes) treatments. Dairy processing induces significant modifications in the structure of milk, from the molecular to the macroscopic length scales. For instance, raw milk is commonly processed by heat treatment and homogenisation to ensure microbial and physical stability, respectively. This processing affects the organisation of the main milk proteins of caseins and whey proteins, and fat globules but preserving the same physical state of liquid milk. In contrast, the addition of rennet to milk results in coagulation providing a different physical (semi)-solid structure, which is the basis of cheese making. The physical state and properties of the dairy matrix dictate the organisation of fat globules. Figure 8.1 shows examples of how the different dairy processing of milk, where the milk proteins and milk fat globules are the main participants in the dairy matrix, affects the physical state and organisation of fat globules. Most of the dairy products are oil-in-water emulsions, except for butter, which is a water-in-oil emulsion. The fat globules in milk are homogenously dispersed in the continuous aqueous phase consisting of proteins, namely caseins and whey proteins, and lactose. The structural properties of milk can be modulated by heat treatment and homogenisation [39]. In yoghurt, the fat droplets are dispersed in a loose gelled protein matrix, obtained by the acid coagulation of caseins by fermentation. The viscosity of yoghurt can be modulated by several factors such as the heat temperature of milk, the addition of stabilisers and the fat content [33]. Similarly, in cheese, a coagulated casein network can be obtained by

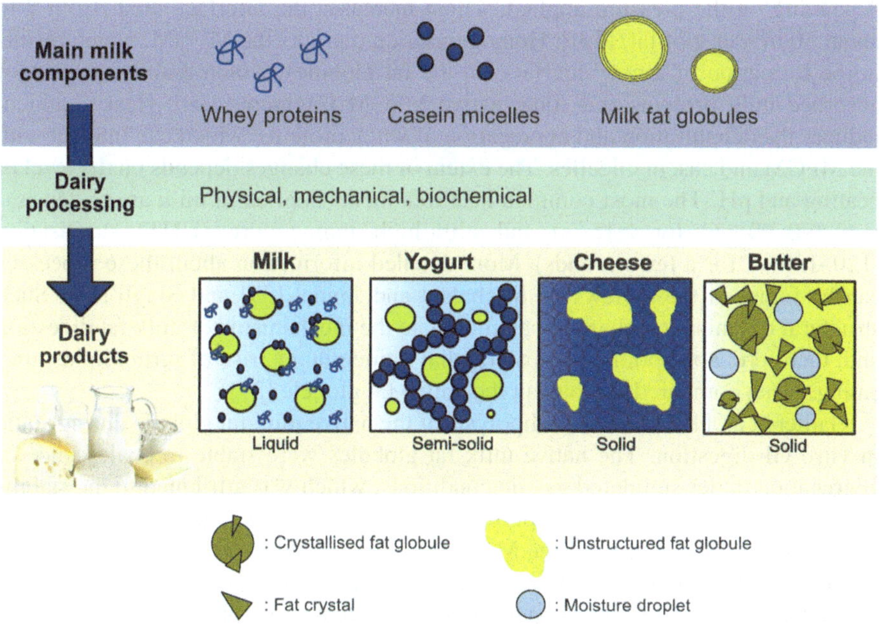

Fig. 8.1 Schematic representation of the main dairy products produced by milk processing, showing the different organisation of milk fat globules in the dairy matrix

the action of rennet, in which the milk fat is embedded without its native structure. The rheological and structural properties of cheese can be modulated by several factors such as the mineral content, curd handling and the maturation process [32]. Butter is obtained by churning cream, which disrupts the MFGM and separates the fat globules leading to a fat phase inversion of partially crystallised TAG, thereby forming the continuous phase of fat containing dispersed water droplets.

The gastrointestinal (GI) digestion of milk fat depends on the accessibility of lipases to the lipid substrate, which will be driven by the surrounding dairy matrix. The size, interfacial composition and organisation of the milk fat globules as well as their interaction within the dairy matrix affects the kinetics of lipid digestion and absorption. The physical state of dairy products has been used in this chapter to separate the distinct behaviour in lipid digestion.

8.2 Liquid Dairy Matrices

The main example of a liquid dairy matrix is milk. The size and organisation of the milk lipid droplets change with processing. The main processes applied to milk are homogenisation, heat treatment and dehydration. The homogenisation of milk reduces the size of the milk fat globule from an average of 4 μm to below 1 μm, depending on the pressure applied, which increases the interface area [from 7 to about 30 m^2 per g of fat, [29]]. Homogenisation disrupts the MFGM, which results in the formation of a new interface on the fat globules, which mainly consists of absorbed milk proteins and some native MFGM fragments [29]. Heat treatment induces the denaturation and aggregation of whey proteins, which can interact with the MFGM and casein micelles. The extent of these changes depends on the level of heating and pH. The most common heat treatments applied in milk are pasteurisation (70–90 °C for ≥15 s) and ultra-high temperature (UHT) sterilisation (120–150 °C for a few seconds). More detailed information about these processes can be found in the reviews by Michalski and Januel [39] and Singh [50]. Such structural organisation changes can modulate the mechanisms of milk fat digestion and, therefore, absorption and postprandial lipidemia, i.e. rise of serum lipids after eating, when compared to lipids in the native fat globule [37].

Gallier et al. [17] studied the lipolysis of the native milk fat globule during static in vitro GI digestion. The native milk fat globules were stable to coalescence or aggregation under simulated gastric conditions, which was attributed to the stabilisation of glycoproteins and phospholipids in the MFGM. In contrast, during intestinal digestion, the integrity of the MFGM was rapidly lost due to the displacement by bile salts leading to destabilisation and disintegration of the globules into lipolytic products. Bile salts have been reported to modify the phospholipid interface, which enhanced the accessibility of pancreatic lipase to the interface [56]. The free FAs release after 1 hour of gastric digestion and 2 hours of intestinal digestion was fast in the first 5 min and thereafter slowed down. Confocal laser scanning microscopy showed that these lipolysis products accumulated at the surface of the fat globules, forming liquid crystalline lamellae, and were solubilised by bile salts forming disk-shaped micelles. This study showed the importance of the MFGM structure in the digestion of milk fat globules. Therefore, any change in the composition and organisation affecting the MFGM structure could influence the digestion of milk fat globules.

There are some studies that have investigated the process-induced changes of milk fat globules in relation to lipid digestion, using mainly in vitro digestion models with or without the gastric digestion step. Berton et al. [6] assessed the effect of the size and interfacial composition of milk fat globules, i.e. the MFGM of the native fat globules vs. milk proteins of the homogenised fat globules, on milk TAG hydrolysis by the presence of human pancreatic lipase, colipase and bile salts in vitro. The smallest native milk fat globules (1.8 μm) resulted in the highest catalytic efficiency of pancreatic lipase compared to average native milk fat globules (3.9 μm) and large native milk fat globules (6.7 μm), which was due to the larger available interfacial surface area (3.6-fold increase in available surface). Homogenisation of the fat globules induced a twofold higher catalytic efficiency compared to the native fat globules. However, this outcome was a moderate increase considering the 28-fold larger available surface suggesting, therefore, that the size was not the most important parameter in the digestion of the homogenised fat globules but the surface composition. Similarly, Ye et al. [58] showed that the overall lipolysis was lower and slower in the raw milk compared to the recombined milk (i.e. anhydrous milk fat mixed with skim milk) with homogenised fat globules. Berton et al. [6] also showed an initial lag phase in the milk TAG hydrolysis curve in the case of the native fat globule milk, which did not occur in the digestion of homogenised milk. This lag phase was attributed to the presence of the trilayer of phospholipids in the MFGM, which could limit the access of the pancreatic lipase of the TAGs. The addition of free FAs, simulating hydrolysis in the gastric phase seemed to avoid the presence of this lag phase [5]. These free FAs could be inserted in the MFGM and favouring the binding of colipase with lipase at the globule surface [42]. However, Ye et al. [58] did not observe this lag phase using only a simulated intestinal digestion. Unlike the native fat globules, the size reduction in homogenised fat globules (from 1.5 down to 0.15 μm) did not show any difference in lipolysis. Similarly, in Ye et al. [58] the two homogenised milks (2.10 and 0.32 μm) did not present any difference in lipid digestion. This could be linked to the physicochemical changes of the fat globules during digestion. Flocculation of the fat globules occurred within the first minutes of intestinal digestion, the proportion of large particles decreased, and small particles were released from the large flocculated particles. Therefore, homogenised droplets can undergo rearrangements leading to milk fat structures with similar physicochemical properties regardless of their initial droplet size, which could explain the weak effect of droplet size on lipolysis. These studies have shown the critical role of the composition and structure of the droplet interface in milk lipid digestion.

Heat treatments, usually combined with homogenisation, are widely used in milk processing. Both processes have been observed to alter the structure, biochemical composition and physicochemical properties of milk. Few studies have shown that this processing impacts the digestive behaviour. Mulet-Cabero et al. [46] applied the heat treatments of pasteurisation (72 °C for 15 s) or UHT treatment (140 °C for 3 s) alone and in combination with homogenisation at pilot-plant scale, which were compared to the native whole milk. The digestion of the samples was performed using a newly developed semi-dynamic adult model, which can simulate the main dynamics of the stomach including gradual acidification, gastric fluid and enzyme secretion, and emptying, and it has been standardised in Mulet-Cabero et al. [44]. This set-up allowed the monitoring of gastric behaviour over time. All milk samples presented protein aggregation and coagula formation within the first 15 min of gastric digestion. The fat globules were entrapped in the coagula structures, as observed by confocal microscopy. Over the digestion time, the homogenised samples creamed regardless of heat treatment, whereas all non-homogenised samples exhibited sedimentation. The heat treatment, in particular UHT treatment, induced more fragmented and softer coagula. This was supported by Ye et al. [57], using a dynamic gastric model, where a heat treatment (90 °C for 20 min) induced clot formation during gastric digestion with a crumbled structure, compared to the firm structure in raw whole milk. Mulet-Cabero et al. [46] showed that the gastric behaviour was critical in the kinetics of lipids delivered, simulating the lipids emptied from the stomach to the duodenum. In general, the homogenised samples presented a delayed lipid emptying due to the formation of a cream layer in the simulated stomach, except in the sample in which homogenisation was combined with UHT treatment.

The latter was attributed to a higher lipid content initially delivered due to the weak coagulum structure. Indeed, the harsh heat treatment in the Ye et al. [57] study induced a faster release of fat globules from the coagula showing, therefore, a positive relationship between the protein breakdown of the coagula and the lipid release. These studies showed that the different colloidal behaviour and rheological properties of the protein coagulum induced by milk processing modulate the simulated delivery of lipids to the duodenum, which could, in turn, influence the absorption rate and bioavailability, and the lipidemia responses. As far as is known, there is only one published study on adult humans assessing the nutritional impact of milk processing, but it relates to the postprandial utilisation of proteins [26], nevertheless they did show that UHT treatment altered the kinetics of protein utilisation. Therefore, the common milk processing of heat treatments and homogenisation should be investigated to assess its impact on the rate and extent of lipid metabolism. There are a few in vitro studies investigating the free FA release during the intestinal digestion of processed milk with a previous static gastric phase. Kopf-Bolanz et al. [25] showed that there was almost a complete fat digestion of the total TAGs from pasteurised and homogenised whole milk. Homogenisation resulted in a faster release of free FA compared to raw milk [55], which was attributed to the higher surface available, thereby enhancing the lipase efficiency. In addition, the authors showed that certain FA were degraded at a faster rate; these were stearic and oleic acids due to their outer position (i.e. S_N1 and S_N3) of the TAG molecule, in contrast to myristic acid. However, the FA profiles were not affected by the processing treatments. However, as a static digestion method was used, the important structural changes occurring in the stomach could not be considered, probably affecting the bioaccessibility kinetics of milk fat globules.

Nevertheless, there are a few animal studies that investigated the effect of the milk fat ultrastructure on lipid metabolism. Michalski et al. [38], using a rat model, measured the ^{13}C abundance in breath from labelled digested TAGs from milk containing the same amount of fat but with a different structure. Non-emulsified and emulsified droplets were produced as coarse droplets of 10 μm covered mainly with phospholipids, and fine droplets of 1 μm covered with caseins, either in a liquid or semi-crystallised state. The general lipid metabolism, as based on the oxidation to CO_2 was lower in the fine droplets coated by caseins, compared to the larger emulsified milk fat particles with different coatings. The non-emulsified and emulsified droplets with phospholipids presented a rapid metabolisation, reaching a peak after 1 hour followed by a rapid decay, in contrast to the constant ^{13}C excretion over 6 hours in the milk fat samples covered by caseins. This is in accordance with the plasma TAG concentration from the study by Michalski et al. [41], showing a delay in fine homogenised fat droplets, compared to the rapid increase with a pronounced peak at 180 min after ingestion of the non-emulsified fat droplets in rats. As possible causes of these findings, the authors pointed out potential differences in gastric emptying, suggesting possible different gastric behaviour and emulsification, and the affinity of lipases to the milk fat structures. To gain more insight into the effect of milk fat structure on lipid digestion, the contents of the respective gastric and small intestinal digestion after ingestion of cream, as concentrated sources of fat, from raw, pasteurised and pasteurised and homogenised whole milk were investigated in rats [16, 18]. In the stomach, the pasteurised sample presented a greater apparent lipolysis, with short- and medium- chain FA being released at a faster rate than long-chain FA [16]. The pasteurised and homogenised sample was digested to a lesser extent, which was attributed to the formation of protein aggregates trapping the fat globules and inhibiting gastric lipase. However, the FA analyses did not take into account possible differences in the gastric emptying rate. The size of the globules in all samples increased under acidic gastric conditions. Protrusions and liquid crystalline-lamellar phases, probably caused by the accumulation of lipolytic products, were visible at the surface of the fat globules in the chyme of raw or pasteurised creams. The formation of liquid-crystalline layers was also observed during small intestinal digestion as well as insoluble crystals possibly consisting of interaction of FA with calcium, regardless of cream treatment [18].

8.3 Solid and Semi-Solid Dairy Matrices

Butter, cheese and yoghurt are the main forms of solid and semi-solid dairy matrices. The latter two are oil-in-water emulsions in which dairy lipids are dispersed in a completely or partially gelled phase. In contrast, butter is a water-in-oil emulsion, in which partially crystallised TAG forms the continuous phase. Cheese is a complex matrix that consists of a casein network with spaces occupied by fat globules, serum and water [32]. Each processing step in cheese making (biochemical, physical, mechanical and ageing treatment) contributes to the different microstructure of each cheese variety having different features, i.e. size and interfacial composition of fat globules, macronutrient structural organisation and rheological properties of the matrix [15]. These characteristics affect the functionality of cheese, including the nutritional properties. This section discusses the impact of solid and semi-solid dairy matrices, mainly cheese since it is more studied, on the digestive behaviour of lipids.

Fang et al. [12] investigated the extent of the breakdown of cheese during static in vitro digestion, in five commercial cheeses having a wide range of textural properties: Camembert, young Cheddar, mature Cheddar, smear-ripened cheese and Mozzarella. Two contrasting digestion behaviours were observed for Camembert and Mozzarella; Camembert presented the highest release of fat (i.e. fat layer isolated after centrifugation based on the amount of fat initially present in the matrix) during GI digestion in contrast to the limited fat release observed in Mozzarella, which was related to different matrix disintegration (i.e. dispersion of cheese into the aqueous phase during digestion). The disintegration of Camembert was fast compared to Mozzarella and reached 78% after 2 min of oral digestion. At the end of gastric digestion, Camembert and Mozzarella reached values of 83% and 27%, respectively. However, during duodenal digestion, the disintegration was rapid and similar for all cheeses at the end of digestion. The different gastric disintegration behaviour was attributed to a higher hardness and cohesiveness of the initial

Mozzarella sample compared to Camembert. Therefore, the physical properties of Mozzarella may limit the accessibility of the protein network to pepsin, and therefore the fat release. In a similar way, the properties of the cheese matrix can modulate the FA release during intestinal digestion. Lamothe et al. [27] showed that Mozzarella presented an overall higher rate and extent of free FA release compared to other studied cheeses.

Cheese composition is an important attribute influencing cheese disintegration during digestion, since it can modulate the textural parameters of cheese. Fang et al. [13] assessed the disintegration and nutrient release of Cheddar (33% fat), light Cheddar (21% fat), Mozzarella (20% fat) and light Mozzarella (14% fat) during static in vitro digestion. The reduction of fat in the cheeses induced a more compact and denser protein matrix resulting in increased hardness and cohesiveness. During digestion, Cheddar cheese presented the highest degree of disintegration (46%) compared to the reduced fat Cheddar (27%). Similarly, Mozzarella showed a higher disintegration (18%) compared to the 15% observed in the light Mozzarella. Similar observations were made regarding fat release during digestion. In the oral phase, the highest fat release was found in Cheddar (62%) compared to light Cheddar (51%) and, Mozzarella (35%) and light Mozzarella (21%). The same trend was observed in the gastric phase in contrast to the similar values obtained after the duodenal digestion. This study showed that cheese disintegration at the end of gastric digestion was correlated to the initial fat content of the cheese. In addition, Fang et al. [13] highlighted the importance of digestion enzymes in cheese disintegration; up to 72% of the overall cheese disintegration was observed at the end of duodenal digestion, compared to 30% when no enzymes were added. It is important to highlight that gastric lipase was not included in the in vitro studies discussed in this section, which is likely to impact the overall lipolysis.

As observed in the previous studies [12, 13, 27], the disintegration of the cheese matrix during GI digestion has been linked to the composition and textural characteristics of cheese. In order to identify the most significant factors influencing cheese disintegration at each digestion step, Guinot et al. [21] assessed the digestion of nine cheeses covering a wide range of texture attributes, compositions and microstructures; young and aged Cheddar, regular and light cream cheese, Parmesan, Feta, Camembert, Mozzarella and sliced processed cheese. The authors showed, using a partial least square regression, that the non-fat solids, moisture in the non-fat substance and phosphorous content were positively correlated whereas the cohesiveness was negatively correlated to the disintegration in the oral phase for 2 min. These four parameters could explain 95% of the variance, and the bolus disintegration in the oral phase was mainly affected by agitation and temperature. At the end of the gastric digestion (120 min), the initial textural parameters of springiness, cohesiveness, adhesiveness and hardness were negatively correlated to the cheese disintegration, explaining 96% of the variance. For instance, aged cheddar presented a higher degradation, which could be explained by its high degree of proteolysis, its capacity to fracture as well as its low cohesiveness and springiness. However, at the end of the small intestinal phase (120 min), these parameters only represented 13% of variance since the digestion behaviour was similar in all the cheeses. Therefore, there was no correlation between textural parameters and free FA release. One would expect that the differences in cheese disintegration in the gastric digestion would affect lipolysis in the small intestine. For example, a cheese that presents low disintegration would remain longer in the stomach, which could delay lipolysis in the duodenum. However, gastric emptying is not simulated in this study and this effect cannot be assessed.

Cheese is an important source of calcium that affects its texture. Ayala-Bribiesca et al. [1] investigated the effect of calcium enrichment in Cheddar cheese on its physical disintegration and lipolysis. $CaCl_2$ was added in the salting step to reach levels of high calcium (29.6 mg Ca/g protein) and very high calcium (46.4 mg Ca/g protein) in Cheddar production. The authors showed that the increase in calcium concentration induced more resistance to degradation during gastric and intestinal digestion due to its matrix hardness. The addition of calcium induced faster lipolysis during intestinal digestion, which could be attributed to the induced microstructure. Also, calcium has been shown to increase lipolysis rates by the removal of the FFAs formed at the droplet interface forming insoluble complexes called calcium soaps, which facilitates lipase accessibility to the substrate [23].

Ayala-Bribiesca et al. [2] further investigated the role of calcium in milk fat lipolysis from cheese during in vitro static digestion. Cheddar-type cheeses were prepared with olein (higher ratio of short-chain FA), stearin (higher ratio of long-chain FA) or regular anhydrous milk fats, and with regular (7136 μg/g cheese) or high (9466 μg/g cheese) calcium levels for salting conditions with NaCl or NaCl + $CaCl_2$, respectively. The cheeses prepared with the stearin fraction were more resistant to disintegration than the other cheeses due to the high melting temperature of the milk fat fraction. Lipolysis was significantly controlled by the type of anhydrous milk fat and the calcium concentration. The cheese prepared with the stearin fraction presented slower lipolysis in the intestinal phase compared to the other cheeses, which was attributed to the higher amount of long-chain FA present in this cheese. Long-chain FA are less soluble and more difficult to cleave by lipase compared to short- and medium-chain FA [43] and they provide a higher melting point (limiting lipase access to a solid physical state of the fat at 37 °C). The high calcium level in the cheeses enhanced the lipolysis rates. The authors observed that calcium interacts with long-chain FA forming calcium soaps, whereas no interaction was detected between short-chain FA and calcium. However, the formation of these calcium-soaps is expected to reduce the absorption of FA. Therefore, Ayala-Bribiesca et al. [3] assessed the effect of the same Cheddar cheeses with different type of fat and calcium levels on the bioavailability of lipids using a rat model. The postprandial lipidemia monitored by TAG concentrations in plasma was statistically different due to the milk fat type ($p = 0.023$) and the calcium level ($p = 0.0054$). The cheese prepared with stearin presented the lowest level of TAG concentration in plasma compared to the regular and olein milk fat cheeses, regardless of the calcium level. Cheeses with a higher level of calcium induced higher and faster TAG peaks in blood, except for the cheeses containing stearin. These results were in line with those obtained with the same experimental cheeses but using an in vitro model [2], and the differences in lipidemia were also attributed to the higher amount of saturated long-chain FA with a higher melting point. The analysis of faecal calcium soaps demonstrated that there was a higher FA excretion (29%) when stearin milk fat cheese was consumed, compared to other cheese types, and the excretion was mainly saturated long-chain FA (C12:0 to C18:0). Similarly, increased calcium in cheese induced the secretion of saturated long-chain FA, in particular, palmitic and steric acid.

A few human studies investigated the effect of cheese matrices on blood lipids. Drouin-Chartier et al. [11] compared young Cheddar and soft cream cheese, which were matched in total energy and macronutrient content. There were no differences in serum TAG in the incremental area under the curve over 8 hours. However, the soft cheese induced a greater increase in TAG concentration at 2 hours compared to the firm cheese. Therefore, this outcome suggests that the physical structure might be more important in the timing and magnitude of the TAG peak value than the overall lipid digestibility. The authors suggested that the smaller lipid droplets and the loose protein gel in the soft cheese could result in an easier access for the enzymes and, subsequent lipolysis. In another clinical study, a traditional fat- and salt- free Norwegian cheese was compared with a Gouda-type cheese. After an 8-week intervention, the consumption of any cheese type did not induce any increase in total and LDL- cholesterol compared to the control group. However, using a stratified analysis, there was a reduction in total cholesterol with Norwegian cheese in those subjects who had a high cholesterol baseline and in the subjects with metabolic syndrome after the Gouda-type cheese consumption.

Yoghurt is obtained by the lactic fermentation of milk resulting in a curd due to the destabilisation of the protein system. The curd can also be obtained by chemical acidification with glucono delta-lactone, instead of biological acidification with starter cultures. Yoghurt exhibits different textures due to milk base heating, starter cultures and yoghurt shearing after fermentation [51] among many other factors, which results in a wide range of physical structures from semi-solid (gels) to liquid (drinking yoghurt). In addition, composition can play an important role in yoghurt texture. For example, Sandoval-Castilla et al. [48] showed different texture attributes and microstructure in yoghurts prepared with fat replacers consisting of whey protein concentrate, microparticulated whey protein and modified tapioca starch. Changes in the texture can modulate lipid digestion, as seen in the case of cheese. However, as far as is known, there is no published study investigating the impact of yoghurt matrices on lipid digestion. In an in vitro study comparing the different physical structures of dairy matrices (liquid, semi-solid and solid), Lamothe et al. [28] compared two types of yoghurt, regular and Greek style, the latter having a higher lipid and protein content. The Greek-style yoghurt presented lower lipolysis than the regular-style yoghurt in the first 30 min of intestinal digestion. However, there were no differences at the end of digestion. In a clinical intervention study, the atherogenic effect of the highly saturated FA level was compared between Camembert cheese and full-fat yoghurt in 159 moderate hypercholesterolemic subjects for 5 weeks. The authors did not find any change in blood pressure and serum lipids after the intervention and between the groups.

8.4 Liquid Vs. Semi-Solid Vs. Solid Dairy Matrices

The initial characteristics and physical structure of the food matrix and its changes within the digestive system need to be considered in order to explain the bioaccessibility and absorption of nutrients. Lamothe et al. [28] compared the digestion of dairy matrices (liquid, semi-solid and solid) using a static in vitro model. The application of different physicochemical conditions (i.e. homogenisation, heat and pH) resulted in different milks, yoghurts and cheeses. Lipolysis was significantly affected by the type of matrix ($p < 0.05$); milk and yoghurt presented a higher and faster release of free FA than cheese, which was explained by a greater accessibility for digestive enzymes. However, a similar release of free FA was observed among milk and yoghurt matrices during the intestinal digestion. The digestion of the cheese matrix led to a higher formation of calcium soaps compared to the other matrices with an equivalent calcium content, regardless of the calcium-to-lipid ratio. This result was attributed to the high concentration of calcium close to the lipid droplets in the cheese matrix, which could enhance the formation of calcium soaps.

Recently, Schmidt et al. [49] investigated lipid digestion in four matrices having a different microstructure and texture but the same nutrient protein:fat and calcium:fat ratios, i.e. Cheddar cheese, a homogenised Cheddar, a micellar casein and cream drink, and a micellar casein and cream gel. The cheese sample had a significantly lower release of free FAs, followed by the homogenised cheese, during the first 30 min of static in vitro intestinal digestion. In contrast, the release of free FAs was higher in the casein drink and gel, presenting similar values between them. The slower digestion of cheese was attributed to the higher resistance of the protein network to breakdown during both gastric and intestinal digestion, which was related to the high viscosity properties both initially and during digestion. Furthermore, larger and coalesced droplets were released from the cheese matrix, thereby impairing lipid accessibility for digestive lipases.

Similar to in vitro studies, there are very few human intervention studies comparing the different physical structures of the dairy matrices, and their influence on postprandial lipidemia. Clemente et al. [8] investigated, in 8 type-2 diabetic patients, the effects on postprandial lipidemia of three meals with a similar composition but the main source of fat had a different physical structure; milk, butter and Mozzarella cheese. The butter-based meal presented a delay in postprandial lipidemia that resulted in a lower incremental TAG area measured for 3 hours, compared to the other two meals. However, there was no significant difference between the meals in the incremental area for the total time measured of 6 hours. The time to peak in plasma TAG was shorter with milk than with Mozzarella ($p < 0.03$) or butter ($p < 0.01$) meals. The authors suggested that the physical structure within a meal may not necessarily influence the overall magnitude of postprandial lipidemia, rather the timing of the TAG peak value. Furthermore, they did not find a relationship between the TAG responses and gastric emptying, measured by ultrasound scan, since the Mozzarella-based meal presented a faster gastric emptying rate, compared to the other two meals.

Some studies have investigated the impact of the physical structure of dairy matrices on serum blood lipids. For example, a randomised crossover study in healthy men was conducted by Tholstrup et al. [53] to investigate the effect of interventions providing 20% of total energy from dairy fat in the form of milk, hard cheese and butter (adjusted for lactose and casein content) for 3 weeks. The consumption of cheese caused a significant but moderate decrease in plasma LDL-cholesterol concentration compared to the butter diet ($p = 0.037$). In contrast, both butter and milk diets resulted in an increase in LDL-cholesterol by 8.7% and 6.4%, respectively. There was no significant difference in the other blood lipids measured. Similarly, Biong et al. [7] showed a reduction in total cholesterol after a Jarlsberg cheese diet compared to a butter diet, in which dairy fat constituted 20% of the energy intake. In another randomised cross-sectional trial by Hjerpsted et al. [22], 13% of energy of the habitual fat intake was replaced with either hard cheese or butter for 6 weeks. After the intervention, cheese intake resulted in a lower serum total, LDL- and HDL- cholesterol, compared to the butter intervention. In addition, there was no increase in total and LDL-cholesterol when cheese was consumed in relation to the control habitual diet. The authors did not find any difference in the faecal fat excretion between the cheese and butter interventions. Similarly, a meta-analysis of randomised controlled trials that compared hard cheese and butter with a similar ratio of polyunsaturated to saturated FA concluded that cheese intake reduced LDL-cholesterol by 6.5%, but no effect on plasma TAGs was found [9]. These studies showed that cheese has the potential to decrease cholesterol compared to butter with the same fat content. In order to obtain more insight into the possible causes of this outcome, Feeney et al. [14] tested the individual cheese components themselves, and totally or partly within a cheese matrix in a randomised parallel intervention using overweight participants. The food matrices were macronutrient-matched and consisted of full fat Cheddar, reduced fat Cheddar + butter and butter + calcium supplementation. After 6 weeks, the authors showed the significant effect of the dairy matrix; dairy fat contained in the cheese resulted in significant lower LDL- and total cholesterol compared with the same constituents consumed as butter, protein and calcium. Some factors have been suggested for the effect of cheese on blood cholesterol concentration; the high content of calcium, fermentation and gastric emptying of cheese.

Calcium has been suggested to improve the blood lipid profile by an increased faecal fat excretion, as one mechanism of action. However, calcium behaves differently in dairy products due to its different organisation in the matrix. In fact, naturally occurring calcium in dairy products and supplementary calcium carbonate did not show the same availability of fat for digestion [31]. Soerensen et al. [52], in a randomised crossover study, compared the effects of the consumption of three isocaloric diets with a similar fat content and composition with a different calcium intake; control diet (500 mg/day without dairy intake) and, milk diet and cheese diet (each on 1700 mg/day including dairy calcium). Compared with the control diet, both milk and cheese diets limited the increase in both total and LDL- cholesterol. The magnitude of the effect for milk tended to be smaller than in the cheese diet, but there were no significant differences. Faecal fat excretion was higher in cheese (5.7 g/day) and milk (5.2 g/day) diets in comparison with the control (3.9 g/day) ($p < 0.001$), but again no significant differences between cheese and milk were found. Faecal fat was correlated, in a multiple linear regression, to changes in total and LDL-cholesterol for all three diets, which might be related to the assumption that faecal fat excretion promotes a decrease in bile acid reabsorption resulting in an increase in bile acids synthesis with plasma cholesterol as a substrate. However, Bendsen et al. [4] did not find significant differences in the bile acid excretion between a high dairy calcium (2,300 mg Ca per day) and low dairy calcium (700 mg Ca per day) diet. In the study by Soerensen et al. [52], the effect of calcium on appetite was also investigated. In general, there were no differences in appetite after the three isocaloric meals. Similarly, Dougkas et al. [10] did not find any difference in the overall energy intake and satiety hormones between milk, yoghurt and cheese consumed as a snack.

As mentioned before, the gastric emptying of dairy matrices plays a crucial role in nutrient digestion and absorption, which has been reviewed by Mulet-Cabero et al. [45]. Mackie et al. [34] compared two isocaloric samples in vivo with a different physical structure. The semi-solid sample was a mixture of cheese and yoghurt whereas the liquid sample was an oil-in-water emulsion stabilised by milk proteins. The gastric behaviour, as shown by magnetic resonance imaging, differed between the treatments; the semi-solid samples presented sedimentation whereas the liquid sample showed creaming within the human stomach. This was linked to the higher fullness rating in the semi-solid sample, compared to the liquid sample. In order to provide some insight into the mechanisms behind this in vivo outcome, in vitro digestion was performed in the same samples using a semi-dynamic gastric model [47]. The authors showed similar colloidal behaviour in the simulated stomach compared to that observed in vivo, which controlled the delivering of nutrients. The sedimentation observed in the semi-solid sample resulted in early emptying of high concentrations of nutrients including lipids, which could possibly explain the higher satiety response observed in humans. By contrast, in the liquid sample, the formation of the cream layer on the top of the simulated stomach caused the delay in lipids. These studies highlight the importance of investigating the gastric behaviour in nutrient digestion kinetics and physiological responses.

8.5 Conclusions

This chapter highlights the importance of the matrix in dairy products on the lipid digestion. The initial physicochemical properties of milk fat in dairy matrices (e.g. size and composition of interfacial surface) needs to be well characterised due to their effects on digestibility. In addition, the interactions of milk fat within the matrix and the rheological properties of the matrix in general are of great relevance. During GI digestion, different factors can control lipid digestion including the release, emptying and absorption of milk fat. Figure 8.2 shows the relevant initial characteristics of the dairy matrix and the main limiting factors of lipid digestion

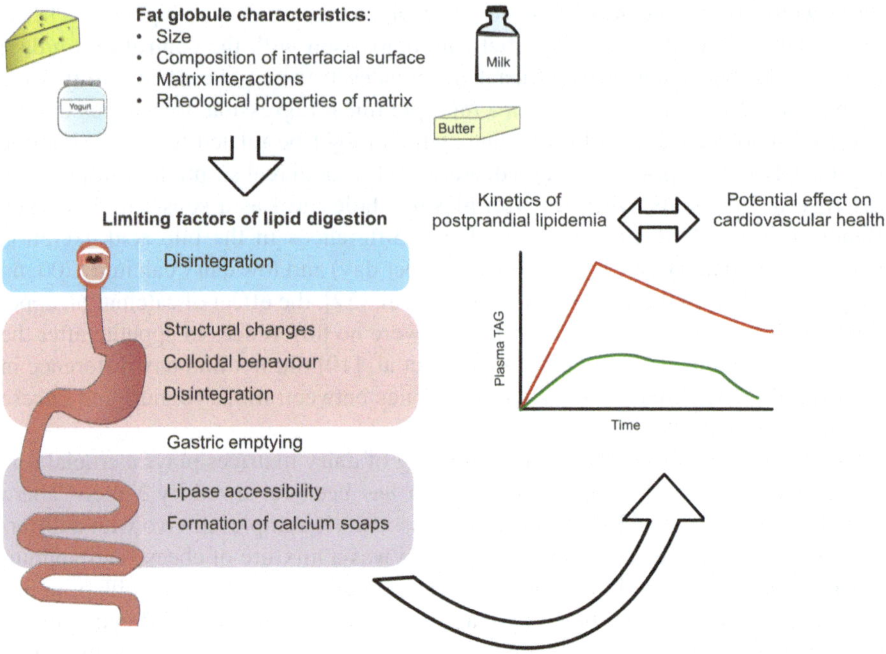

Fig. 8.2 Summary of the relevant characteristics of milk fat in the dairy matrix and the limiting factors of fat digestion in the upper gastrointestinal tract

within the GI tract. These factors can modulate the postprandial lipidemia which, in turn, influence physiological responses. Elevated postprandial lipidemia has resulted in a risk factor for CVD [24]. As alluded to in this chapter, data obtained in humans are to date very limited. Human intervention studies are needed to elucidate the role of the dairy matrix in milk fat digestion. In this respect, investigating the kinetics of postprandial lipidemia in relation to gastric emptying and gastric behaviour is needed. This should be undertaken using adequate methodologies that can follow the emptying of the gastric chyme. Moreover, the long-term effects on the lipid metabolism profile and cardiovascular risks of different dairy matrices need to be better investigated in order to provide an understanding of the nutritional quality of dairy products.

References

1. Ayala-Bribiesca E, Lussier M, Chabot D, Turgeon SL, Britten M (2016) Effect of calcium enrichment of Cheddar cheese on its structure, in vitro digestion and lipid bioaccessibility. Int Dairy J 53:1–9
2. Ayala-Bribiesca E, Turgeon SL, Britten M (2017) Effect of calcium on fatty acid bioaccessibility during in vitro digestion of Cheddar-type cheeses prepared with different milk fat fractions. J Dairy Sci 100(4):2454–2470

3. Ayala-Bribiesca E, Turgeon SL, Pilon G, Marette A, Britten M (2019) Reprint of "Postprandial lipaemia and fecal fat excretion in rats is affected by the calcium content and type of milk fat present in Cheddar-type cheeses". Food Res Int 118:65–71
4. Bendsen NT, Hother A, Jensen S, Lorenzen JK, Astrup A (2008) Effect of dairy calcium on fecal fat excretion: a randomized crossover trial. Int J Obes 32(12):1816–1824
5. Bernbäck S, Bläckberg L, Hernell O (1989) Fatty acids generated by gastric lipase promote human milk triacylglycerol digestion by pancreatic colipase-dependent lipase. Biochimica et Biophysica Acta (BBA)-Lipids and Lipid Metabolism 1001(3):286–293
6. Berton A, Rouvellac S, Robert B, Rousseau F, Lopez C, Crenon I (2012) Effect of the size and interface composition of milk fat globules on their in vitro digestion by the human pancreatic lipase: native versus homogenized milk fat globules. Food Hydrocoll 29(1):123–134
7. Biong AS, Müller H, Seljeflot I, Veierød MB, Pedersen JI (2004) A comparison of the effects of cheese and butter on serum lipids, haemostatic variables and homocysteine. Br J Nutr 92(5):791–797
8. Clemente G, Mancini M, Nazzaro F, Lasorella G, Rivieccio A, Palumbo AM, Rivellese AA, Ferrara L, Giacco R (2003) Effects of different dairy products on postprandial lipaemia. Nutr Metab Cardiovasc Dis 13(6):377–383
9. de Goede J, Geleijnse JM, Ding EL, Soedamah-Muthu SS (2015) Effect of cheese consumption on blood lipids: a systematic review and meta-analysis of randomized controlled trials. Nutr Rev 73(5):259–275
10. Dougkas A, Minihane AM, Givens DI, Reynolds CK, Yaqoob P (2012) Differential effects of dairy snacks on appetite, but not overall energy intake. Br J Nutr 108(12):2274–2285
11. Drouin-Chartier J-P, Tremblay AJ, Maltais-Giguère J, Charest A, Guinot L, Rioux L-E, Labrie S, Britten M, Lamarche B, Turgeon SL (2017) Differential impact of the cheese matrix on the postprandial lipid response: a randomized, crossover, controlled trial. Am J Clin Nutr 106(6):1358–1365
12. Fang X, Rioux L-E, Labrie S, Turgeon SL (2016) Commercial cheeses with different texture have different disintegration and protein/peptide release rates during simulated in vitro digestion. Int Dairy J 56:169–178
13. Fang X, Rioux L-E, Labrie S, Turgeon SL (2016) Disintegration and nutrients release from cheese with different textural properties during in vitro digestion. Food Res Int 88:276–283
14. Feeney EL, Barron R, Dible V, Hamilton Z, Power Y, Tanner L, Flynn C, Bouchier P, Beresford T, Noronha N, Gibney ER (2018) Dairy matrix effects: response to consumption of dairy fat differs when eaten within the cheese matrix—a randomized controlled trial. Am J Clin Nutr 108(4):667–674
15. Fox PF, Guinee TP, Cogan TM, McSweeney PLH (2017) Cheese: structure, rheology and texture. In: Fundamentals of cheese science. Springer US, Boston, pp 475–532
16. Gallier S, Cui J, Olson TD, Rutherfurd SM, Ye A, Moughan PJ, Singh H (2013) In vivo digestion of bovine milk fat globules: effect of processing and interfacial structural changes. I. Gastric digestion. Food Chem 141(3):3273–3281
17. Gallier S, Ye A, Singh H (2012) Structural changes of bovine milk fat globules during in vitro digestion. J Dairy Sci 95(7):3579–3592
18. Gallier S, Zhu XQ, Rutherfurd SM, Ye A, Moughan PJ, Singh H (2013) In vivo digestion of bovine milk fat globules: effect of processing and interfacial structural changes. II. Upper digestive tract digestion. Food Chem 141(3):3215–3223
19. Givens DI (2017) Saturated fats, dairy foods and health: a curious paradox? Nutr Bull 42(3):274–282
20. Griffin B (2017) Serum low-density lipoprotein as a dietary responsive biomarker of cardiovascular disease risk: consensus and confusion. Nutr Bull 42(3):266–273
21. Guinot L, Rioux L-E, Labrie S, Britten M, Turgeon SL (2019) Identification of texture parameters influencing commercial cheese matrix disintegration and lipid digestion using an in vitro static digestion model. Food Res Int 121:269–277

22. Hjerpsted J, Leedo E, Tholstrup T (2011) Cheese intake in large amounts lowers LDL-cholesterol concentrations compared with butter intake of equal fat content. Am J Clin Nutr 94(6):1479–1484
23. Hu M, Li Y, Decker EA, McClements DJ (2010) Role of calcium and calcium-binding agents on the lipase digestibility of emulsified lipids using an in vitro digestion model. Food Hydrocoll 24(8):719–725
24. Kolovou G, Ooi TC (2013) Postprandial lipaemia and vascular disease. Curr Opin Cardiol 28(4):446–451
25. Kopf-Bolanz KA, Schwander F, Gijs M, Vergères G, Portmann R, Egger L (2012) Validation of an in vitro digestive system for studying macronutrient decomposition in humans. J Nutr 142(2):245–250
26. Lacroix M, Bon C, Bos C, Léonil J, Benamouzig R, Luengo C, Fauquant J, Tomé D, Gaudichon C (2008) Ultra high temperature treatment, but not pasteurization, affects the postprandial kinetics of Milk proteins in humans. J Nutr 138(12):2342–2347
27. Lamothe S, Corbeil M-M, Turgeon SL, Britten M (2012) Influence of cheese matrix on lipid digestion in a simulated gastro-intestinal environment. Food Funct 3(7):724–731
28. Lamothe S, Rémillard N, Tremblay J, Britten M (2017) Influence of dairy matrices on nutrient release in a simulated gastrointestinal environment. Food Res Int 92:138–146
29. Lopez C (2005) Focus on the supramolecular structure of milk fat in dairy products. Reprod Nutr Dev 45(4):497–511
30. Lopez C (2011) Milk fat globules enveloped by their biological membrane: unique colloidal assemblies with a specific composition and structure. Curr Opin Colloid Interface Sci 16(5):391–404
31. Lorenzen JK, Nielsen S, Holst JJ, Tetens I, Rehfeld JF, Astrup A (2007) Effect of dairy calcium or supplementary calcium intake on postprandial fat metabolism, appetite, and subsequent energy intake. Am J Clin Nutr 85(3):678–687
32. Lucey J, Johnson M, Horne D (2003) Invited review: perspectives on the basis of the rheology and texture properties of cheese. J Dairy Sci 86(9):2725–2743
33. Lucey JA, Munro PA, Singh H (1998) Rheological properties and microstructure of acid milk gels as affected by fat content and heat treatment. J Food Sci 63(4):660–664
34. Mackie AR, Rafiee H, Malcolm P, Salt L, van Aken G (2013) Specific food structures supress appetite through reduced gastric emptying rate. Am J Physiol Gastrointest Liver Physiol 304(11):G1038–G1043
35. Månsson HL (2008) Fatty acids in bovine milk fat. Food Nutr Res 52
36. Fox PF (2003) Milk proteins: general and historical aspects. In Advanced Dairy Chemistry—1 Proteins (pp. 1-48). Springer, Boston, MA
37. Michalski M-C (2009) Specific molecular and colloidal structures of milk fat affecting lipolysis, absorption and postprandial lipaemia. Eur J Lipid Sci Technol 111(5):413–431
38. Michalski M-C, Briard V, Desage M, Geloen A (2005) The dispersion state of milk fat influences triglyceride metabolism in the rat. Eur J Nutr 44(7):436–444
39. Michalski M-C, Januel C (2006) Does homogenization affect the human health properties of cow's milk? Trends Food Sci Technol 17(8):423–437
40. Michalski M-C, Michel F, Geneste C (2002) Appearance of submicronic particles in the milk fat globule size distribution upon mechanical treatments. Lait 82(2):193–208
41. Michalski M-C, Soares AF, Lopez C, Leconte N, Briard V, Geloen A (2006) The supramolecular structure of milk fat influences plasma triacylglycerols and fatty acid profile in the rat. Eur J Nutr 45(4):215–224
42. Momsen MM, Dahim M, Brockman HL (1997) Lateral packing of the pancreatic lipase cofactor, colipase, with phosphatidylcholine and substrates. Biochemistry 36(33):10073–10081
43. Mu H, Porsgaard T (2005) The metabolism of structured triacylglycerols. Prog Lipid Res 44(6):430–448
44. Mulet-Cabero A-I, Egger L, Portmann R, Ménard O, Marze S, Minekus M, Le Feunteun S, Sarkar A, Grundy MML, Carrière F, Golding M, Dupont D, Recio I, Brodkorb A, Mackie A

(2020) A standardised semi-dynamic in vitro digestion method suitable for food – an international consensus. Food Funct 11(2):1702–1720
45. Mulet-Cabero A-I, Mackie AR, Brodkorb A, Wilde PJ (2020) Dairy structures and physiological responses: a matter of gastric digestion. Crit Rev Food Sci Nutr:1–16
46. Mulet-Cabero A-I, Mackie AR, Wilde PJ, Fenelon MA, Brodkorb A (2019) Structural mechanism and kinetics of in vitro gastric digestion are affected by process-induced changes in bovine milk. Food Hydrocoll 86:172–183
47. Mulet-Cabero A-I, Rigby NM, Brodkorb A, Mackie AR (2017) Dairy food structures influence the rates of nutrient digestion through different in vitro gastric behaviour. Food Hydrocoll 67:63–73
48. Sandoval-Castilla O, Lobato-Calleros C, Aguirre-Mandujano E, Vernon-Carter E (2004) Microstructure and texture of yogurt as influenced by fat replacers. Int Dairy J 14(2):151–159
49. Schmidt, J. M., Kjølbæk, L., Jensen, K. J., Rouy, E., Bertram, H. C., Larsen, T., Raben, A., Astrup, A., & Hammershøj, M. (2020). Influence of type of dairy matrix micro-and macrostructure on in vitro lipid digestion. *Food & Function*
50. Singh H (2004) Heat stability of milk. Int J Dairy Technol 57(2–3):111–119
51. Sodini I, Remeuf F, Haddad S, Corrieu G (2004) The relative effect of milk base, starter, and process on yogurt texture: a review. Crit Rev Food Sci Nutr 44(2):113–137
52. Soerensen KV, Thorning TK, Astrup A, Kristensen M, Lorenzen JK (2014) Effect of dairy calcium from cheese and milk on fecal fat excretion, blood lipids, and appetite in young men. Am J Clin Nutr 99(5):984–991
53. Tholstrup T, Høy C-E, Andersen LN, Christensen RD, Sandström B (2004) Does fat in milk, butter and cheese affect blood lipids and cholesterol differently? J Am Coll Nutr 23(2):169–176
54. Thorning TK, Bertram HC, Bonjour J-P, De Groot L, Dupont D, Feeney E, Ipsen R, Lecerf JM, Mackie A, McKinley MC (2017) Whole dairy matrix or single nutrients in assessment of health effects: current evidence and knowledge gaps. Am J Clin Nutr 105(5):1033–1045
55. Tunick MH, Ren DX, Van Hekken DL, Bonnaillie L, Paul M, Kwoczak R, Tomasula PM (2016) Effect of heat and homogenization on in vitro digestion of milk. J Dairy Sci 99(6):4124–4139
56. Wickham M, Garrood M, Leney J, Wilson PD, Fillery-Travis A (1998) Modification of a phospholipid stabilized emulsion interface by bile salt: effect on pancreatic lipase activity. J Lipid Res 39(3):623–632
57. Ye A, Cui J, Dalgleish D, Singh H (2016) The formation and breakdown of structured clots from whole milk during gastric digestion. Food Funct 7(10):4259–4266
58. Ye A, Cui J, Singh H (2010) Effect of the fat globule membrane on in vitro digestion of milk fat globules with pancreatic lipase. Int Dairy J 20(12):822–829

Chapter 9
Interactions Among Macronutrients and Their Effect on Lypolisis

A. Heredia, A. Asensio-Grau, J. Calvo-Lerma, and A. Andrés

9.1 Introduction

The digestion and absorption of dietary lipids is a very complex mechanism involving insoluble substances, neutral and amphiphilic lipids, and different lipases that act in the stomach and small intestine [2]. Therefore, the intestinal absorption of the products of lipolysis (mainly free fatty acids), as well as free cholesterol and fat-soluble vitamins, enormously depends on the efficiency of the sequential action of lipolytic enzymes on the different substrates.

During digestion, the acidity of the stomach (pH 1–5) is the property that can affect the most the emulsification of the fat, since in these conditions, ionization is limited and the emulsion can even be broken. However, some scientific studies have shown that lipids can undergo emulsification in gastric phase, since many foods contain compounds with emulsifying and/or stabilizing properties [13]. Additionally, contractions and peristaltic waves taking place in the stomach also contribute to the emulsification of lipids by means of propelling the gastric content, which causes the dispersion of fats and the disintegration of other solid particles [2, 28]. Later in the intestinal phase, the gastric content is mixed with the pancreatic and biliary juices, which leads to very important changes in the properties of the medium. On the one hand, pH raises until neutral values that allows the ionization of the acid groups; and on the other, new lipid species from the biliary juices (bile salts, phospholipids and free cholesterol), with important emulsifying properties, are incorporated as well. The bile secreted by the gallbladder helps form the oil/water dispersion of fat globules (fat micelles), making them accessible for the digestive enzymes. After the

A. Heredia · A. Asensio-Grau · J. Calvo-Lerma · A. Andrés (✉)
Instituto de Ingeniería de Alimentos para el Desarrollo, Universitat Politècnica de València, Valencia, Spain
e-mail: aandres@tal.upv.es

M. M.-L. Grundy, P. J. Wilde (eds.), *Bioaccessibility and Digestibility of Lipids from Food*, https://doi.org/10.1007/978-3-030-56909-9_9

formation of fat micelles, pancreatic enzymes placed on the water interface, will hydrolyze the triglycerides from fatty acids that will be absorbed by the enterocytes located in the small intestine.

The increasing prevalence of diseases related to the dietary lipids intake (obesity, hypercholesterolemia, cardiovascular diseases, metabolic syndrome, etc.) as well as the treatment of diseases that occur with maldigestion and/or malabsorption of fats (chronic pancreatitis, cystic fibrosis, gallbladder disease, sphincter of Oddi dysfunction, etc.), have aroused great interest in the scientific community for studying the factors and mechanisms governing lipids digestion [22, 23, 37]. Indeed, gastrointestinal environment could vary within different individuals depending on their age, gender and/or diet, being these differences crucial under specific digestive disorders [33]. The scientific contributions to this subject come from different areas of knowledge, traditionally from the biomedical and pharmaceutical field. Although in recent years, the study of digestion processes has also gained great attention within the frame of Food Science and Technology.

The fate of lipids digestion depends on different factors: (a) those related to the lipid structure and molecular composition, (b) lipids organization within the food matrix structure, (c) presence and amount of other compounds in the digestion medium (proteins, carbohydrates, amphipathic substances, etc.) and (d) the volume and composition of the digestive fluids secretions, gastric and pancreatic (both depending but especially the latter, on age, health status, amount and type of fat ingested in the diet, etc.). The study of food digestion from food science approach is integral, and aims at analyzing the influence of all these factors on the digestibility and bioavailability of nutrients. The resulting knowledge would be particularly relevant for the development and/or adaptation of food manufacturing processes and food formulations to specific populations groups. The development of in vitro digestion methodologies that mimic the digestion processes is a useful approach to study the influence of the factors involved in macronutrients digestibility, since they allow the reproduction of the physiological gastrointestinal process in a controlled and reproducible form. In this way, it is possible to know out the state of the reaction at each point of the process, and attribute the results only to the conditions of the experiment. In vitro models are especially valuable for screening purposes and for a systematic approach to analyze ingredients interactions that can occur during food digestion.

Ingredients interactions and their effects on food quality is a complex subject. Moreover, if nutrients digestibility and/or bioavailability is included in the quality concept. These interactions can occur not only during manufacturing, storage and food preparation, but also along the digestion process, being able to modulate the absorption and use of nutrients that is the ultimate goal of food intake. The type and magnitude of the interactions during manufacturing and storage of food systems depends on both the composition and structure of the matrix, and the variables of the environment or the processes to which they are subjected. In the formulation and processing of foods, the molecular interactions are quite well known, which allows acting either by promoting or by avoiding them according to the final objective pursued. However, the concept of interaction in food systems is much broader and might be understood with an approach that also encompasses the interactions taking place during the digestion processes.

The vast majority of foods are multiphase mixed dispersed systems in which the organization of the different components, the formation of the different phases and their physical state give rise to food matrices with specific structures and properties. Functional properties of macronutrients are a result of interactions between them and other food compounds such as water. Then, altering the food composition of the small molecular species can contribute to significant changes in matrix food properties. Besides the food composition, processing implying high pressure and temperature also has a key role on promoting ingredients interactions (i.e. protein-lipid interactions in extrusion process). In the same way, the exposure to digestion environment (mechanical forces, pH, enzymes, etc.) promotes food de-structuring, new molecular interactions and the formation of new structures. In this sense, novel scientific knowledge is needed in order to minimize or maximize interactions among food ingredients and/or with some physiological compounds (gastrointestinal fluids) during digestion, depending on the personalized final goal of nutrition. From a practical point of view, the study of lipid interactions and their effect on lipolysis seem to be the most appealing, for their direct impact in health-related outcomes. The field is quite unexplored; however, some evidences have started to grow in the literature. From this perspective, this chapter aims at pointing out the most significant interactions known up to now to affect lipids digestion.

9.2 Nutrients' Interactions and Their Effect on Lipolysis

Binary, or even ternary, interactions occurring among macro and micronutrients in foods could affect the bioaccessibility of the components taking part. Particularly, interactions of lipids with carbohydrates (fibre, starch, cellulose) and/or proteins (fibrilar, globular) are especially relevant in terms of lipids digestibility. These molecular interactions can be natively present in raw materials (e.g. legumes, cereals or seeds, among others) or can be generated during processing or storage that leads to modifications in the food structure (e.g. pasta, bread, dairy products or processed meat). The synchronic intake of individual foods in the same dish or meal can also result in new additional interaction. Recently, studies of food digestion are paying special attention to macronutrients interactions; although the vast majority of them have been performed in model systems, or ideal emulsions, in which the complexity and multifactor nature of foods matrices are simplified in order to drive solid conclusions. This section attempts to summarize the main scientific evidences found in both model-systems and real food matrices, related to the impact of macronutrients interactions on lipids digestion.

9.2.1 Knowledge Established Throughout Model-Systems: Molecular Foundations

Concerning the presence of proteins and their effect on lipolysis, Turgeon and Rioux [38] conceived protein networks as a surrogate of the food matrix in which fat globules are embedded. Therefore, lipolysis would be conditioned by the rate and extent of proteolysis: as long as protein network is broken down, lipids are released from the food matrix, becoming accessible to lipases [10]. In addition, some differences have been found regarding protein structure (fibrillar or globular) and origin (animal or vegetal). In fact, plant proteins might exert a greater preventing effect against lipolysis compared to animal ones. Wheat gliadins are more insoluble and hydrophobic than diary proteins, so they form a more compact interfacial film than caseins, which is resistant to displacement by bile salts. This film hinders the diffusion of lipase towards the surface of the fat globules resulting in a lower extent of digestion. Similarly, soybean proteins have been reported to be more resistant than β-lactoglobulin to interfacial displacement by bile salts at the oil-water interface, also leading to decreased lipid digestion [32]. On the other hand, lipids and proteins from fish seem to be more easily digested than mammalians ones because of the lower collagen presence in fish muscle than in mammalian muscle tissues [19]. Pepsin seems to have better affinity for myofibrillar and sarcoplasmic proteins than for connective ones, which are more fibrous and difficult to hydrolyse. This affinity also results in lesser lipolysis extent and rate in different meat products.

Finally, Nieva-Echevarría et al. [26] reported that the effect of proteins on lipolysis is different depending on digestion time and bile salts concentration. Apparently, as long as peptides or amino acids are released from digestion, they might favour lipolysis at the first step by enhancing the emulsification capacity of less polar glycerides (di and triglycerides). However, as the proportion of more polar molecules (monoglycerides and fatty acids) with a higher emulsification capacity increases, the action of lipases at the fat globules interface seems to be hampered, leading to a lower hydrolysis of monoglycerides and therefore reducing the complete lipolysis. This effect is especially relevant when the bile salts concentration in the digestive medium is drastically lessened. As it is well known, bile salts are a natural biosurfactant that is present in the gut lumen and have an important role in lipolysis by regulating the composition of lipid-water interface, where hydrolysis occurs [12]. Bile salts solubilize newly formed lipolytic products into bile-phospholipid mixed micelles, thus removing them from the interface together with other surfactants, such as proteins, that may also be present in the adsorption layer, displacing bound lipases. Complementarily, Nieva-Echevarría et al. [27] reported that bile salts concentration below 18.75 g/L greatly favoured the hydrolysis not only in triglycerides, but also in partial glycerides, especially diglycerides. The lower extent of lipolysis observed at higher concentrations (30–60 g/L) by Nieva-Echevarría et al. [27] agrees with previous studies [6, 21]. Under high concentrations, the absence of sufficient colipase results in a repulsion of lipases at the interface because of biliary molecules completely covering it.

With respect to chains of polysaccharides, they protect the oil-in-water emulsions against lipolysis by forming a thick hydrated layer that impedes lipases physical contact with the fat globule [25]. In fact, polysaccharides are used to thicken foods in industrial processing as they have little propensity towards emulsification, given their lack of surface activity [9]. Moreover, thermal treatments occurring in baking or frying, have been reported to induce amylose-lipid complexes generation. These complexes seem to reduce the access of digestive enzymes to their substrates, limiting both lipid and starch digestibility [34]. Overall, carbohydrates could affect lipid digestion in the small intestine through a variety of physicochemical mechanisms, such as increasing viscosity of the digestion medium, alteration of droplet disruption or coalescence kinetics, binding of bile salts, phospholipids or enzymes [16, 20].

9.2.2 Interactions of Lipids with Other Components in Real Foods: Structure Coming into Play

The above-described foundations refer to lipid, protein and carbohydrate molecules interacting with each other in model systems, without considering the effect of the physical structure in which they are contained. This could mean that interactions among each other are expressed with no restrictions, and that the observed effects are clear and with no interferences. However, the fact is that macronutrient molecules in real foods are assembled with each other conforming the 3D structure a food matrix is, increasing the level of complexity around the concept of nutrient interactions. Structural properties of the food matrices play a crucial role in the release of molecules to the digestion medium, where the lipolysis reaction really occurs. In other words, food structure coming into play implies that interactions among nutrients take place once these have been released to a medium where reaction can occur.

Food matrix acts as a nutrient-container during storage and as a nutrient-release system during digestion, and depending on it, the release and presence of macronutrients in digestion medium will be conditioned [38]. Thus, lipid digestion may depend on the breakdown of these matrix materials and lipid release before they can be exposed to lipases, as previously suggested in Sect. 9.2.1 [8]. Besides, many of these non-lipid components may interfere with lipid digestion by altering the viscosity of the digestion media, by competing with lipase for the oil-water interface of the emulsified lipids, or by interfering with enzyme activity [36].

Guo et al. [16] performed a review, which compiled the main studies assessing the role of food matrix as a determinant of lipolysis. Despite many researches have been conducted to characterise digestion in different foods, the methodologies that have been followed are uneven, so establishing comparisons with the obtained results up to date has to be done with caution and the possibility for that is overall limited. In addition, research on food digestion has addressed different objectives and, while it is true that different studies tangentially deal with the common topics, full experimental designs focused on assessing lipid digestibility in real food matrices are very scarce.

Among the few studies of digestion performed in real foods and focused on lipolysis, it is worth mentioning the one carried out by Calvo-Lerma et al. [7] in which a total of 52 foods belonging to 10 different categories (dairy products, fish, meat, sweets, cereal-based products, nuts, fruits, free oils and fats, eggs and potato-based products) were digested with the aim of elucidating the relevance of macronutrients interactions, food structure and host-intestinal conditions, on lipolysis. As it is shown in Fig. 9.1, the main conclusion achieved was that protein-lipid and carbohydrate-lipid interactions only significantly ($p < 0.001$) affect lipolysis extent under altered intestinal conditions of both reduced intestinal pH (6) and bile secretion (1 mM) such as given in some pathologies. This effect was not reported under healthy standard intestinal conditions characterized by intestinal pH of 7 and bile salts concentration of 10 mM. Moreover, the relevance of interactions on lipolysis extent under suboptimal intestinal conditions was, at the same time, dependent on fat content in foods. Thus, lipolysis decreased along with an increase in protein or starch content in those foods with a low (1–8%) or medium lipid content (9–20%). In foods with high lipid content (21–40%), protein-lipid interaction was found negligible; while starch-lipid interaction positively affected lipolysis ($p < 0.001$). As shown in the Fig. 9.1a, food products as olive oil, cured cheese or salmon with high lipid content, achieved similar lipolysis extent, around 80-%, regardless of their protein content. In contrast, in foods with low and medium content of lipids and high protein content, lipolysis was highly affected by protein-lipid interaction, as in beef steak or cured ham.

Similarly, part of the effect of the starch-lipid interaction on lipolysis (Fig. 9.1b) can be exemplified as follows: lipolysis achieved in breads, with high-starch and

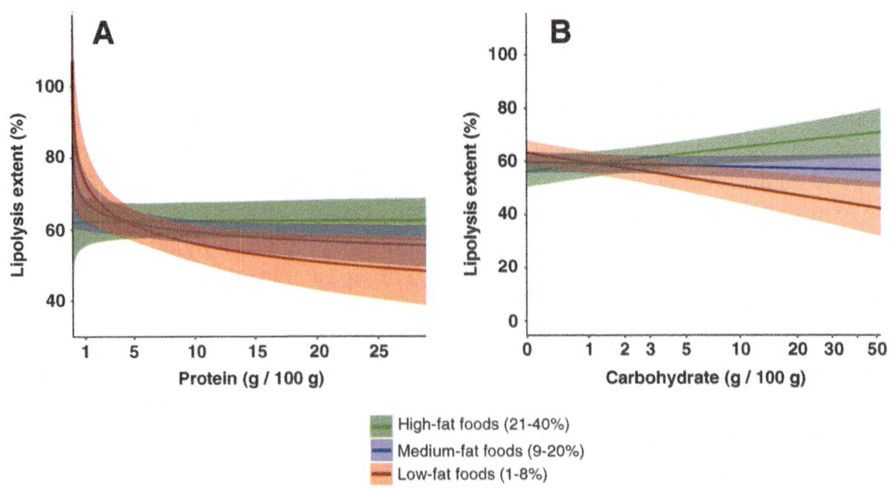

Fig. 9.1 Nutrient interaction and the effect on lipolysis: protein-lipid interaction (**a**), and starch-lipid interaction (**b**). (Figure adapted from Calvo-Lerma et al. [7])

low-fat content, was found to be lower than in biscuits of lower starch but medium fat content. In French fries, chips and nuts, foods with high starch and fat contents, a positive interaction of these macronutrients was found on lipolysis.

Likewise in real foods, the significance of these interactions on lipolysis, are modulated by the lipid organization in the food matrix. Calvo-Lerma et al. [7] tries to know out the complementary role of the food structure on lipolysis by the application of Michalski's classification [24] (Fig. 9.2). Thus, the 52 studied foods were firstly classified as: (1) foods with complex solid structure, (2) foods with a continuous aqueous phase or (3) foods with a continuous lipid phase. Secondly, foods with complex solid structure were divided in two subgroups, those with lipids inclusion in a protein matrix (e.g. cheese) and those with lipids inclusion in carbohydrate and protein matrix (e.g. cookies). Lipids in foods with a continuous aqueous phase can be structured as oil in water emulsion (e.g. milk) or as intracellular lipid droplets and membrane structures (tissues) (e.g. meat, egg yolk, vegetables). Finally, different subgroups of foods with a continuous lipid phase can be established: free fat (e.g. oil, lard), foods with particles dispersed in solid fat (e.g. chocolate) and water in oil emulsions (e.g. butter or margarine). According to Calvo-Lerma et al. [7], It can be affirmed that the way how fat is integrated in the food matrix can also significantly govern lipids digestion, being lipids in solid particles embedded in continuous liquid phase (e.g. chocolate) the worst digested regardless to the intestinal conditions. In addition, it might be pointed out that, similarly to the relevance of macronutrients' interactions, the lipid organization in the food structure seems to be much more relevant under altered intestinal conditions (pH 6 and bile salts concentration 1 mM) (Fig. 9.2a) than in healthy gastrointestinal environment (pH 7 and bile salts concentration of 10 mM) (Fig. 9.2b). This finding may have a close relationship with the fact that suboptimal intestinal conditions of reduced pH and bile salts concentration could have several implications on lipolysis. The role of bile salts concentration has been previously exposed in previous Sect. 9.2.1.

Foods presenting *oil-in-water emulsion structures*, which are mainly milk and dairy products such as yoghurt, are the type of structure that can reach the highest lipolysis extents. Moreover, foods with oil-in-water emulsion present with a short range of final lipolysis extents, thus small variability can be attributed to the different food matrices. However, as observed in the figure, yoghurt and other derivatives reach less lipolysis than milk in its natural form. Moving to another type of structure, such as *lipid inclusion in a protein matrix*, we find two very different foods: meat products and cheese. Considering meat, a wide range in lipolysis extents can be observed, from 45% to 100%. An additional consideration must be taken to understand this evidence, which is the degree of de-structuration of the matrix that is present in processed meat (sausage, hamburger, coldmeats) in contrast to meats in their native form (beef steak, chicken drumstick). As previously explained, degradation of the food matrix is a key factor determining lipolysis, and in the case of fresh meat, this is barely inexistent. This makes that the rupture of the matrix has to be achieved during digestion in order to lipids get released and become accessible to the enzymes. On the contrary, paying attention to processed meat, the de-structuration has already taken place before digestion, and higher lipolysis extents

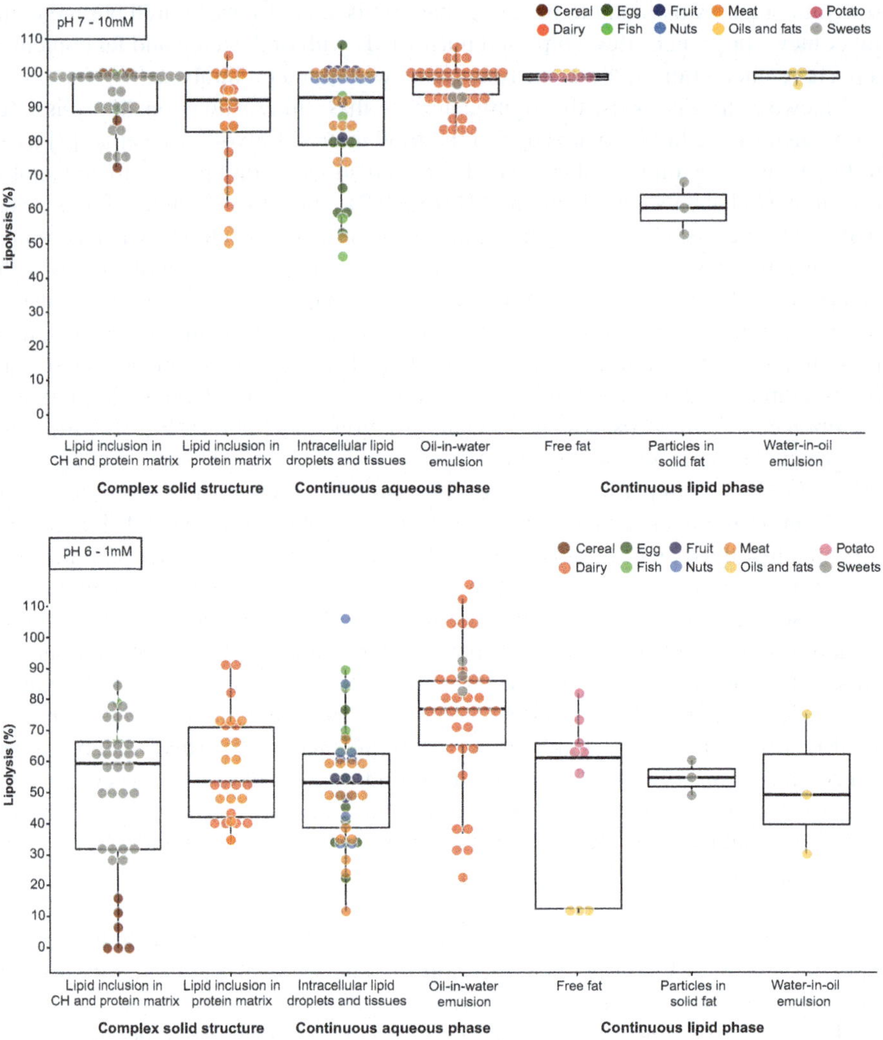

Fig. 9.2 Boxplots representing lipolysis extent of the assessed food groups according to the lipid structure in the food matrix, under the digestion conditions of (**a**) intestinal pH 6 and bile salts concentration 1 mM; and (**b**) intestinal pH 7 and bile salts concentration 10 mM. Cereal: white bread and dark bread; Egg: poached egg, hard egg, omelette; Fruit: avocado, olive; Meat: chicken drumstick, pork loin, beef steak, cured ham, cooked ham, luncheon meat, sausage, hamburger; Potato: French fries and chips; Dairy: full-fat cow milk, semi-skimmed cow milk, goat milk, milk-shake, natural yoghurt, yoghurt with sugar, liquid yoghurt, custard, crème caramel, fresh cow cheese, fresh goat cheese, cured cheese, cream cheese, soft cheese; Fish: salmon, canned tuna; Nuts: walnuts, peanuts, fried peanuts; Oils and fats: olive oil and butter; Sweets: regular biscuit, chocolate-chips biscuit, chocolate-stuffed biscuit, breakfast cereal, croissant, doughnut, waffle, milk-chocolate, chocolate spread. (Figure adapted from Calvo-Lerma et al. [7])

are more easily achieved. In the case of cheese, as also previously discussed, the origin of fat could be the determinant of the success of lipolysis, despite most of the cheese matrices present with a similar structure.

To set another example of foods presenting the same lipid structure within the matrix, but with different lipolysis extents, *the intracellular lipid droplets in tissues* should be discussed. This is an even more heterogeneous category in which such diverse foods as nuts, egg and fish are gathered. As shown in the figure, egg presents with very different lipolysis extents depending on the cooking technique: poached egg contains lipids in a liquid yolk, which is easily broken down and mixed with digestive fluids, allowing for high lipolysis extents close to 100%. If the egg is boiled, the yolk acquires a solid structure, which makes the matrix degradation harder and thus, the release of lipids for its consequent hydrolysation. Finally, in the form of omelette, the protein networks trap lipid molecules inside, and proteolysis has to occur before lipids can be made accessible to lipases. In the case of fish (salmon, concretely) the very high content of protein, which is in the form of fibrilles, entail an even stronger network, and lipid release from the matrix could become even more difficult. In addition, the high ratio protein to lipid, could cause an excess of products of proteolysis in the fat droplet interface as proteolysis progresses, which could compromise the removal role of products of lipolysis that bile slats play to facilitate lipase action. In contrast, in the case of nuts, high lipolysis extents were reached, probably because the matrix was highly minced during the oral stage (as part of the experimental design) and the components of the matrix (protein, lipid and carbohydrate) were no longer belonging to a cohesive structure, and thus were easily accessible to the digestion agents.

Therefore, despite some foods present the same lipid structure within the food matrix, there are specific food matrix properties in each food determining the fate of digestion of the lipids they contain. Indeed, there are processed foods presenting the same formulation in ingredients, which are the bakery products, and the same lipid structure within the following matrix, which is *lipid inclusion in carbohydrate and protein matrix*, that show different lipolysis extent. The possible explanation could be that within bakery products two types of structure can be differentiated: those in which the dough is raised during cooking, such as puffy pastries, and those that remain with a compact structure, such as biscuits. In the first type of bakery, the food matrix seems to be more easily broken down during digestion, so higher lipolysis extents are reached, compared to the hard structure of the second type, in which degradation of the matrix requires additional mechanic agitation.

9.3 Case-Studies: The Role of Nutrient Interactions Along with Food Structure on Lipolysis

This section gathers illustrative examples of digestion studies performed in rich-fat foods belonging to different categories and concerning the role of nutrients- interactions and food structure on lipolysis.

9.3.1 Milk and Dairy Products

Milk and dairy products are oil-in-water emulsions in which lipids are dispersed in an aqueous liquid phase (milk, cream), in a partially gelled phase (yoghurt, cheeses) or in a dry medium rich in proteins (powders). In cheese, fat is either dispersed as native milk fat globules (soft cheeses), present as fat globules more or less aggregated or coalesced, dispersed as small fat globules covered by proteins after high shear stress homogenisation (blue cheeses, some fresh cheeses), or in the form of free fat domains covered by milk polar lipids (hard-type cheeses).

Studies have shown that peptides and free fatty acids are more easily released from milks and yogurts (which are liquid and semi-solid matrices, respectively) vs. cheeses (which are solid matrices) owing to easier breakdown of the food matrix and greater accessibility by digestive enzymes in the former two. For cheese, the structural characteristics and hardness are determinants of lipid digestion. Fang et al. [11] compared the disintegration and lipolysis of cheeses with different textural properties Fang et al. [11], and found that regular Cheddar cheese had a significantly higher extent of lipid digestion than light Mozzarella cheese. The larger fat globules in Cheddar cheese acted as weak points in its microstructure and texture, which led to a greater decrease in hardness and greater disintegration during *in vitro* digestion. By contrast, light Mozzarella consisted of a denser fibrous protein matrix due to the stretching step that occurs during its fabrication. Its protein matrix showed a lower rate of disintegration and protein hydrolysis, thereby providing enhanced protection to dispersed fat, and hence a lower extent of digestion Fang et al. [11]. Therefore, in milk and dairy products, the main factors related to the food matrix conditioning lipolysis are the structure acquired during processing, which has to be disintegrated.

9.3.2 Egg

In egg, lipids are present in the yolk, while the white part contains mainly the protein. Lipids are dispersed in the form of lipoproteins, e.g. high density (HDL) and low density (LDL) lipoproteins. Egg admits many ways of cooking resulting in very different food matrix, which finally impact, on macronutrients digestibility. Asensio-Grau et al. [4] studied how influence the cooking procedure on protein and lipid digestibility of eggs. The cooking method generated a great impact on matrix structure of egg resulting in different structural changes in egg proteins and lipids. In spite of the same nutritional composition, the cooked eggs (poached egg, boiled egg and omelette) have different matrix degradation during digestion affecting on lipid bioavailability. The intensity of the thermal treatment during cooking, in terms of time, determined the protein denaturation to form insoluble aggregates, which interfere in lipid digestion. In fact, also the denatured egg yolk protein can be adsorbed at the oil:water interface avoiding the bile salts access and decreasing lipolysis. Omelette and boiled egg resulted in a lower matrix degradation and hence lower lipid and protein digestibility compared to poached egg. Overall, it seems that in egg the structure imparted by the cooking technique is the most determinant factor of lipolysis.

9.3.3 Meat Products

In meat, triacylglicerides (TAG) are mainly present in the adipocytes, the cells that form the adipose tissue. TAG can also be found in the muscles in the form of isolated adipocytes or droplets and within cell muscles. In processed meat, these fats are partly crystallised at ambient temperature, and sometimes even at body temperature, due to the presence of more than 40% of long-chain saturated fatty acids, palmitic acid being mainly located in sn-2 position in lard.

Asensio-Grau et al. [3] studied the lipid digestibility in meat products and also the effect of the food structure on the lipolysis. The results from the study unveiled that lipid hydrolysis depend on the food matrix structure. Different meat products were studied and compared, on the one the hand natural meat matrices and on the other hand different processed meats that presented an unstructured matrix. The unstructured matrices showed a higher matrix degradation and hence lipid digestibility compare to the natural matrices. Food structure is the main factor controlling digestion. Natural meat products presented the fat globules inside of a solid protein matrix. In order to make lipids accessible for the enzymes, proteolysis has to occur. In case of unstructured meats, the structural changes occurred before the food intake and thus improve the lipid accessibility. Processed meat do not have a structure in which proteins are bonded to each other and thus lipids are more accessible to lipases than those lipids in structured meats.

9.3.4 Nuts

In nuts, lipids are stored in organites called oleosomes, which are constituted by a hydrophobic core rich in TAG surrounded by a monolayer of phospholipids and proteins. The oleosins have strong steric hindrance. This organization ensures the oleosomes very high stability against thermal or detergents injuries.

To this respect, Grundy et al. [14] focused the research on the impact of almonds' food structure on lipolysis. In almonds, oil bodies, which are the oil-bearing structures that may represent over 50% of the nut weight, are located within thin-walled cells. So, the release of oil bodies from almond cells or the diffusion of lipase towards these cells is necessary for lipolysis. Approximately 10% of lipids in raw and roasted almonds have been observed to be released from almonds upon mastication with the cells remaining largely intact. The rate and extent of lipid digestion of almond cells were much lower than that of the isolated oil bodies (22 vs. 69%

hydrolysis after 1 h digestion) during *in vitro* intestinal digestion [14]. *In vitro* digestion studies are highly consistent with in vivo human trials. Therefore, the digestion and bioaccessibility of lipids in almonds was regulated by the structure and properties of the cell walls surrounding the oil bodies.

Complementarily, Paz-Yépez et al. [30] also studied the influence of the particle size of peanuts and walnuts on lipid digestibility. Matrix degradation and hence lipolysis were affected by the particle size in each type of nut. The particle size of the nuts provides information on the importance of chewing on the enzyme's accessibility to the macronutrients. The physical transformation occurred in nuts with smaller particle size increased the surface area of intra-cellular nutrients improved hence the availability of lipids for hydrolysis by the digestive enzymes.

9.3.5 Rich-Fat Cereals and Seeds

Some authors have described the role of plant-material structure on bioactive compounds delivery along gastrointestinal tract and on macronutrients digestibility. Among cereals, studies have paid special attention on oat because its highest fat content, around 7%, and the relevance of lipolysis rate and extent, and β-glucans presence, on hydrophobic phytochemicals and cholesterol bioavailability [1, 35]. In mature oat grains, oil-bodies are mainly accumulated in the starchy endosperm forming a continuous oil matrix that interacts with both protein and starch; meanwhile they are also present in the aleurone layer, embryo and scutellum as oleosins [18].

Oil bodies must be extracted from the food matrix during digestion to be hydrolyze by the lipases action. In this sense, disruption undergoes during chewing, and salivation, are crucial to lipids release from the food matrix. Even though chewing were effective in terms of oil bodies extraction, the presence of fibre, mainly β-glucans, seems to hinder the extent and rate of lipolysis during the further stages of digestion, gastric and intestinal ones. Grundy et al. [15] attempts to explain mechanistic involving this phenomenon by adding different oat-based products, such as flakes, bran or flour, to an emulsion (control). Indeed, the presence of flakes significantly reduced the lipid digestion of the emulsion.

Based on these results, β-glucan's role on lipid digestion seems to be more dependent on the binding or entrapment of the digestive products, occurring differently during the digestion of flakes, bran or flour, than on the total amount of β-glucans or the increase of the viscosity of the digestive fluids due to presence of the dietary fiber. The hydration capacity of polysaccharides also conditions the role plays by oat compounds on lipolysis [39] The same phenomenon is expected to occur in sesame, chia or flax seeds digestibility, which are especially rich in lipids (up to 35%) but also in proteins and fibre. In fact, oil extraction from seeds is a common process to obtain a healthy lipidic ingredient of easier digestion than the whole seeds. Milling, soaking, germination of fermentation, among other processes, can be also performed to induce compositional and structural changes that might result in better digestibility and enhanced bioaccesibility of bioactive compounds of seeds [17, 29]. It is important to point out the lack of scientific studies aiming at elucidating the structural factors controlling plant-materials digestion.

9.4 Effect of Co-digestion on Lipolysis: Meal Factors

The previous sections have followed a step-wise approach to explain the following concept: interactions among lipids and other components (protein, carbohydrate, bile salts) taking place in real foods individually consumed. The food matrix can be conceived as a limiting element for these reactions to occur, as nutrients have first to get released to be available to interact with the rest of the compounds. Thus, while in Sect. 9.2.1 the molecular foundations seem to be well-established and general statements can be made (e.g., vegetal protein can prevent lipolysis more than animal protein), in Sect. 9.2.2 the complexity of the food matrix reveals that more concrete phenomena might be occurring in function of the specific food being studied. In this sense, the extent and rate to which breakdown of the matrix is effective will condition the amount and type of nutrients available in the digestion medium, to interact with each other and with enzymes, bile salts and other components of the digestive fluids.

The level of complexity increases even more when two or more foods are co-digested. Co-digestion of different foods makes the way to the influence of other factors known as "meal factors". The mixture of different foods in the same dish or meal, results in an oral bolus – and subsequent chyme – with different structural properties, thus conditioning release of the nutrients they are made of. In addition, depending on the type and amount of co-digested foods, a new and wide range of possible interactions appears, which may result in different products or digestion extension than when digested alone.

While there are some strong evidences about "meal factors", these are still limited as refers to its impact to lipolysis. For example, the negative interaction between calcium and iron occurring as a luminal event is very well established, as both minerals compete for divalent metal transporter at the apical membrane of intestinal cells. Another example dealing with iron is the strong inhibitory effect that phytates have on its eventual uptake. The positive effect of an interaction with iron has been described as well, which deals with ascorbic acid, especially nonheme iron, which comes mainly from plant foods. Ascorbic acid facilitates iron absorption by forming a chelate with ferric iron at acid pH that remains soluble at the alkaline pH of the duodenum.

In contrast, little attention has been paid, up to date, to study interactions of macromolecules, such as carbohydrate, protein and lipid on hydrolysis and absorption. A possible reason for the scarce studies could be the complexity of food matrices entail *per se* – which are still vaguely described – and the unlimited number of interactions that could occur among different components. However, as a matter of fact, combining different foods in one intake is the most common eating habit, as normally foods are prepared as recipes, dishes, sandwiches, etc., and consumed altogether. The gastronomy definition of "Dish" is "*a specific food preparation which contains different variety of food*". The composition of a dish depends on the combined individual foods, being carbohydrates the main macronutrient in pasta dishes and protein in meat and fish dishes. Consequently, the importance of co-ingestion, co-digestion and meal-factors should be addressed and considered a priority. In this sense, there are a few studies [5, 31] exploring lipolysis extent when combining different matrices in the same digestion. Preliminary results of in vitro co-digestion experiments show the existence of interactions that do not always respond to the same trend. Figure 9.3 shows the results of lipolysis measured after the in vitro digestion of different foods, alone or in combination. High protein foodstuffs, low fat (beef) or high fat (salmon), co-digested with High Carbohydrate food, low fat (boiled potato) or high fat (French fries) resulted in a lower or higher lipolysis extent respectively, than when digested alone. In these examples, both the presence of carbohydrates and the presence of oil (embedded in potato chips) seem to determine the final result of the digestion of the total fat of the system. In a similar way as potato based dishes, pasta is another of the basic ingredients of recipes to which lipid-rich sauces are added, such as Bolognese sauce (cheese, meat and fried tomato) or pesto sauce (oil, cheese, basil and pine nuts). Although the extent of lipolysis in this type of sauces is much greater than that obtained in products such as meat (beef: A) or fish (salmon: B), a decrease in the lipid digestion is also observed when they are in vitro co-digested together with the pasta. These are only

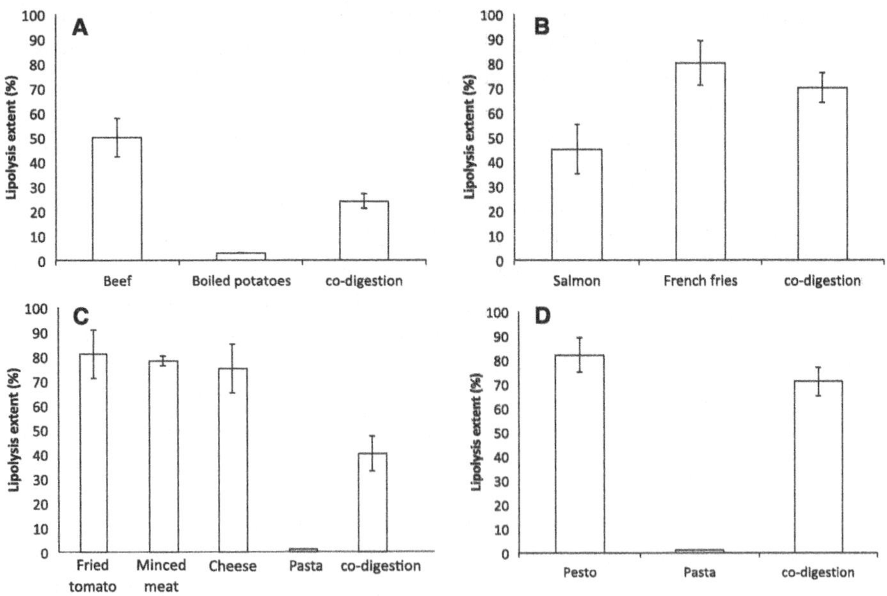

Fig. 9.3 In vitro digestion of different foods, separately and co-digested: Impact on lipolysis extent. (**a**) beef steak and boiled potato; (**b**) salmon and French fries; (**c**) pasta with minced meat, cheese and fried tomato; (**d**) pasta with pesto sauce

preliminary results pointing out the need to explore more in deep what finally happens during digestion of a meal beyond the final nutritional value of the daily intake.

9.5 Final Remarks and Conclusions

Foods are extremely complex systems, in which the macronutrient composition determines the structure of the food matrix, and in function of the food matrix, different interactions can occur, resulting in different lipolysis extents when subjected to in vitro digestion.

The knowledge we have available up to date point at an effect caused by nutrient interactions related to the food matrix structure, and to the specific type of constituents (nutrients) of the matrix, thus suggesting very specific situations depending on the concrete type of food. However, some general conclusions can be drawn: foods with lipids that are in the form of oil-in-water are less susceptible to be affected by interactions with other nutrients, although this structure is attributed to milk and dairy products, in which food matrix is less cohesive. In contrast, foods with more complex structures, such as intracellular lipid droplets in tissues, will result in different types of interactions among lipid with carbohydrate and protein, leading to different lipolysis extents. Additionally, from a general overview, research points that foods with low lipid content and high carbohydrate or protein content have reduced lipolysis extents, while foods with moderate to high lipid contents are not as much affected by the interactions with the other macronutrients. These interactions seem to be also present in co-digestion of different foods, but this scenario is even more complex than the individual digestion of foods alone, and despite some cases have been studied, the virtually infinite number of combinations of different types of food may lead to other type of findings.

Overall, nutrient interactions occur in the digestion medium after release from the food matrix, so the existing evidence in model systems can be assumed. Studies up to date, however, indicate that the relevant aspect that should be well established and characterised is how these nutrients are released from the matrix, which is something specific from each food (and the way it is processed and cooked) and imparted by the digestion conditions, and what really conditions how interactions and digestion eventually occur. What is more, the complexity of the system could even increase when combining different foods in the same digestion.

To conclude, the information presented in this chapter can be conceived as a first approach to the challenge of studying the huge complexity a food matrix entails, and its implication on lipid digestion. Further and broad research is encouraged.

References

1. Alvarez-Sala A, Garcia-Llatas G, Cilla A, Barberá R, Sánchez-Siles LM, Lagarda MJ (2016) Impact of lipid components and emulsifiers on plant sterols bioaccessibility from milk-based fruit beverages. J Agric Food Chem 64(28):5686–5691. https://doi.org/10.1021/acs.jafc.6b02028
2. Armand M (2007) Lipases and lipolysis in the human digestive tract: where do we stand? Curr Opin Clin Nutr Metab Care 10(2):156–164. https://doi.org/10.1097/MCO.0b013e3280177687
3. Asensio-Grau A, Calvo-Lerma J, Heredia A, Andrés A (2018) Fat digestibility in meat products: influence of food structure and gastrointestinal conditions. Int J Food Sci Nutr 0(0):1–10. https://doi.org/10.1080/09637486.2018.1542665
4. Asensio-Grau A, Peinado I, Heredia A, Andrés A (2018) Effect of cooking methods and intestinal conditions on lipolysis, proteolysis and xanthophylls bioaccessibility of eggs. J Funct Foods 46(May):579–586. https://doi.org/10.1016/j.jff.2018.05.025
5. Asensio-Grau A, Frassineti S, Heredia A, Andrés A (2018) Lipid digestibility in single or combined food. In: 4th International & 5th National Student Congress of Food Science and Technology, p 2341–2240 (74). Avecta Valencia (Spain)
6. Borgström B, Erlanson C (1973) Pancreatic lipase and co-lipase: interactions and effects of bile salts and other detergents. Eur J Biochem 37(1):60–68
7. Calvo-Lerma J, Fornés-Ferrer V, Heredia A, Andrés A (2018) In vitro digestion of lipids in real foods: influence of lipid organization within the food matrix and interactions with nonlipid components. J Food Sci 83(10):2629–2637. https://doi.org/10.1111/1750-3841.14343
8. Chen L, Remondetto GE, Subirade M (2006) Food protein-based materials as nutraceutical delivery systems. Trends Food Sci Technol 17(5):272–283. https://doi.org/10.1016/j.tifs.2005.12.011
9. Dickinson E (2009) Hydrocolloids as emulsifiers and emulsion stabilizers. Food Hydrocoll 23(6):1473–1482. https://doi.org/10.1016/j.foodhyd.2008.08.005
10. Dickinson E (2012) Emulsion gels: the structuring of soft solids with protein-stabilized oil droplets. Food Hydrocoll 28(1):224–241. https://doi.org/10.1016/j.foodhyd.2011.12.017
11. Fang X, Rioux LE, Labrie S, Turgeon SL (2016) Disintegration and nutrients release from cheese with different textural properties during in vitro digestion. Food Res Int 88:276–283. https://doi.org/10.1016/j.foodres.2016.04.008
12. Golding M, Wooster TJ (2010) The influence of emulsion structure and stability on lipid digestion. Curr Opin Colloid Interface Sci 15(1–2), 90–101
13. Golding M, Wooster TJ, Day L, Xu M, Lundin L, Keogh J, Cliftonx P (2011) Impact of gastric structuring on the lipolysis of emulsified lipids. Soft Matter 7(7):3513–3523. https://doi.org/10.1039/c0sm01227k
14. Grundy MML, Carrière F, Mackie AR, Gray DA, Butterworth PJ, Ellis PR (2016) The role of plant cell wall encapsulation and porosity in regulating lipolysis during the digestion of almond seeds. Food Func 7(1):69–78. https://doi.org/10.1039/c5fo00758e
15. Grundy MML, McClements DJ, Ballance S, Wilde PJ (2018) Influence of oat components on lipid digestion using an in vitro model: impact of viscosity and depletion flocculation mechanism. Food Hydrocoll 83(May):253–264. https://doi.org/10.1016/j.foodhyd.2018.05.018
16. Guo Q, Ye A, Bellissimo N, Singh H, Rousseau D (2017) Modulating fat digestion through food structure design. Prog Lipid Res 68(October):109–118. https://doi.org/10.1016/j.plipres.2017.10.001
17. Guzmán-Ortiz FA, San Martín-Martínez E, Valverde ME, Rodríguez-Aza Y, Berríos JDJ, Mora-Escobedo R (2017) Profile analysis and correlation across phenolic compounds, isoflavones and antioxidant capacity during germination of soybeans (Glycine max L.). CyTA 15(4):516–524. https://doi.org/10.1080/19476337.2017.1302995
18. Heneen WK, Karlsson G, Brismar K, Gummeson PO, Marttila S, Leonova S et al (2008) Fusion of oil bodies in endosperm of oat grains. Planta 228(4):589–599. https://doi.org/10.1007/s00425-008-0761-x

19. Kong F, Tang J, Lin M, Rasco B (2008) Thermal effects on chicken and salmon muscles: tenderness, cook loss, area shrinkage, collagen solubility and microstructure. LWT-Food Sci Technol 41(7):1210–1222
20. Lairon D, Play B, Jourdheuil-Rahmani D (2007) Digestible and indigestible carbohydrates: interactions with postprandial lipid metabolism. J Nutr Biochem 18(4):217–227. https://doi.org/10.1016/j.jnutbio.2006.08.001
21. Li Y, Hu M, McClements DJ (2011) Factors affecting lipase digestibility of emulsified lipids using an in vitro digestion model: proposal for a standardised pH-stat method. Food Chem 126(2):498–505. https://doi.org/10.1016/j.foodchem.2010.11.027
22. Lo CM, Tso P (2009) Physicochemical basis of the digestion and absorption of triacylglycerol. In: Designing functional foods: measuring and controlling food structure breakdown and nutrient absorption. Woodhead Publishing, p 94–125. https://doi.org/10.1533/9781845696603.1.94
23. Maldonado-Valderrama J, Wilde P, Macierzanka A, MacKie A (2011) The role of bile salts in digestion. Adv Colloid Interf Sci 165(1):36–46. https://doi.org/10.1016/j.cis.2010.12.002
24. Michalski MC, Genot C, Gayet C, Lopez C, Fine F, Joffre F,... & steering committee of RMT LISTRAL. (2013). Multiscale structures of lipids in foods as parameters affecting fatty acid bioavailability and lipid metabolism. Prog Lipid Res 52(4), 354–373
25. Nakamura A, Yoshida R, Maeda H, Corredig M (2006) Soy soluble polysaccharide stabilization at oil-water interfaces. Food Hydrocoll 20(2–3 SPEC. ISS.):277–283. https://doi.org/10.1016/j.foodhyd.2005.02.018
26. Nieva-Echevarría B, Goicoechea E, Guillén MD (2017) Effect of the presence of protein on lipolysis and lipid oxidation occurring during in vitro digestion of highly unsaturated oils. Food Chem 235:21–33. https://doi.org/10.1016/j.foodchem.2017.05.028
27. Nieva-Echevarría B, Goicoechea E, Manzanos MJ, Guillén MD (2016) A study by 1H NMR on the influence of some factors affecting lipid in vitro digestion. Food Chem 211:17–26. https://doi.org/10.1016/j.foodchem.2016.05.021
28. Pafumi Y, Lairon D, De La Porte PL, Juhel C, Storch J, Hamosh M, Armand M (2002) Mechanisms of inhibition of triacylglycerol hydrolysis by human gastric lipase. J Biol Chem 277(31):28070–28079. https://doi.org/10.1074/jbc.M202839200
29. Pająk P, Socha R, Broniek J, Królikowska K, Fortuna T (2019) Antioxidant properties, phenolic and mineral composition of germinated chia, golden flax, evening primrose, phacelia and fenugreek. Food Chem 275(September 2018):69–76. https://doi.org/10.1016/j.foodchem.2018.09.081
30. Paz-Yépez C, Peinado I, Heredia A, Andrés A (2018) Influence of particle size and intestinal conditions on in vitro lipid and protein digestibility of walnuts and peanuts. Food Res Int 119(June 2018):951–959. https://doi.org/10.1016/j.foodres.2018.11.014
31. Paz-Yépez C, Asensio-Grau A, Calvo-Lerma J, Heredia Gutiérrez A, Andrés A (2019) Study of lipolysis under in vitro digestion of different pasta dishes. In: 6th International Conference on Food digestion, p 378/201. Food Research International (ELSEVIER), Granada (Spain)
32. Qiu C, Zhao M, Decker EA, McClements DJ (2015) Influence of protein type on oxidation and digestibility of fish oil-in-water emulsions: Gliadin, caseinate, and whey protein. Food Chem 175:249–257. https://doi.org/10.1016/j.foodchem.2014.11.112
33. Shani-Levi C, Alvito P, Andrés A, Assunção R, Barberá R, Blanquet-Diot S et al (2017) Extending in vitro digestion models to specific human populations: perspectives, practical tools and bio-relevant information. Trends Food Sci Technol 60:52–63. https://doi.org/10.1016/j.tifs.2016.10.017
34. Singh J, Dartois A, Kaur L (2010) Starch digestibility in food matrix: a review. Trends Food Sci Technol 21(4):168–180
35. De Smet E, Mensink RP, Plat J (2012) Effects of plant sterols and stanols on intestinal cholesterol metabolism: suggested mechanisms from past to present. Mol Nutr Food Res 56(7):1058–1072. https://doi.org/10.1002/mnfr.201100722
36. Taylor JD, Linman MJ, Wilkop T, Cheng Q (2009) Regenerable tethered bilayer lipid membrane arrays for multiplexed label-free analysis of lipid-protein interactions on

poly(dimethylsiloxane) microchips using SPR imaging. Anal Chem 81(3):1146–1153. https://doi.org/10.1021/ac8023137
37. Torcello-Gómez A, Maldonado-Valderrama J, de Vicente J, Cabrerizo-Vílchez MA, Gálvez-Ruiz MJ, Martín-Rodríguez A (2011) Investigating the effect of surfactants on lipase interfacial behaviour in the presence of bile salts. Food Hydrocoll 25(4):809–816. https://doi.org/10.1016/j.foodhyd.2010.09.007
38. Turgeon SL, Rioux LE (2011) Food matrix impact on macronutrients nutritional properties. Food Hydrocoll 25(8):1915–1924. https://doi.org/10.1016/j.foodhyd.2011.02.026
39. Veverka M, Dubaj T, Veverková E, Šimon P (2018) Natural oil emulsions stabilized by β-glucan gel. Colloids Surf A Physicochem Eng Asp 537(October 2017):390–398. https://doi.org/10.1016/j.colsurfa.2017.10.043

Part IV
Other Forms of Lipids

Chapter 10
Lipid Digestion and Bioaccessibility of Lipid-Soluble Compounds

Ali Rashidinejad, Sébastien Marze, and Harjinder Singh

10.1 Introduction

Food digestion within the gastrointestinal tract involves several processes, which are mainly mechanical and chemical, although some are physical and biological. The complexity of the interactions between food components and the gastrointestinal environment makes it difficult to understand the exact role that each component plays to favour health or disease. For decades, nutrients and micronutrients were studied only through their absorption and biological fate, often in isolation from the food. However, it is now recognized that the nature and the structure of a food play important roles in the production and liberation of nutrients and, in some cases, even dictate the subsequent kinetics of absorption. In particular, efficient lipid digestion and efficient liberation from the food within the gastrointestinal tract are known to be essential for the optimal assimilation of lipid-soluble molecules (e.g. fat-soluble vitamins, carotenoids, polyphenols etc.). However, this relationship is not fully understood, as some of the factors involved have not been studied sufficiently. Thus, the goal of this chapter is to review our current knowledge about these factors, making the link between in vitro gastrointestinal digestion and in vivo absorption studies with respect to the structure of the fat-soluble materials. To this end, a few definitions are given, and then the main lipid-soluble molecules found in food, together with what is currently known about their fate within the

A. Rashidinejad (✉) · H. Singh
Riddet Institute, Massey University, Palmerston North, New Zealand
e-mail: A.Rashidinejad@massey.ac.nz

S. Marze
Biopolymères Interactions Assemblages, INRA, Nantes, France

M. M.-L. Grundy, P. J. Wilde (eds.), *Bioaccessibility and Digestibility of Lipids from Food*, https://doi.org/10.1007/978-3-030-56909-9_10

gastrointestinal tract and after absorption, are described. Finally, in the last part of the chapter, the role of lipid digestion is highlighted, by presenting research findings obtained from both in vitro and in vivo experiments, as well as from numerical simulations.

10.2 Definitions for Lipid-Soluble Molecules in the Framework of Lipid Digestion

The term 'lipid solubility' or 'lipophilicity' is generally referred to as the ability of a compound/substance/chemical to dissolve in lipids (fats and oils). Such compounds, which are known as 'lipophilic compounds', also dissolve in non-polar solvents (e.g. hexane, toluene etc.), but do not dissolve in polar/hydrophilic (water-loving) solvents. However, the solubility of a lipophilic compound in an aqueous-based medium depends on several different factors, such as temperature, pH, ionic strength and the presence of solubilizing compounds (e.g. bile salts). These factors are crucial when the digestion of lipid-soluble molecules/compounds is considered.

The bioaccessibility of a lipophilic compound is defined as the amount incorporated in mixed micelles relative to the amount in the food. Mixed micelles form in the small intestine (duodenum), and are composed of (1) bile secreted by the gallbladder and the liver (bringing bile salts, phospholipids and cholesterol) and; (2) free fatty acids, monoglycerides and lysophospholipids, which are produced by hydrolysis of the lipids (triglycerides and phospholipids respectively). Several steps are needed before this incorporation is achieved: (1) mechanical and chemical breakdown of food structures, releasing nutrients and lipophilic compounds; (2) hydrolysis of some esterified lipophilic compounds by specific or non-specific enzymes; (3) transfer of the lipophilic compounds from the lipid phase to the mixed micelles. These steps are discussed in Sect. 10.4.

With respect to the lipid solubility of fat-soluble compounds and their behaviour after digestion in the gastrointestinal tract, bioavailability is an important step in ensuring their bioefficacy. Several different stages are associated with the bioavailability of hydrophobic/fat-soluble compounds. They include the liberation of the specific compound from the lipid structure in the digestive tract, its absorption from the intestinal lumen, its distribution and metabolism in the body and the elimination of its excess from the body. Obviously, a good understanding of the metabolic and absorptive fate of fat-soluble compounds can help with the promotion of their effect on human health. Various factors such as the food matrix (especially fat content and fat type), protein transporters, molecular structure, bioaccessibility and metabolizing enzymes can affect the bioavailability of fat-soluble compounds in the body [110]. In Sect. 10.3, we discuss different types of lipid-soluble compounds, their structure, their solubility properties in the human digestive system and their metabolism, absorption, bioavailability and bioaccessibility.

10.3 Lipid-Soluble Molecules: Structure, Metabolism, Bioaccessibility, Absorption and Bioavailability

10.3.1 Lipid-Soluble Vitamins

Four of the lipid-soluble vitamins known in humans are needed to maintain good health; i.e. vitamins A, D, E and K. The main characteristics of these vitamins and their vitamers (provitamins) are presented in Table 10.1 [13]. Unlike water-soluble vitamins, which are substantially excreted from the body and thus their concentration in the urine is a robust predictor of their consumption or deficiency, fat-soluble vitamins are absorbed through the digestive system with the help of different types of lipids/fats [49, 88]. Thus, the deficiency and the toxicity of fat-soluble vitamins are more complex than those of water-soluble vitamins. Although the deficiency of fat-soluble vitamins because of their malabsorption can be significant in several cases, the accumulation of some of these vitamins in the body (e.g. vitamins A and D) can result in hypervitaminosis [88]. Fortunately, cooking does not cause a significant loss of fat-soluble vitamins, but the human body needs to have access to a source of these vitamins every day, and can store excess amounts in the liver and adipose tissues.

Table 10.1 Main characteristics of fat-soluble vitamins and their vitamers

Common name	Main molecular species found in the Western diet	RDA[a]	Mean/median daily intake[b]
Pre-formed vitamin A	Retinyl palmitate	900 REA[c]	598–682 μg
Provitamin A carotenoids	β-Carotene	–	2.15–2.62 mg
	α-Carotene	–	0.39 mg
	β-Cryptoxanthin	–	0.12–0.14 mg
Vitamin E	d-α-Tocopherol	15 mg	9.8–10.3 mg
	d-γ-Tocopherol	–	–
Vitamin D	Cholecalciferol	5 μg	2.9 μg
Vitamin K	Phylloquinone	120 μg	70–80 μg
	Menaquinone	–	21 μg
Non-provitamin A	Lycopene	–	6.6–12.7 mg
Carotenoids	Lutein/Zeaxanthin	–	2.0–2.3 mg[d]
Phytosterols	Sitosterol	–	167–437 mg
	Stigmasterol	–	
	Campesterol	–	

Modified from Borel [13]

[a]US RDA (recommended dietary allowance), or adequate intake (vitamin D), for the adult male population

[b]Mean/median daily intake for the adult male population

[c]*REA* retinol activity equivalents

[d]Lutein + zeaxanthin

In the small intestine, fats and fat-associated compounds can move across the mucosal membranes, enter the bloodstream and, finally, enter the liver, their storage site. Therefore, if the amount of fat in a diet is inadequate, the absorption of the fat-soluble vitamins in the body may be impaired [107].

10.3.1.1 Vitamin A and Carotenoids

Vitamin A, which exists in foods in two principal forms (i.e. retinol and the carotenes), can be found in foods such as the liver (and some other meat products) of different animals, ghee and butter (and other dairy foods), vegetables with orange and dark colours (e.g. sweet potato, carrot, pumpkin, broccoli, spinach etc.), eggs and different fruits (e.g. apricots, papayas, mangos etc.) [130]. Structurally speaking, vitamin A contains the class composed of retinol, retinal, retinoic acid, carotenoids and lycopene (carotenes are also known as provitamin A, with the main one being β-carotene). Several important functions for vitamin A in the human body are known, such as its role in growth and development (in the form of retinoic acid, a hormone-like growth factor for different cells), vision (in the form of retinal) and maintenance of the immune system [124, 143]. Figure 10.1a shows the different chemical structures of vitamin A and its provitamins [123].

As the pure alcohol form of vitamin A has low stability, this vitamin is found in a retinyl ester form in different animal tissues; this is also the commercially produced form of vitamin A [35]. Retinyl palmitate, the major form of vitamin A in animal-based food products, is an ester that can be converted to retinol in the small intestine. Retinol, which is also the storage form of vitamin A, can be converted to an active aldehyde form, i.e. retinal, when required [35]. The carotenoid family, such as α-, β- and γ-carotenes, as well as the xanthophyll β-cryptoxanthin are the provitamins of vitamin A in both herbivores and omnivores that have the enzyme β-carotene 15,15′-dioxygenase to cleave the carotene molecule (i.e. β-carotene) and to convert it to retinol in the intestinal mucosa [140].

The dietary uptake and the transport of vitamin A are shown in Fig. 10.1b [26]. Some specific pancreatic and intestinal enzymes can hydrolyse the dietary retinyl esters to retinol, which is then taken up by the cells in the intestinal mucosa (i.e. enterocytes) [9]. Fat plays several important and substantial functions in this regard: it is a stimulator for the enzymes that are responsible for the hydrolysis of retinyl esters; it is also an important factor for the formation of the micelle that is required for the solubilization of carotenoids and retinol in the intestinal lumen; it promotes the formation of chylomicrons [59]. As a hydrophobic molecule, in enterocytes, retinol is bound by a specific compound called 'cellular retinol binding protein, CRBP II', which is able to bind most of the existing retinol in intestinal cells [31]. Almost half of the provitamin A carotenoids are oxidized, converted to retinal and, finally, reduced to retinol; the remainder are absorbed in mucosa cells in their intact form [10]. For more efficient absorption, because of its fat-loving nature, retinol is first esterified with long chain fatty acids. Then, the retinyl esters, along with the

(a)

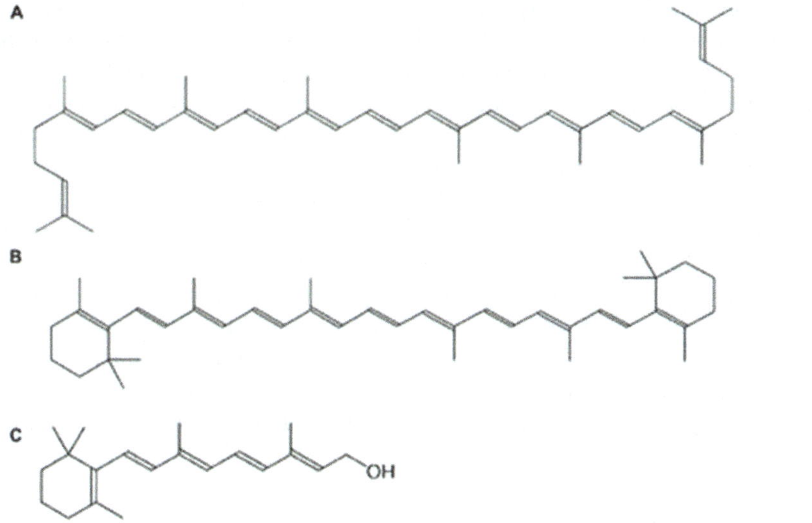

(b)

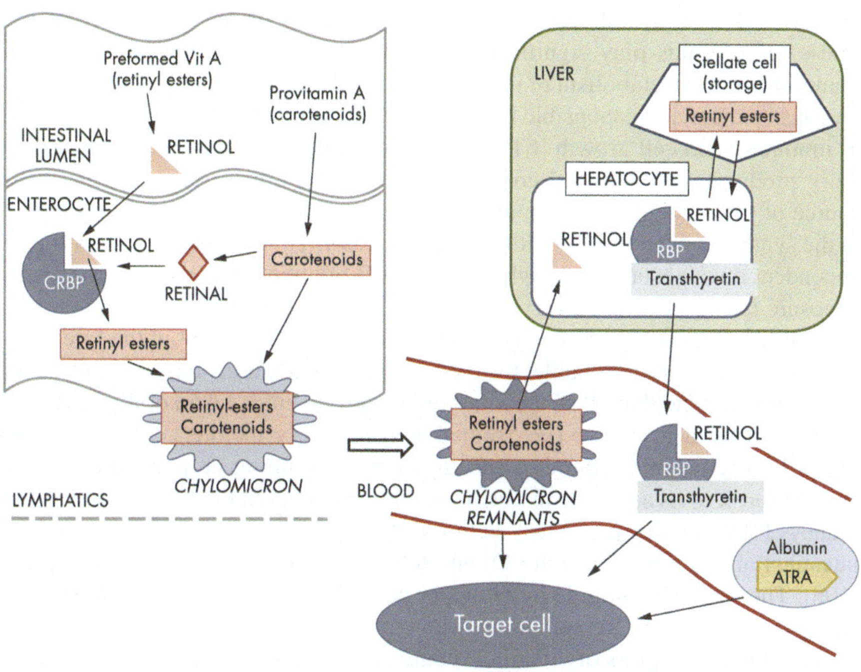

Fig. 10.1 (**a**) Chemical structure of vitamin A and its precursors: (A) lycopene; (B) β-carotene; (C) all-trans-retinol [123]. (**b**) Dietary uptake and transport of vitamin A. (Redrawn from Conaway et al. [26], with permission)

original/intact carotenoids, are incorporated into chylomicrons with other fat-soluble compounds (e.g. triglycerides, cholesterol, cholesterol esters etc.) [60].

After the hydrolysis by lipoprotein lipase of chylomicrons in the bloodstream, the chylomicron remnants containing retinyl esters [31] are taken up by hepatocytes through receptor-mediated endocytosis [28]. If there is excess retinol in the body, it becomes re-esterified and remains in the liver stellate cells (as the main storage site for vitamin A). However, some smaller quantities of retinyl esters and/or carotenoids may also be transferred by chylomicrons to extrahepatic tissues for storage and further utilization [31]. Once the retinyl esters in the liver stellate cells are hydrolysed, retinol can be transported back to hepatocytes so that it can be bound by retinol binding protein (RBP), which is a specific transport protein for this lipid-soluble molecule [120]. Subsequently, the retinol–RBP complex can enter the bloodstream and combine with transthyretin, which is another protein that is synthesized in the liver [41]. Such binding of the retinol–RBP complex to transthyretin results in the prevention of the clearance of retinol–RBP by the kidneys [132].

10.3.1.2 Vitamin D

Vitamin D, as shown in Fig. 10.2a, exists in two forms in humans, i.e. D_3 (cholecalciferol) and D_2 (ergocalciferol), and involves a class of fat-soluble secosteroids. These secosteroids play significant roles in increasing the intestinal absorption, homeostasis and metabolism of important minerals, such as calcium, phosphate and magnesium, and are responsible for several other biological effects in the body, such as immunity and cell growth [64]. Only a few foods (e.g. milk, vitamin-D-fortified dairy products, oily fish and cod liver) contain vitamin D. The principal natural source of this vitamin is a chemical reaction that occurs under the skin and results in the synthesis of cholecalciferol from cholesterol. This synthesis process is highly dependent on exposure to sunlight, and to UVB radiation in particular. However, as exposure to the sun can be very variable in different populations around the world and during different seasons and climates, recommendations about the safe amount of exposure are very unreliable; the dietary recommendations for vitamin D characteristically assume that all of the vitamin D requirement is taken orally [24].

Regardless of its source (i.e. from skin synthesis or the diet), vitamin D is biologically inactive and requires a specific enzyme (binding protein) to hydroxylate and convert it to its active form. Such conversion occurs in the liver as well as in the kidneys [100]. It was previously believed that there was a passive process for the intestinal absorption of vitamin D, but new data show it to be a complex process [108]. The fate of vitamin D in the upper part of the human's gastrointestinal tract has been reviewed by Reboul [108], who concluded that the binding of micelles to the membrane receptors of vitamin D could be an important initiator with respect to the entry of this fat-soluble vitamin into the enterocytes [108].

As shown in Fig. 10.2b, several different proteins are involved in the transport of vitamin D across the enterocytes, some of which and their mechanisms of action (and pathways) are as yet unknown [108]. According to this process, mixed micelles

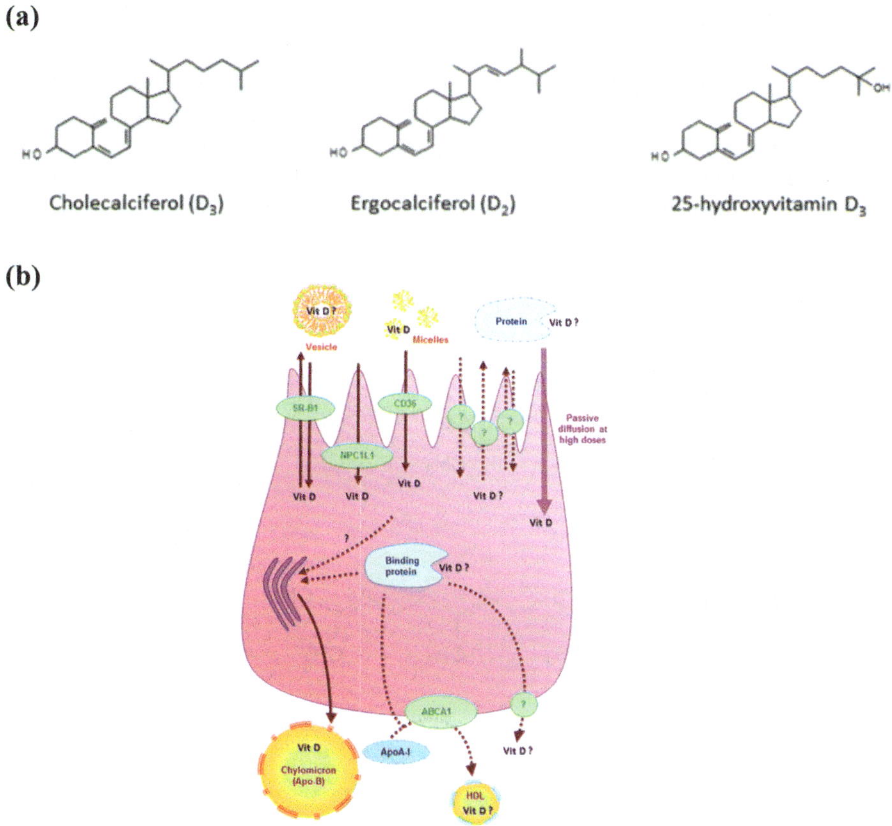

Fig. 10.2 (**a**) Chemical structures of the main molecular forms of vitamin D. (**b**) Proteins that are involved in the transport of vitamin D through the enterocytes. (Redrawn from Reboul [108], with permission)

(or possibly vesicles and/or proteins) initially aid the take-up of vitamin D, by some apical membrane transporters, including SR-BI, NPC1L1 and CD36. Generally, part of the dietary vitamin D may enter the cell through passive diffusion; this is more relevant when pharmacological doses of vitamin D are administered. Then, apical membrane transporters (such as SR-BI) can aid in the efflux of a fraction of the vitamin D back to the intestinal lumen, whereas another fraction is transported to the chylomicron incorporation site. Finally, the free form of vitamin D is secreted in the lymph and into chylomicrons; in other words, it enters the apolipoprotein-B-dependent route. However, this is only one of the possible secretion pathways for this vitamin and some non-apolipoprotein-B-dependent routes (e.g. via ABCA1 and HDL) may also exist [108]. As mentioned earlier, although the intracellular transport of vitamin D involves its binding to some specific proteins, the whole mechanism is yet to be clearly identified, especially in relation to the transporters, the associated pathways and the molecular mechanisms underlying this phenomenon.

10.3.1.3 Vitamin E

Vitamin E is found in different foods, such as oil from different sources (e.g. wheat germ, hazelnut, canola etc.), meat products (e.g. fish, oysters and chicken) and dairy products (e.g. milk, cheese, butter and ghee), and includes a group of various fat-soluble compounds. This group includes four tocopherols as well as four tocotrienols; their chemical structures are presented in Fig. 10.3 [50]. Vitamin E expresses

a

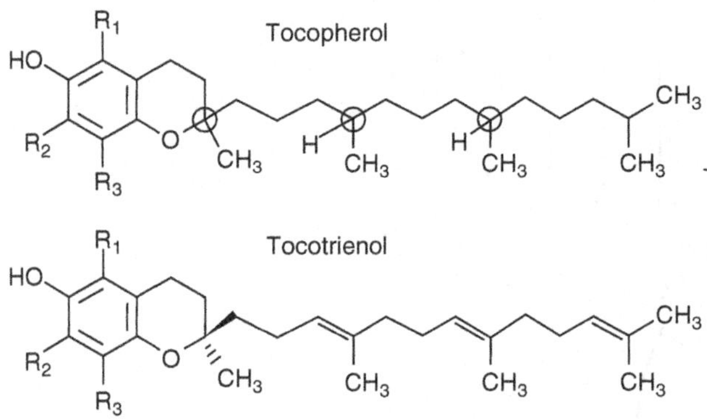

	R_1	R_2	R_3
α	CH_3	CH_3	CH_3
γ	H	CH_3	CH_3
β	CH_3	H	CH_3
δ	H	H	CH_3

b

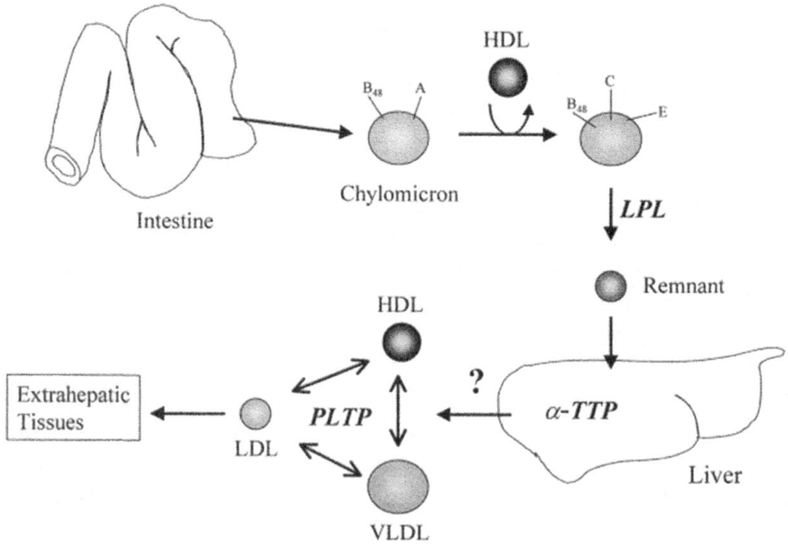

Fig. 10.3 (**a**) Chemical structures of tocopherols and tocotrienols. The circles mark the three chiral centres in the tocopherols, which include 2, 4′ and 8′. (**b**) Absorption, transport and distribution of tocopherols in the human body. LPL: lipoprotein lipase, PLTP: plasma phospholipid transfer protein, HDL: high density lipoprotein, LDL: low density lipoprotein, VLDL: very low density lipoprotein. (Redrawn from Manolescu et al. [86] and Hosomi et al. [67] respectively (free access))

some antioxidant functions in cell membranes in the human body and there are theories that it may be involved in controlling gene expression, regulating enzyme activity (e.g. protein kinase C) and regulating the transduction of cell signals. Problems of vitamin E deficiency that have been reported to date are rarely related to a diet that is low in vitamin E; rather they are associated with an underlying cause that is related to impaired dietary fat digestion [5, 17].

Both known sources of vitamin E (i.e. tocopherols and tocotrienols) are present in α, β, γ and δ forms, depending on the number of methyl groups and the position at which they are located on the chromanol ring [17]. All of the vitamers of vitamin E possess a chromane double ring, along with a hydroxyl group. This feature allows them to donate a hydrogen atom, by which free radicals are reduced. They also contain a hydrophobic side chain that enables them to penetrate into biological membranes.

The metabolism of tocopherols (including the stereoisomers of synthetic α-tocopherol) and tocotrienols starts with their absorption from the intestinal lumen. Vitamin E is emulsified together with the fat-soluble components present in the food. Later, the lipolysis and emulsification of the newly formed lipid droplets can lead to the production of mixed micelles. Through passive diffusion, these micelles are then absorbed in the brush border membrane of Golgi of the mucosa cells, which can aid in the re-assembly of tocopherols together with triglycerides, phospholipids, apolipoproteins and cholesterol to chylomicrons [67]. After absorption, the micelles are incorporated into chylomicrons, secreted into the portal vein and transferred to the liver, where they can be stored. It is estimated that 51–86% of vitamin-E-related compounds (i.e. tocopherols and tocotrienols) are absorbed from the intestinal lumen. The portion of vitamin E (and its vitamers) that is not absorbed is excreted via the faeces. Moreover, excretion of vitamin E by the liver via bile and into the intestinal lumen is also possible. It is subsequently either reabsorbed or excreted (via the faeces), whereas all of the vitamers of vitamin E are excreted via the urine [5, 97]. Figure 10.3b is a schematic view of the absorption, transport and distribution of tocopherols in the human body [67].

Once α-tocopherol has reached the liver, it is taken up by a specific transferring protein called 'α-TTP'. α-TTP is known to be the key intracellular transport protein for vitamin E. With the aid of this protein, the secretion of α-tocopherol into the bloodstream is mediated via a non-Golgi-dependent pathway [70]. However, the other forms of vitamin E can be degraded to a compound called '2′-carboxethyl-6-hydroxychromane (CEHC)'. During this process, the phytic tail of the molecule is first truncated and then sulphated or glycuronidated, which results in the molecule being water soluble and excreted via the urine. The same degradation process applies to α-tocopherol, and produces '2,5,7,8-tetramethyl-2-(2′-carboxyethyl)-6-hydroxychromane (α-CEHC)'. However, this process is slower than that for other forms of vitamin E, because of the protection mechanism of α-TTP. Thus, the consumption of a large amount of α-tocopherol will result in an increased amount of α-CEHC in the urine, which is an efficient mechanism for disposing of excess vitamin E from the human body [5, 97].

10.3.1.4 Vitamin K

Vitamin K is mostly found in leafy green vegetables (e.g. lettuce, Swiss chard, spinach etc.), Brassica vegetables (e.g. kale, broccoli, cabbage, cauliflower and brussel sprouts), some fruits (e.g. kiwifruit, avocados, grapes etc.) and vegetable oils (in particular, soybean oil). It includes a group of structurally similar hydrophobic compounds that the human body requires for the complete synthesis and function of specific proteins, which act as the prerequisites for blood coagulation. In addition, these proteins are also required for controlling the binding of elements, such as calcium, in bones and other tissues [33, 114]. A lack of vitamin K is often associated with an impaired blood coagulation system and subsequent uncontrolled bleeding. Various clinical studies have shown that a deficiency of vitamin K can result in weakened bones in the skeletal system and, thus, can lead to osteoporosis, as well as to the promotion of calcification in the arteries and soft tissues of the human body [40, 78, 114].

Vitamin K belongs to a family containing '2-methyl-1,4-naphthoquinone' derivatives (Fig. 10.4a), including vitamin K_1 and vitamin K_2, as the natural vitamers, and synthetic vitamers. As shown in Fig. 10.4b, the dietary sources of these vitamers (e.g. vitamin K_1 and vitamin K_2) are different. For example, vitamin K_1 is found in leafy greens whereas vitamin K_2 is mostly sourced from foods such as natto (the most important source) and some cheeses. Vitamin K_1 can be converted into vitamin K_2 by the bacteria in the human gut. In addition, some homologues of vitamin K_2 (most notably, MK-7 to MK-11) may also be produced by the gut bacteria, typically through lengthening of the isoprenoid side chain of vitamin K_2.

Although vitamin K is well absorbed from a typical diet, the total vitamin K levels are dependent on the status of the digestive health [57, 79]. For healthy humans (i.e. with healthy intestines), vitamin K deficiency is considered to be quite rare, regardless of the fact that the K_1 form is abundant in a normal diet containing leafy and salad vegetables, herbs etc. This is different in the case of newborns, who have not been fed foods that are rich in vitamin K and whose gut is not yet well developed; in most countries, newborns usually receive a vitamin K injection or oral vitamin K drops upon birth, to prevent bleeding or the development of haemorrhagic disease [57, 79].

Similar to the other fat-soluble vitamins, vitamin K and its vitamers are absorbed by a pathway similar to that for fat/triacylglycerols [53]. In foods containing vitamin K, the vitamin is initially dissolved within small lipid droplets, which contain triacylglycerols. When the food is ingested, the lipid droplets and lipophilic compounds undergo alterations in their composition, size and dimensions because of the changes in the conditions of the digestive tract (e.g. pH, ionic strength, mechanical forces, the activity of digestive enzymes and molecular interactions) [87, 93]. The triacylglycerols are digested by intestinal lipase, which results in the liberation of free fatty acids and monoacylglycerols, which in turn can associate with bile salts and phospholipids and can form mixed micelles [103]. Vitamin K becomes incorporated into the hydrophobic interior of the mixed micelles. These vitamin-K-loaded mixed micelles can be absorbed by an active or passive transport mechanism, but

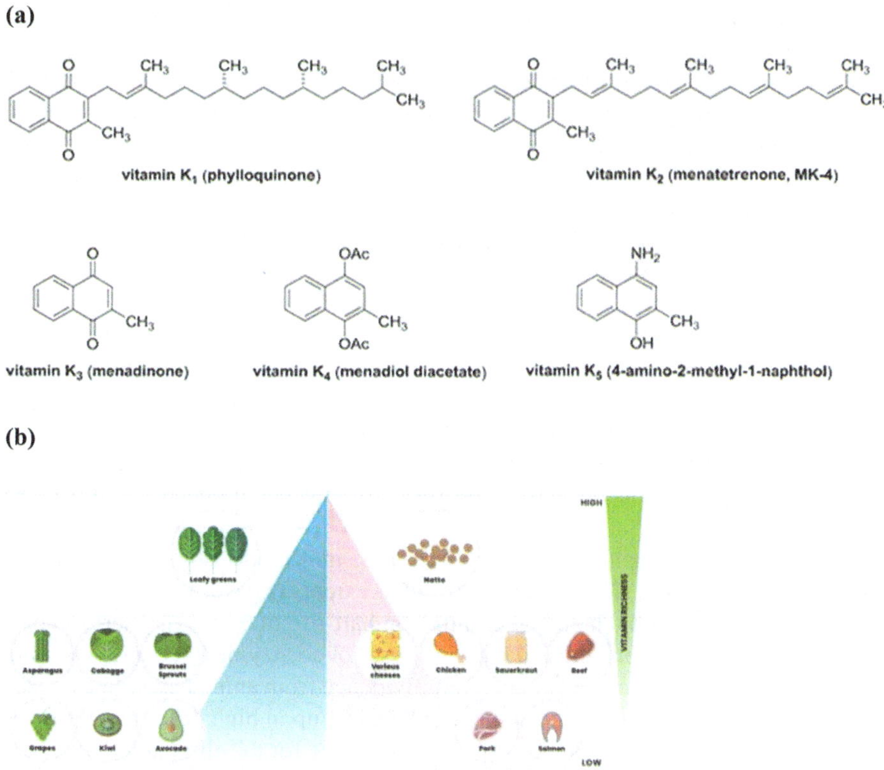

Fig. 10.4 (**a**) Chemical structures of different vitamers of vitamin K, including natural vitamers (vitamin K_1 and vitamin K_2) and synthetic vitamers (vitamin K_3, vitamin K_4 and vitamin K_5). (**b**) Dietary sources of vitamin K_1 (left) and vitamin K_2 (right). (Modified from Halder et al. [57])

they first need to be transported through the mucus layer to reach the epithelial cells. The absorbed fatty acids and glycerols along with vitamin E are reassembled into triacylglycerols and are then packed into chylomicrons (lipoproteins) [71]. After they have been excreted from the epithelial cells (the basolateral side), the chylomicrons eventually reach the bloodstream and pass through the lymphatic system [94].

10.3.2 Coenzyme Q10

Coenzyme Q10 is found in various foods, including meat products (e.g. beef, chicken, fish etc.), oils (e.g. olive, grapeseed, sunflower etc.), nuts (e.g. peanut, walnut, almond, pistachio etc.), vegetables (e.g. broccoli, cauliflower, spinach etc.) and fruits (e.g. avocado, grapes, blackcurrant etc.). It is also known as coenzyme Q, ubiquinone and ubidecarenone and abbreviations such as CoQ10, Q10 and CoQ are widely used. Coenzyme Q10 comprises a 1,4-benzoquinone, in which 'Q' is the chemical group of quinone and '10' is the number of isoprenyl subunits in the tail of the molecule [42, 62, 106].

Coenzyme Q10 plays a number of different roles in all respiratory eukaryotic cells. As it is a component of the electron transport chain, coenzyme Q10 contributes to aerobic cellular respiration and thus has a vital effect on energy generation in the form of ATP [38, 43]. It is also considered to be a potent and important antioxidant in the human body [7].

Similar to other hydrophobic compounds/vitamins, the absorption of coenzyme Q10 in the human digestive system occurs by the same process as that of lipids/triglycerides. In particular, the mechanism of uptake of coenzyme Q10 is similar to that of vitamin E (discussed in Sect. 10.3.1.3). Firstly, emulsification and micelle formation in the small intestine are facilitated by the secretion of pancreatic enzymes and bile [7]. Of course, as in the case of other fat-soluble compounds, the presence of various lipids can stimulate the excretion of bile acids and thus greatly enhance the absorption of coenzyme Q10. In this regard, it has been found that coenzyme Q10 from exogenous sources is best absorbed from the small intestine if it is taken with a meal containing fat [101, 136].

Because of its hydrophobic nature and its large molecular weight, dietary coenzyme Q10 is absorbed slowly and to a limited extent. It has a time at maximum concentration of around 6 h and an elimination half-life of 33 h [7]. There appears to be a rational correlation between ingested doses of coenzyme Q10 (up to a certain point) and its plasma levels in humans [22]. Evidence from animal experiments also confirms that this fat-soluble coenzyme can be taken up in high doses by all tissues (e.g. heart and brain mitochondria) [4]. The main route for the elimination of excess coenzyme Q10 is through bile excretion and the faeces [102].

Recently, López-Lluch et al. [82] demonstrated that the bioavailability of coenzyme Q10 in different supplemental formulations was significantly dependent on the individual, the carrier lipids and the solubilization of the coenzyme Q10 products. Even when the composition of a particular formulation was highly effective, individuals showed different responses to the administered doses. Such responses depended on some unknown factors, such as lifestyle, body mass index, sex, weight and age [82].

10.3.3 Lipophilic Polyphenols

Polyphenols (also known as polyhydroxyphenols, polyphenolic compounds and phenolic compounds), which are the most abundant antioxidants in the human diet (especially in fruits, herbs and vegetables), are a structural class of mainly natural and organic chemical compounds that are categorized by the presence of phenol structural units. Every polyphenolic compound presents different and unique chemical, physical and biological (i.e. metabolic and therapeutic health benefits, toxicity etc.) properties, because of differences in the number and properties of these phenol structures. Some examples of hydrophobic polyphenolic compounds are curcumin, quercetin, rutin, hesperidin, catechin, resveratrol, naringin, phenolic acids (caffeic acid, formic acid, gallic acid etc.) and some tannins [56, 105].

Several epidemiological studies have confirmed the health-promoting effects of polyphenols/antioxidants as a result of the consumption of foods that are rich in these natural compounds. In particular, they reduce the risk of chronic diseases, such as cardiovascular diseases, diabetes and cancer, through preventing the impairment caused by oxidative stress in certain biomolecules (e.g. nucleic acids and proteins) [23]. Although some polyphenols with large molecular weight (e.g. proanthocyanidins) are poorly absorbed in humans, phenolic acids (e.g. caffeic acid) can be absorbed easily through the gut barrier. After intestinal absorption, there is conjugation of the polyphenols to sulphate, glucuronide and methyl groups in the inner tissues and the gut mucosa. Therefore, non-conjugated polyphenols are not expected to be present in the plasma, which is a good functionality for facilitating the excretion of these phenolic compounds and limiting their possible toxicity [116].

Although polyphenols show antioxidant activity in vitro, it is their subsequent metabolism and absorption in the digestive tract that governs their biological characteristics, including antioxidant activity [125]. Only the fraction of the polyphenols that are released from the food matrix during digestion (in the small and/or large intestine) are considered to be part of the digesta. Food polyphenols in their native form exist mainly as polymers, esters and glycosides, which cannot be absorbed as such, and therefore need to be hydrolysed by endogenous enzymes and/or microflora enzymes in the digestive tract [142].

Regardless of their hydrophobicity, it has been shown that about 48% and 42% of polyphenols can be digested in the small intestine and the large intestine respectively. The nature of the food matrix itself (e.g. the presence of fat in the case of hydrophobic polyphenols) may also affect their bioavailability as they can react with some constituents in the food matrix [85]. Some other factors, such as pH, the gastrointestinal environment and the presence of bile salts, significantly affect the metabolism and bioavailability of polyphenols [85].

An overview of the general bioavailability of polyphenols, including the corresponding critical steps that can occur from ingestion to excretion, is shown in Fig. 10.5 [11]. Several different proteins and transporters are known to be involved in polyphenol bioavailability. Polyphenols, such as flavonoids, are absorbed in the human gastrointestinal tract, and metabolites and unchanged compounds are excreted in the urine and the faeces [27]. The gut bacteria can provide enzymes to split the heterocyclic ring of polyphenols and degrade them to other compounds such as phenyl acids, which may then be conjugated, absorbed, metabolized further by the bacteria or excreted [2]. Soon after ingestion, the polyphenols enter the bloodstream/plasma.

As mentioned earlier, if polyphenols cannot be absorbed in the small intestine, they move into the colon, where the microflora can metabolize them extensively and convert them into a wide array of phenolic acids with lower molecular weight [37, 47]. The fate of undigested polyphenols has been reported in many studies [36, 80, 109]. The intestinal bacteria can hydrolyse glycosides to aglycones, which can

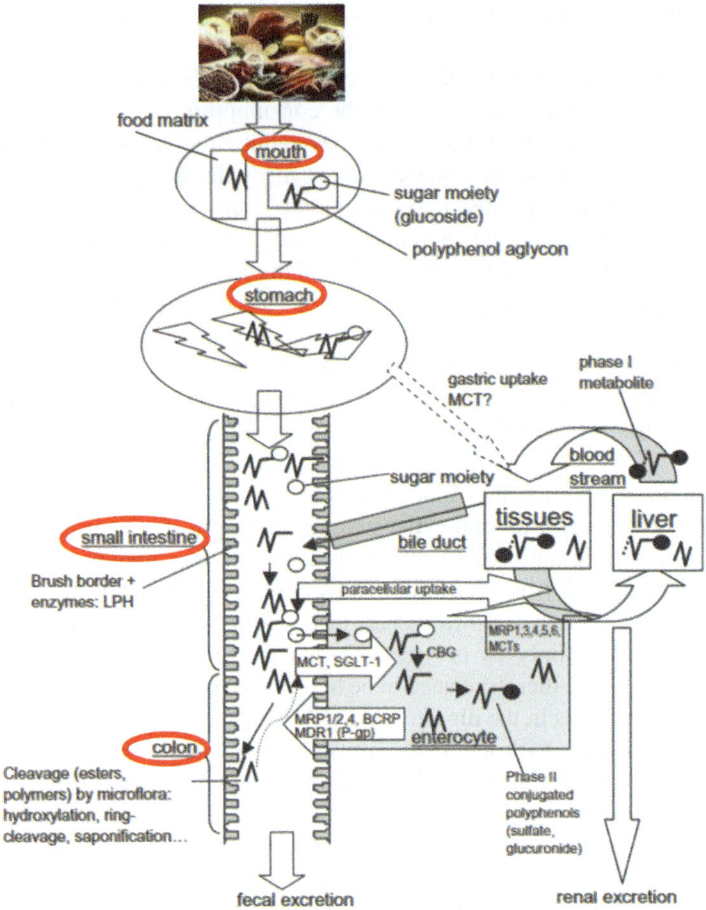

Fig. 10.5 Overview of the bioavailability of polyphenols, showing critical steps that occur during release, digestion, uptake, transport and excretion. Abbreviations; BRCP: breast cancer resistance protein, CBG: cytosolic β-glucosidase, LPH: lactase-phlorizin hydrolase, MCT: monocarboxylic acid transporter, MRP: multidrug resistance proteins, Pgp: P-glycoprotein, SGLT1: sodium-glucose linked transporter 1, MDR1: multidrug resistance protein. (Modified from Bohn [11])

subsequently be transformed into different acids by the action of esterases, β-glucosidase and β-rhamnosidase. The enzymes within the gut microflora can catalyse the catabolism products of flavonoids into simple units such as gallic acid and catechin. Moreover, these enzymes are able to carry out actions such as hydrolysis, demethylation, decarboxylation and dehydroxylation. A range of compounds can be formed from these actions, depending on the structure of the specific phenolic compound being digested [80]. For example, flavonols will result in the production of hydroxyphenylacetic acid, whereas flavanones and flavones can be catabolized to hydroxyphenylpropionic acid. However, flavanols can be degraded into both hydroxyphenylpropionic acid and phenylvalerolactone acid [80, 125]. All of these metabolites, formed from the degradation of phenolic compounds, will ultimately result in the synthesis of benzoic acid. These metabolites can be absorbed and enter the blood circulation, bind to serum albumin and finally be transported to the liver, where they might undergo processes, such as hydroxylation, O-methylation and demethylation, or be conjugated to glucuronide and sulphated derivatives. A part of these may be secreted, together with bile, back into the small intestine and again undergo hydrolysis and subsequently either be absorbed and enter the circulation or be excreted via the faeces [80, 125].

10.3.4 Sterols

Sterols or steroid alcohols, which are found in both plant (including fungi) and animal foods, include a key class of organic molecules and a subgroup of steroids (with a hydroxyl group at the 3-position of the A ring). The sterols from plant and animal sources are called 'phytosterols' and 'zoosterols' respectively. These organic compounds can also be produced by some specific types of bacteria, although the sterols produced in this way may possess different functions [45, 137]. The basic chemical structures of sterol and its different derivatives are shown in Fig. 10.6 [117].

Cholesterol from animal sources is the main sterol that makes a vital contribution to the structure of cell membranes and has an important function as a precursor to steroid hormones and fat-soluble vitamins. Campesterol, stigmasterol and sitosterol are the most familiar phytosterols, whereas ergosterol is a well-known sterol that is found in different types of fungi [76]. Sterols are in fact amphipathic lipids with a

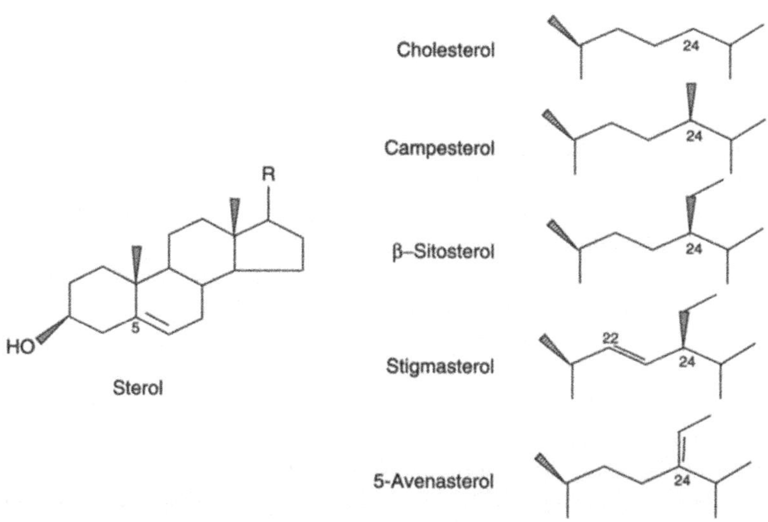

Fig. 10.6 Chemical structure of sterols. (Redrawn from Segura et al. [117], with permission)

quite flat molecular structure, in which the hydroxyl group on the A ring is polar and the rest of the aliphatic chain is non-polar [45].

The absorption of sterol in the small intestine is a selective process; phytosterols and ergosterol are absorbed poorly or may not be absorbed at all, whereas cholesterol can be absorbed efficiently [126]. Intestinal cholesterol absorption is the main means for the entry of cholesterol into the blood, which can therefore affect LDL plasma levels [74, 75]. However, the molecular mechanisms of cholesterol absorption are yet to be fully understood. A transporter called 'ATP-binding cassette (ABC) transporter' or 'ABCG5/ABCG8' has been identified to be the apical cholesterol export pump. This mechanism promotes almost complete efflux of phytosterols and partial efflux of cholesterol from enterocytes into the intestinal lumen. This may explain the selective process of cholesterol absorption in the human body with respect to phytosterols and other non-cholesterol sterols, as discussed above [74]. Additionally, there is a putative protein for the import of cholesterol, but it is yet to be fully characterized [74].

10.4 Role of Lipid Digestion in the Bioaccessibility and Bioavailability of Lipophilic Compounds

Obviously, lipid-soluble molecules are not alone in the gastrointestinal tract, and they interact with all other nutrients and food compounds during digestion. In this section, we review the general factors that control their bioavailability, highlighting the role of lipids in enhancing their bioaccessibility.

10.4.1 The SLAMENGHI Factors

Many factors contribute to the bioaccessibility of lipophilic compounds during digestion. A general classification was developed to evaluate micronutrient bioavailability; it was first applied to iron [138] and then to carotenoids [20]. In this scheme, bioavailability is explained by the so-called SLAMENGHI factors: Species of the compound; molecular Linkage (e.g. esterified); Amount consumed in a meal; Matrix in which the compound is incorporated; Effectors of absorption and bioconversion (e.g. other nutrients, digestive juice compounds); Nutrient status of the host; Genetic factors (e.g. for different ethnic groups); Host-related factors (e.g. health status); Interactions between all factors. The first five factors (the physicochemical factors) can be applied to explain the bioaccessibility of most lipophilic compounds and have been reviewed for the specific case of emulsions [92]. In this section, the physicochemical factors that control the bioaccessibility of lipophilic compounds in foods are discussed, with some illustrative examples regarding their effect on their bioavailability in humans.

10.4.1.1 Amount and Type of Lipophilic Compound

As most of the in vitro digestion protocols are static, the amounts of lipophilic compound and bile salt (and their ratio) are fixed, and this could limit bioaccessibility when the lipophilic compound is in excess, as these compounds are usually incorporated in bile salt mixed micelles. Therefore, in a bioaccessibility study, this ratio should be based on physiological values, i.e. the normal intake of the lipophilic compound and the amount of secreted bile salt, neglecting bile salt recirculation. Several studies have shown that the amount of lipophilic compound (or, reciprocally, the amount of bile salt) modifies its bioaccessibility. In three studies, increasing the amount of β-carotene in an emulsion decreased its percentage bioaccessibility, whereas increasing the amount of bile salt increased the percentage bioaccessibility of β-carotene [128, 144]. All other studies showed that increasing the amount of bile salt increased the bioaccessibility of carotenoids [30, 68]. These trends are usually linear within a specific range and then plateau. This confirms that bioaccessibility is limited by the lipophilic compound/bile salt ratio. In vivo, the percentage bioavailability was found to decrease when increasing doses of phenolic compounds were fed to humans, because the bioavailability increased only sublinearly with the dose [46, 84, 121]. Although many chemical and biological processes are involved in vivo, bioaccessibility studies could be useful for optimizing the amount fed as mixed micelle incorporation is a prerequisite to absorption.

It is well known that bioavailability and bioaccessibility vary widely among lipophilic compounds, because of their specific molecular properties [12, 91]. Less is known about the effects of the molecular linkage of specific compounds. One human study reported variable lutein bioavailability for free lutein compared with lutein di-ester; free lutein was more bioavailable only when completely dissolved in oil [16]. This trend was also seen in an in vitro study of the bioaccessibility of lutein and lutein esters from milk or yoghurt; free lutein was more bioaccessible than lutein esters [145]. Free carotenoids from cajá pulp beverages were also found to be more bioaccessible (in both percentage and amount) than carotenoid esters [32]. Similarly, mainly non-esterified zeaxanthin was found in mixed micelles after the in vitro digestion of various vegetables containing zeaxanthin esters, in the presence of carboxyl ester lipase, which hydrolysed the esters. Moreover, mainly non-esterified zeaxanthin was taken up by Caco-2 cells [55]. With this knowledge, the main limiting factor for absorption appears to be ester hydrolysis. However, more work is needed to understand the interplay between ester hydrolysis and micellization of both esterified and free lipophilic compounds.

10.4.1.2 Amount and Type of Lipid

There is considerable literature on the effect of the type of lipid on the bioaccessibility of lipophilic compounds. Most in vitro studies on β-carotene have shown that a higher final bioaccessibility is reached at a given intestinal digestion time (usually 2 h) for long chain triglycerides than for medium chain triglycerides [95, 118, 152].

However, Nguyen et al. [98] reported the complete kinetics of the bioaccessibility of β-carotene from droplets of various oils and found an inverse trend after 40 min of intestinal digestion. At the end of droplet digestion (spanning from 40 min to 4 h), the final bioaccessibility value was similar for all oils [98]. For various vegetable salad carotenoids (carotenes, lycopene), it has been reported that the bioaccessibility increases with the chain length of the added triglyceride, but not with its degree of unsaturation. No significant trend was observed for lutein + zeaxanthin [69]. In other studies, no significant trend or an inverse trend was observed for these xanthophylls [54]. For lutein, one significant inverse trend of higher bioaccessibility with medium chain triglycerides than with long chain triglycerides [149] and one significant trend of higher bioaccessibility with long chain triglycerides than with medium chain triglycerides [99] were reported. These inverse trends might originate from the duration of the intestinal digestion phase in vitro, which is usually too short (2 h) for the slowly digested long chain triglycerides (such as those in fish oils). This aspect is discussed further in the next section (Sect. 10.4.1.3). For lycopene, bioaccessibility was also found to be higher in long chain triglycerides [25]. All other lipophilic compounds that have been studied (curcumin, cholecalciferol, α-tocopherol, eugenol) follow the same trend of higher bioaccessibility with long chain triglycerides than with medium chain triglycerides [3, 83, 154].

This trend has also been reported in bioavailability studies on humans. The fasting blood serum concentration of cholecalciferol was found to be higher when humans were fed peanut oil than when they were fed medium chain triglycerides [65]. The concentrations of retinyl esters and β-carotene were found to be higher in the blood chylomicrons when humans were fed long chain triglycerides (olive oil or sunflower oil) than when they were fed medium chain triglycerides (butter) [14, 115]. The concentration of carotenoids was found to be higher in the triacylglycerol-rich fraction of blood when humans were fed canola oil, compared with butter [55].

The amount of lipid also plays an important role in enhancing the bioaccessibility of lipophilic compounds. However, all in vitro studies have shown that this is true only up to an optimal amount, beyond which the bioaccessibility decreases [3, 25, 69]. This is again probably due to the static nature of in vitro digestion protocols, in which triglyceride digestion products compete with lipophilic compounds, especially with long chain triglycerides, for mixed micelle incorporation.

In humans, the concentration of carotenoids in the blood chylomicrons was found to be higher from a vegetable salad with 28 g of added fat, compared with 6 g of added fat or with a fat-free salad [19]. The same trend was reported for all carotenoids in the triacylglycerol-rich fraction of blood after the consumption of a vegetable salad with 20 g of added fat or 8 g of added fat [55]. For a vegetable salad with 0, 2, 4, 6, 8 or 32 g of added fat, there was an increasing trend in the concentration of lipophilic compounds in the blood chylomicrons. This trend was linear for α-carotene, lycopene, retinyl palmitate and phylloquinone, and was non-linear for lutein, tocopherols and β-carotene [139]. Thus, in the normal range investigated, there seems to be no significant competition effect between triglyceride digestion products and lipophilic compounds in vivo.

10.4.1.3 Droplet Size

When food and/or digesta contain emulsified lipid droplets, most in vitro digestion studies have shown that the bioaccessibility of lipophilic compounds increases significantly when the mean droplet size decreases [68, 113, 147]. This occurs because, at a given lipid concentration, more surface area is available for mass transfer from the lipid phase to the aqueous micellar phase. However, for curcumin, this trend was found to be significant in only one study [118], with two other studies reporting no significant effect of the droplet size [3, 154].

In vivo, a human study found no significant effect of the droplet size on the bioavailability of vitamin A and vitamin E [12]. In rats, no significant effect was found for ergocalciferol [113] or ubiquinone [21], and only one significant trend was found for the maximum plasma concentration of astaxanthin [1]. This absence of an effect of the droplet size is probably due to a change in the droplet size distribution in the gastrointestinal tract. For example, Borel et al. [15] found that the mean droplet size was not significantly different in the duodenum. It is thus essential to measure the droplet size distribution throughout digestion in vitro, to ensure that the initial size difference is retained during the intestinal phase, where the release takes place. With a stable droplet size distribution, an effect is nevertheless expected for the absorption of lipophilic compounds in vivo, as a cellular study with stable emulsions of different mean droplet sizes showed that the uptake of 5-demethyltangeretin was much more efficient for mean droplet diameters of 134 and 250 nm, compared with 406 nm [151].

10.4.1.4 Molecular Interactions

All the effects described above may occur simultaneously, influencing each other. These are the so-called Interactions in the SLAMENGHI acronym. Molecular interactions, which are involved in the Matrix and in the Effectors of absorption factors, should be discussed separately. These are the interactions between the different nutrients and compounds from the food, and also with the digestive juice compounds from the host. For example, it is known that divalent minerals interact strongly with lipids, bile salts and many lipophilic compounds, reducing their bioaccessibility/bioavailability [29]. Nutrient–nutrient interactions at the lipid droplet interface have been discussed by Marze [92]. Briefly, these nutrients (mainly lipids, proteins and polysaccharides) associate or compete at the droplet interface. Depending on the exact interaction, this may result in (1) droplet destabilization by flocculation, coalescence or creaming, and/or; (2) inhibition of the interfacial activity of lipases and/or bile salts. These effects influence lipid digestion and in turn the bioaccessibility of lipophilic compounds.

Molecular interactions in the bulk have also been described and have been summarized by Marze [89]. In this context, the most studied compound appears to be dietary fibre. Some fibres were initially found to bind bile salts in vitro [77, 122], whereas all lipids in the mixed micelles were in the bound form [131]. The mechanism was similar to that exerted by synthetic bile salt sequestrates. Bioaccessibility studies were conducted to evaluate the effect of various dietary fibres. Sugar beet pectin was found to reduce the bioaccessibility of β-carotene [146]. In other studies with citrus pectin, the reduction was found to be larger with a lower degree of methyl esterification [133, 135], but the inverse trend was observed in the presence of phosphatidylcholine [134]. The bioaccessibility of β-carotene was also decreased when a nanoemulsion was encapsulated in alginate beads [150], or when a nanoemulsion was entrapped in a κ-carrageenan gel [119], and decreased more with increasing dietary fibre concentration. A decrease was also found for the bioaccessibility of curcumin when a nanoemulsion was encapsulated in either alginate or κ-carrageenan beads [150].

In vivo, some soluble dietary fibres were found to reduce the absorption of cholesterol and lipid in the rat [39]. In a meta-analysis, a significant effect was confirmed in humans for fasting plasma cholesterol concentrations but not for lipid concentrations [18]. In two studies of carotenoid absorption in humans, it was found that some soluble dietary fibres did not change the plasma cholesterol and lipid concentrations significantly, but nevertheless reduced the plasma carotenoid concentration significantly, except that of canthaxanthin [111, 112]. Although the in vitro results clearly show that soluble dietary fibres sequester bile salts and/or lipids, their effect in vivo is still not fully understood, and the few studies involving lipophilic compounds are thus the subject of speculation.

10.4.2 Relation Between Lipid Digestion and Lipophilic Compound Bioaccessibility

10.4.2.1 Simulation of Lipophilic Compound Bioaccessibility

A numerical simulation was developed to study the interplay between the digestion of triglycerides and the bioaccessibility of lipophilic vitamins [90, 91]. This was based on multi-agent modelling, which represents each molecular species as an autonomous particle with specific properties and rules. One static oil droplet with triglyceride and lipophilic vitamin particles and a continuous aqueous phase with lipase or lipase–bile salt particles was modelled, with an interface where the products of digestion competed with lipase/bile salts. Final digestion products (monoglyceride and fatty acid particles) were produced after a given number of contacts with lipase at the interface, inversely proportional to the experimental lipolysis rates for various triglycerides. Final digestion products and lipophilic vitamins were solubilized in simple bile salt micelles after a given number of contacts with bile salt at the interface, inversely proportional to the experimental solubilization ratios (mass of solubilizate per mass of bile salt) for various fat-soluble vitamins and lipophilic drugs. All particles were set to diffuse according to their molecular mass and the viscosity of the oil phase. Lipase–bile salt particles were moved randomly to reflect their faster movement because of the low viscosity of water and some convective flows in the aqueous phase. Bile salt micelles were set to (1) saturate at the solubilization ratio (static in vitro case) or; (2) recycle by discharging their solubilizates (in vivo case).

The results showed that the bioaccessibility kinetics of lipophilic vitamins always followed the bioaccessibility kinetics of the digestion products, and were slower with slowly digested triglycerides. The final lipophilic vitamin bioaccessibility was always lower in the saturated case than in the recycled case. It decreased with the triglyceride chain length in the saturated case, whereas it increased with the triglyceride chain length in the recycled case. This was explained by incomplete triglyceride digestion in the saturated case and complete triglyceride digestion in the recycled case. Whenever triglyceride digestion was complete or when a comparison was made at a given degree of triglyceride digestion, the bioaccessibility of lipophilic vitamins was always higher for longer chain triglycerides.

To validate these results experimentally, a semi-dynamic microfluidic digestion system was developed [98]. Isolated droplets of triglycerides containing a lipophilic vitamin could be generated, immobilized and digested under continuous digestive fluid flow and renewal, reflecting the recycled case of the simulation. The release kinetics of the digestion products and of the lipophilic vitamin could be monitored by measuring the droplet size and the vitamin autofluorescence using confocal fluorescence microscopy. The microfluidic digestion results were in very good qualitative agreement with the simulation results in the recycled case, except for the final lipophilic vitamin bioaccessibility, which was always underestimated in the first version of the simulation [90]. A second version improved this discrepancy by accounting for additional factors, the formation of mixed micelles (digestion products–bile salt) being the most significant factor [91].

These findings suggest that triglyceride digestion should be complete when comparing lipophilic compound bioaccessibility for various triglycerides at a given intestinal digestion time (usually 2 h). As the amount of bile salt is fixed in static in vitro protocols, this can be achieved by decreasing the amount of triglyceride. An alternative is to compare lipophilic compound bioaccessibility for various triglycerides at a given degree of triglyceride digestion. This may require a longer digestion duration for long chain triglycerides, which should be kept within the realistic limit of 4 h (average small intestine transit time).

10.4.2.2 Simulation of Mixed Micelle Incorporation

The last step in making a lipophilic compound bioaccessible is its incorporation into mixed micelles. It is important to note that various terms, such as solubilization, micellization and, of course, bioaccessibility (which is relative to the initial amount), are used for this concept. Mixed micelle is a generic term referring to nanoscale particles composed of bile constituents (bile salts, phospholipids, cholesterol) and digested lipids (fatty acids, monoglycerides, lysophospholipids). It generally refers to a spherical particle, although it has been shown repeatedly that various structures are formed in vitro and in vivo [63, 72]. The major structures are the spherical micelle and vesicle, but ellipsoidal, disk-like and cylindrical structures (among others) have also been reported [44, 92], depending on the relative amounts of the constituents. This is an important aspect, as these particles are likely to transport lipids and lipophilic compounds differently in the small intestine.

For the incorporation of lipophilic compounds, the seminal works in the fasted state of Kossena et al. [72] showed that various particles formed as the amount of fatty acid/monoglyceride in the bile mixture was varied. Each particle was found to have a specific incorporation capacity for lipophilic drugs, increasing with the fatty acid/monoglyceride amount, and depending on their chain length and degree of unsaturation [72]. Subsequent in vivo experiments demonstrated that the absorption of lipophilic drugs in the rat also varied greatly depending on the type of particle infused into the duodenum, increasing proportionally with the solubilization capacity [73]. Such a relationship is currently under systematic investigation for food lipophilic compounds in the fed state (Marze, project AssemBiles). There is a substantial amount of data on the solubilization capacity for lipophilic drugs in the fasted state (as compiled by Wiedmann and Kamel [141]), but very few results for food lipophilic compounds in the fed state.

The incorporation of lipophilic compounds into mixed micelles has been stressed to be a key step towards absorption. However, mostly macroscopic data are available to quantify this incorporation, although it is controlled by interactions at the molecular scale. As studying these interactions precisely is challenging experimentally, computer simulations have also been developed, mostly using molecular dynamics. Initially, bile salt, bile salt–phospholipid or bile salt–phospholipid–cholesterol assemblies were modelled to characterize their structures [44]. Recently, other lipids such as fatty acids were included in the simulations [127], in order to study the incorporation of lipophilic compounds for more realistic assemblies. For this purpose, only two articles have been published: lipophilic drug incorporation in bile salt–phospholipid assemblies in the fasted state [66]; lipophilic drug incorporation in bile salt–phospholipid–fatty acid–monoglyceride assemblies in both the fasted state and the fed state [8]. The results in the fed state confirmed that, when the fatty acid–monoglyceride concentration increased, the assembly structure changed from micellar to lamellar, with a progressive increase in particle size. These changes were accompanied by an increased incorporation of danazol (a lipophilic drug). Danazol became aggregated above the incorporation capacity, and hence was no longer in its solubilized form. Figure 10.7 illustrates these findings for several conditions. This type of simulation should be applied to the incorporation of food lipophilic compounds into mixed micelles, in order to study the universality of this behaviour.

10.4.3 Lipid Digestion in Natural Foods

Lipids in natural foods generally occur in the form of complex structures in which triglyceride particles are coated with a solubilizing, stabilizing layer or multi-layer of membrane phospholipids and proteins. Breaking down the surrounding

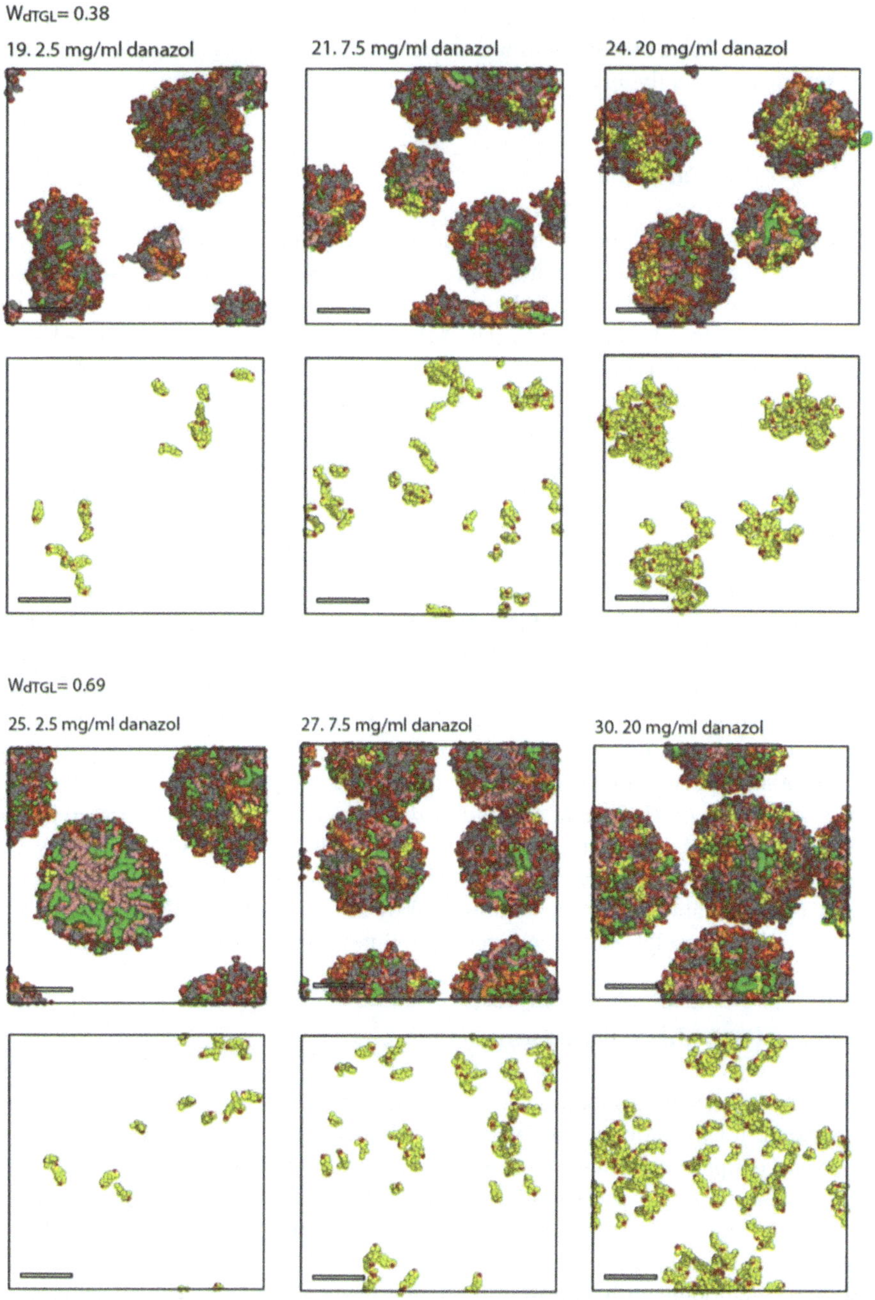

Fig. 10.7 Localization of danazol within the final frames of selected simulations, with changing danazol concentration. Danazol is coloured yellow, glycodeoxycholate is grey, lysophosphocholine is orange, oleic acid is pink, glyceryl monooleate is green and oxygen atoms are red. The box indicates the periodic boundary and the length of the scale bar is 3.0 nm. Water atoms have been omitted. (Reprinted with permission from Birru et al. [8])

structures and releasing the lipid droplets from the cells, seed bodies or whatever locating matrix has a profound influence on our ability to digest the lipid and use its components efficiently and effectively.

10.4.3.1 Milk Fat Globules

Milk is an oil-in-water emulsion containing milk fat globules (MFGs), which are composed of a core that is rich in triglycerides and is stabilized by the milk fat globule membrane (MFGM). MFGs have a mean diameter around 4 μm with a size range from 0.2 to 10 μm [61, 81]. The MFGM has a thickness of about 4–10 nm (trilayer) and is composed of polar lipids, specific membrane proteins, glycoproteins, cholesterol and some other minor components [81].

To date, the digestion behaviour of different types of milk, including raw milk, has been studied in vitro and the effect of digestion on the microstructure, proteolysis and lipolysis has been described [6, 52, 148]. Gallier et al. [52] reported that, because of a high zeta-potential, under in vitro gastric conditions, no coalescence or aggregation of MFGs was observed. As the phospholipids that are present in the structure of MFGs are not digested in the stomach, the integrity of these globules is retained during gastric digestion [52, 58]. However, under in vitro intestinal conditions, the lipolytic products are solubilized when a lamellar phase of these products becomes accumulated at the surface of MFGs, and bile salt micelles are formed [52]. Bile salts can displace some of the membrane on MFGs, which allows the pancreatic lipase–colipase complex to anchor at the surface of the MFG, because the intestinal lipolysis of MFGs does not present a lag phase and is efficient [52]. The components of MFGs that are solubilized by bile salts are then absorbed by the intestinal cells. Short and medium chain fatty acids are quickly absorbed by the enterocytes and are transported by the portal vein for orientation towards β-oxidation, whereas long chain fatty acids are absorbed by the small intestinal enterocytes and are transported through the lymphatics as chylomicrons [52, 96].

10.4.3.2 Oil Bodies (Oleosomes)

Oil bodies, which are also known as oleosomes, sphaerosomes or spherosomes, are responsible for the storage and synthesis of lipids in plant cells (about 98% of an oleosome is lipid), especially oil seeds. Oil bodies provide the required energy for the seed during germination and have a size in the range from 0.2 to 2.5 μm [129]. In fact, oleosomes are small cell organelles containing a single membrane. These organelles arise from the endoplasmic reticulum and are bounded by a single (but half unit) membrane with a phospholipid monolayer [129, 153]. The phospholipids are composed of polar heads and hydrophobic tails towards the cytosol and the inner side respectively. The membrane of oleosomes is stabilized by oleosins (proteins), which is the reason for the name 'oleosomes'. However, only 2% of oleosomes are composed of the membrane proteins; some of these proteins have enzymatic functions, by which they contribute to the synthesis of lipids [34, 153]. There are three basic structural domains in every oleosin: (1) an amphipathic domain, which is present at both the N-terminal and the C-terminal; (2) a highly hydrophobic domain, which penetrates into the triacylglycerol matrix. The amphipathic domain resides on the surface of the oil bodies, resulting in their stability via steric hindrance and electronegative repulsion [48, 104].

Some research on the in vitro digestion of oleosomes and oil bodies has been carried out. One interesting example is the work of Gallier and Singh [51], who studied the physicochemical and structural changes of almond oil bodies in an emulsion formulation under simulated gastrointestinal digestion conditions. The emulsion containing almond oil bodies showed similar behaviour to that of a protein-stabilized emulsion, where flocculation of the oil bodies occurred under simulated gastric digestion conditions. Throughout the gastric digestion, the surface of the oil bodies was covered by proteins, peptides and phospholipids. Under simulated intestinal digestion conditions, the interfacial peptides and phospholipids were displaced by bile salts, so that the flocs were disrupted. The oil body membrane had a negative impact on the efficiency of the gastric digestion, which in turn led to the accumulation of long chain fatty acids (the main lipolytic products) at the surface of the oil bodies and, correspondingly, limited the activity of pancreatic lipase [51].

10.5 Conclusions

The structure, metabolism, absorption, bioavailability and bioaccessibility of lipid-soluble compounds, as well as the role of digestion with respect to the relationship between lipid digestion and the bioaccessibility of lipophilic compounds, have been discussed. To the best of our knowledge, the main aspects that still need to be studied in order to understand the role of gastrointestinal digestion in the bioavailability of lipophilic compounds are related to the molecular properties. Firstly, the role of the molecular form and linkage of the compounds is not always clear. For example, the hydrolysis kinetics of esters is known for only a few compounds, and the effect of dietary fibre linkage to some polyphenols is mostly unknown. Secondly, knowledge of the gastrointestinal metabolites of lipophilic compounds, how they are produced and how they are absorbed is mostly lacking, except for a few model carotenoids and polyphenols. This hinders our understanding of the fate of the parent molecules from ingestion to excretion. Thirdly, the interactions between nutrients other than lipids and lipophilic compounds, as well as the organization of the mixed micelles in the real situation of multiple nutrients, are not well described. Indeed, competition between molecules for incorporation seems to be more frequent than currently thought. Finally, the effect of natural or designed food structures (at all scales) on the liberation of some lipophilic compounds should be accounted for, as it constitutes a major barrier to digestion and absorption in some cases.

References

1. Affandi M, Julianto T, Majeed ABA (2012) Enhanced oral bioavailability of astaxanthin with droplet size reduction. Food Sci Technol Res 18(4):549–554
2. Aherne SA, O'Brien NM (2002) Dietary flavonols: chemistry, food content, and metabolism. Nutrition 18(1):75–81
3. Ahmed K, Li Y, McClements DJ, Xiao H (2012) Nanoemulsion- and emulsion-based delivery systems for curcumin: encapsulation and release properties. Food Chem 132(2):799–807
4. Alessandrì MG, Scalori V, Giovannini L, Mian M, Bertelli AA (1988) Plasma and tissue concentrations of coenzyme Q10 in the rat after intravenous administration by a microsphere delivery system or in a new type of solution. Int J Tissue React 10(2):99–102
5. Azzi A (2018) Many tocopherols, one vitamin E. Mol Asp Med 61:92–103
6. Berton A, Rouvellac S, Robert B, Rousseau F, Lopez C, Crenon I (2012) Effect of the size and interface composition of milk fat globules on their in vitro digestion by the human pancreatic lipase: native versus homogenized milk fat globules. Food Hydrocoll 29(1):123–134
7. Bhagavan HN, Chopra RK (2006) Coenzyme Q10: absorption, tissue uptake, metabolism and pharmacokinetics. Free Radic Res 40(5):445–453
8. Birru WA, Warren DB, Han SF, Benameur H, Porter CJH, Pouton CW, Chalmers DK (2017) Computational models of the gastrointestinal environment. 2. Phase behavior and drug solubilization capacity of a type I lipid-based drug formulation after digestion. Mol Pharm 14(3):580–592
9. Blaner WS, Obunike JC, Kurlandsky SB, Al-Haideri M, Piantedosi R, Deckelbaum RJ, Goldberg IJ (1994) Lipoprotein lipase hydrolysis of retinyl ester. Possible implications for retinoid uptake by cells. J Biol Chem 269(24):16559–16565
10. Blomstrand R, Werner B (1967) Studies on the intestinal absorption of radioactive β-carotene and vitamin A in man: conversion of β-carotene into vitamin A. Scand J Clin Lab Invest 19(4):339–345
11. Bohn T (2014) Dietary factors affecting polyphenol bioavailability. Nutr Rev 72(7):429–452
12. Borel P (2003) Factors affecting intestinal absorption of highly lipophilic food microconstituents (fat-soluble vitamins, carotenoids and phytosterols). Clin Chem Lab Med 41(8):979–994
13. Borel P (2009) Nutrigenetic effect on intestinal absorption of fat-soluble microconstituents (vitamins A, E, D and K, carotenoids and phytosterols). In: Barbosa-Cánovas G, Mortimer A, Lineback D, Spiess W, Buckle K, Colonna P (eds) Global issues in food science and technology. Academic Press, San Diego, pp 31–42
14. Borel P, Tyssandier V, Mekki N, Grolier P, Rochette Y, Alexandre-Gouabau MC, Azais-Braesco V (1998) Chylomicron beta-carotene and retinyl palmitate responses are dramatically diminished when men ingest beta-carotene with medium-chain rather than long-chain triglycerides. J Nutr 128(8):1361–1367
15. Borel P, Pasquier B, Armand M, Tyssandier V, Grolier P, Alexandre-Gouabau MC, Azais-Braesco V (2001) Processing of vitamin A and E in the human gastrointestinal tract. Am J Physiol 280(1):G95–G103
16. Bowen PE, Herbst-Espinosa SM, Hussain EA, Stacewicz-Sapuntzakis M (2002) Esterification does not impair lutein bioavailability in humans. J Nutr 132(12):3668–3673
17. Brigelius-Flohe R, Traber MG (1999) Vitamin E: function and metabolism. FASEB J 13(10):1145–1155
18. Brown L, Rosner B, Willett WW, Sacks FM (1999) Cholesterol-lowering effects of dietary fiber: a meta-analysis. Am J Clin Nutr 69(1):30–42
19. Brown MJ, Ferruzzi MG, Nguyen ML, Cooper DA, Eldridge AL, Schwartz SJ, White WS (2004) Carotenoid bioavailability is higher from salads ingested with full-fat than with fat-reduced salad dressings as measured with electrochemical detection. Am J Clin Nutr 80(2):396–403
20. Castenmiller JJM, West CE (1998) Bioavailability and bioconversion of carotenoids. Annu Rev Nutr 18:19–38

21. Cho HT, Salvia-Trujillo L, Kim J, Park Y, Xiao H, McClements D (2014) Droplet size and composition of nutraceutical nanoemulsions influences bioavailability of long chain fatty acids and coenzyme Q10. Food Chem 156:117–122
22. Chopra RK, Goldman R, Sinatra ST, Bhagavan HN (1998) Relative bioavailability of coenzyme Q10 formulations in human subjects. Int J Vitam Nutr Res 68(2):109–113
23. Cilla A, De Palma G, Lagarda MJ, Barberá R, Farré R, Clemente G, Romero F (2009) Impact of fruit beverage consumption on the antioxidant status in healthy women. Ann Nutr Metab 54(1):35–42
24. Glerup H, Mikkelsen K, Poulsen L, Hass E, Overbeck S, Thomsen J et al (2000) Commonly recommended daily intake of vitamin D is not sufficient if sunlight exposure is limited. J Intern Med 247(2):260–268
25. Colle IJP, Van Buggenhout S, Lemmens L, Van Loey AM, Hendrickx ME (2012) The type and quantity of lipids present during digestion influence the in vitro bioaccessibility of lycopene from raw tomato pulp. Food Res Int 45(1):250–255
26. Conaway HH, Henning P, Lerner UH (2013) Vitamin A metabolism, action, and role in skeletal homeostasis. Endocr Rev 34(6):766–797
27. Cook N, Samman S (1996) Flavonoids—chemistry, metabolism, cardioprotective effects, and dietary sources. J Nutr Biochem 7(2):66–76
28. Cooper AD (1997) Hepatic uptake of chylomicron remnants. J Lipid Res 38(11):2173–2192
29. Corte-Real J, Bohn T (2018) Interaction of divalent minerals with liposoluble nutrients and phytochemicals during digestion and influences on their bioavailability – a review. Food Chem 252:285–293
30. Corte-Real J, Desmarchelier C, Borel P, Richling E, Hoffmann L, Bohn T (2018) Magnesium affects spinach carotenoid bioaccessibility in vitro depending on intestinal bile and pancreatic enzyme concentrations. Food Chem 239:751–759
31. D'Ambrosio DN, Clugston RD, Blaner WS (2011) Vitamin A metabolism: an update. Nutrients 3(1):63–103
32. da Costa GA, Mercadante AZ (2018) In vitro bioaccessibility of free and esterified carotenoids in cajá frozen pulp-based beverages. J Food Compos Anal 68:53–59
33. Nutrition Data (2013) Nutrition facts, calories in food, labels, nutritional information and analysis. Available online at: http://www.nutritiondata.com/ (accessed 15 February 2019).
34. De Chirico S, di Bari V, Foster T, Gray D (2018) Enhancing the recovery of oilseed rape seed oil bodies (oleosomes) using bicarbonate-based soaking and grinding media. Food Chem 241:419–426
35. DeMan JM (1999) Principles of food chemistry, 3rd edn. Aspen Publishers, Gaithersburg
36. Déprez S, Brezillon C, Rabot S, Philippe C, Mila I, Lapierre C, Scalbert A (2000) Polymeric proanthocyanidins are catabolized by human colonic microflora into low-molecular-weight phenolic acids. J Nutr 130(11):2733–2738
37. Deprez S, Mila I, Huneau J-F, Tome D, Scalbert A (2001) Transport of proanthocyanidin dimer, trimer, and polymer across monolayers of human intestinal epithelial Caco-2 cells. Antioxid Redox Signal 3(6):957–967
38. Dutton PL, Ohnishi T, Darrouzet E, Leonard MA, Sharp RE, Cibney B, Moser CC (2000) Coenzyme Q oxidation reduction reactions in mitochondrial electron transport. In: Kagan VE, Quinn PI (eds) Coenzyme Q: molecular mechanisms in health and disease. CRC Press, Boca Raton, pp 65–82
39. Ebihara K, Schneeman BO (1989) Interaction of bile-acids, phospholipids, cholesterol and triglyceride with dietary-fibers in the small-intestine of rats. J Nutr 119(8):1100–1106
40. Eby CS (2016) Bleeding and vitamin K deficiency. In: Teruya J (ed) Management of bleeding patients. Springer International Publishing, Cham, pp 145–150
41. Episkopou V, Maeda S, Nishiguchi S, Shimada K, Gaitanaris GA, Gottesman ME, Robertson EJ (1993) Disruption of the transthyretin gene results in mice with depressed levels of plasma retinol and thyroid hormone. Proc Natl Acad Sci U S A 90(6):2375–2379

42. Erlanson C, Borgström B (1968) The identity of vitamin A esterase activity of rat pancreatic juice. Biochim Biophys Acta 167(3):629–631
43. Ernster L, Dallner G (1995) Biochemical, physiological and medical aspects of ubiquinone function. Biochim Biophys Acta 1271(1):195–204
44. Euston SR (2017) Molecular simulation of biosurfactants with relevance to food systems. Curr Opin Colloid Interface Sci 28:110–119
45. Fahy E, Subramaniam S, Brown HA, Glass CK, Merrill AH Jr, Murphy RC, Dennis EA (2005) A comprehensive classification system for lipids. Eur J Lipid Sci Technol 107(5):337–364
46. Feliciano RP, Mills CE, Istas G, Heiss C, Rodriguez-Mateos A (2017) Absorption, metabolism and excretion of cranberry (poly) phenols in humans: a dose response study and assessment of inter-individual variability. Nutrients 9(3). https://doi.org/10.3390/nu9030268
47. Fraga CG, Schramm DD, Schmitz HH, Hammerstone JF, Sullards MC, Zhu QY, Keen CL (2002) Procyanidin dimer B2 [epicatechin-(4β-8)-epicatechin] in human plasma after the consumption of a flavanol-rich cocoa. Am J Clin Nutr 76(4):798–804
48. Frandsen GI, Mundy J, Tzen JT (2001) Oil bodies and their associated proteins, oleosin and caleosin. Physiol Plant 112(3):301–307
49. Fukuwatari T, Shibata K (2008) Urinary water-soluble vitamins and their metabolite contents as nutritional markers for evaluating vitamin intakes in young Japanese women. J Nutr Sci Vitaminol 54(3):223–229
50. Galli F, Azzi A, Birringer M, Cook-Mills JM, Eggersdorfer M, Frank J, Özer NK (2017) Vitamin E: emerging aspects and new directions. Free Radic Biol Med 102:16–36
51. Gallier S, Singh H (2012) Behavior of almond oil bodies during in vitro gastric and intestinal digestion. Food Funct 3(5):547–555
52. Gallier S, Ye A, Singh H (2012) Structural changes of bovine milk fat globules during in vitro digestion. J Dairy Sci 95(7):3579–3592
53. Giammanco A, Cefalù AB, Noto D, Averna MR (2015) The pathophysiology of intestinal lipoprotein production. Front Physiol 6(61). https://doi.org/10.3389/fphys.2015.00061
54. Gleize B, Tourniaire F, Depezay L, Bott R, Nowicki M, Albino L, Borel P (2013) Effect of type of TAG fatty acids on lutein and zeaxanthin bioavailability. Br J Nutr 110(1):1–10
55. Goltz SR, Campbell WW, Chitchumroonchokchai C, Failla ML, Ferruzzi MG (2012) Meal triacylglycerol profile modulates postprandial absorption of carotenoids in humans. Mol Nutr Food Res 56(6):866–877
56. Gomez M, Martinez MM (2018) Fruit and vegetable by-products as novel ingredients to improve the nutritional quality of baked goods. Crit Rev Food Sci Nutr 58(13):2119–2135
57. Halder M, Petsophonsakul P, Akbulut AC, Pavlic A, Bohan F, Anderson E, Schurgers L (2019) Vitamin K: double bonds beyond coagulation insights into differences between vitamin K1 and K2 in health and disease. Int J Mol Sci 20(4). https://doi.org/10.3390/ijms20040896
58. Hamosh M, Peterson JA, Henderson TR, Scallan CD, Kiwan R, Ceriani RL, Hamosh P (1999) Protective function of human milk: the milk fat globule. Semin Perinatol 23(3):242–249
59. Harrison EH (2012) Mechanisms involved in the intestinal absorption of dietary vitamin A and provitamin A carotenoids. Biochim Biophys Acta 1821(1):70–77
60. Harrison EH, Hussain MM (2001) Mechanisms involved in the intestinal digestion and absorption of dietary vitamin A. J Nutr 131(5):1405–1408
61. Heid HW, Keenan TW (2005) Intracellular origin and secretion of milk fat globules. Eur J Cell Biol 84(2–3):245–258
62. Hernández-Camacho JD, Bernier M, López-Lluch G, Navas P (2018) Coenzyme Q10 supplementation in aging and disease. Front Physiol 9(44). https://doi.org/10.3389/fphys.2018.00044
63. Hernell O, Staggers JE, Carey MC (1990) Physical-chemical behavior of dietary and biliary lipids during intestinal digestion and absorption. 2. Phase analysis and aggregation states of luminal lipids during duodenal fat digestion in healthy adult human beings. Biochemistry 29(8):2041–2056
64. Holick MF (2004) Sunlight and vitamin D for bone health and prevention of autoimmune diseases, cancers, and cardiovascular disease. Am J Clin Nutr 80(6):1678S–1688S

65. Holmberg I, Aksnes L, Berlin T, Lindback B, Zemgals J, Lindeke B (1990) Absorption of a pharmacological dose of vitamin-d3 from 2 different lipid vehicles in man – comparison of peanut oil and a medium chain triglyceride. Biopharm Drug Dispos 11(9):807–815
66. Holmboe M, Larsson P, Anwar J, Bergstrom CAS (2016) Partitioning into colloidal structures of fasted state intestinal fluid studied by molecular dynamics simulations. Langmuir 32(48):12732–12740
67. Hosomi A, Arita M, Sato Y, Kiyose C, Ueda T, Igarashi O, Inoue K (1997) Affinity for α-tocopherol transfer protein as a determinant of the biological activities of vitamin E analogs. FEBS Lett 409(1):105–108
68. Hou ZQ, Liu YW, Lei F, Gao YX (2014) Investigation into the in vitro release properties of beta-carotene in emulsions stabilized by different emulsifiers. LWT Food Sci Technol 59(2):867–873
69. Huo T, Ferruzzi MG, Schwartz SJ, Failla ML (2007) Impact of fatty acyl composition and quantity of triglycerides on bioaccessibility of dietary carotenoids. J Agric Food Chem 55(22):8950–8957
70. Kaempf-Rotzoll DE, Traber MG, Arai H (2003) Vitamin E and transfer proteins. Curr Opin Lipidol 14(3):249–254
71. Kayden HJ, Traber MG (1993) Absorption, lipoprotein transport, and regulation of plasma concentrations of vitamin E in humans. J Lipid Res 34(3):343–358
72. Kossena GA, Boyd BJ, Porter CJH, Charman WN (2003) Separation and characterization of the colloidal phases produced on digestion of common formulation lipids and assessment of their impact on the apparent solubility of selected poorly water-soluble drugs. J Pharm Sci 92(3):634–648
73. Kossena GA, Charman WN, Boyd BJ, Porter CIH (2005) Influence of the intermediate digestion phases of common formulation lipids on the absorption of a poorly water-soluble drug. J Pharm Sci 94(3):481–492
74. Kramer W, Glombik H, Petry S, Heuer H, Schäfer H-L, Wendler W, Weyland C (2000) Identification of binding proteins for cholesterol absorption inhibitors as components of the intestinal cholesterol transporter. FEBS Lett 487(2):293–297
75. Kramer W, Girbig F, Corsiero D, Burger K, Fahrenholz F, Jung C, Müller G (2003) Intestinal cholesterol absorption: identification of different binding proteins for cholesterol and cholesterol absorption inhibitors in the enterocyte brush border membrane. Biochim Biophys Acta 1633(1):13–26
76. Kreis W, Müller-Uri F (2018) Biochemistry of sterols, cardiac glycosides, brassinosteroids, phytoecdysteroids and steroid saponins. Annu Plant Rev 40:304–363
77. Kritchevsky D, Story JA (1974) Binding of bile-salts in-vitro by nonnutritive fiber. J Nutr 104(4):458–462
78. Kuroda T, Uenishi K, Ohta H, Shiraki M (2019) Multiple vitamin deficiencies additively increase the risk of incident fractures in Japanese postmenopausal women. Osteoporos Int 30(3):593–599
79. Kurosu M (2017) Vitamin K2 biosynthesis: drug targets for new antibacterials. In: Gordeladze JO (ed) Vitamin K2–vital for health and wellbeing. IntechOpen, London
80. Lee HC, Jenner AM, Low CS, Lee YK (2006) Effect of tea phenolics and their aromatic fecal bacterial metabolites on intestinal microbiota. Res Microbiol 157(9):876–884
81. Lopez C (2011) Milk fat globules enveloped by their biological membrane: unique colloidal assemblies with a specific composition and structure. Curr Opin Colloid Interface Sci 16(5):391–404
82. López-Lluch G, del Pozo-Cruz J, Sánchez-Cuesta A, Cortés-Rodríguez AB, Navas P (2019) Bioavailability of coenzyme Q10 supplements depends on carrier lipids and solubilization. Nutrition 57:133–140
83. Majeed H, Antoniou J, Hategekimana J, Sharif HR, Haider J, Liu F, Zhong F (2016) Influence of carrier oil type, particle size on in vitro lipid digestion and eugenol release in emulsion and nanoemulsions. Food Hydrocoll 52:415–422

84. Manach C, Morand C, Gil-Izquierdo A, Bouteloup-Demange C, Remesy C (2003) Bioavailability in humans of the flavanones hesperidin and narirutin after the ingestion of two doses of orange juice. Eur J Clin Nutr 57(2):235–242
85. Manach C, Scalbert A, Morand C, Rémésy C, Jiménez L (2004) Polyphenols: food sources and bioavailability. Am J Clin Nutr 79(5):727–747
86. Manolescu B, Atanasiu V, Cercasov C, Stoian I, Oprea E, Buşu C (2008) So many options but one choice: the human body prefers α-tocopherol. A matter of stereochemistry. J Med Life 1(4):376–382
87. Mao L, Miao S (2015) Structuring food emulsions to improve nutrient delivery during digestion. Food Eng Rev 7(4):439–451
88. Maqbool A, Stallings VA (2008) Update on fat-soluble vitamins in cystic fibrosis. Curr Opin Pulm Med 14(6):574–581
89. Marze S (2013) Bioaccessibility of nutrients and micronutrients from dispersed food systems: impact of the multiscale bulk and interfacial structures. Crit Rev Food Sci Nutr 53(1):76–108
90. Marze S (2014) A coarse-grained simulation to study the digestion and bioaccessibility of lipophilic nutrients and micronutrients in emulsion. Food Funct 5(1):129–139
91. Marze S (2015) Bioaccessibility of lipophilic micro-constituents from a lipid emulsion. Food Funct 6(10):3218–3227
92. Marze S, Gaillard C, Roblin P (2015) In vitro digestion of emulsions: high spatiotemporal resolution using synchrotron SAXS. Soft Matter 11(26):5365–5373
93. McClements DJ, Xiao H (2012) Potential biological fate of ingested nanoemulsions: influence of particle characteristics. Food Funct 3(3):202–220
94. McClements DJ, Li F, Xiao H (2015) The nutraceutical bioavailability classification scheme: classifying nutraceuticals according to factors limiting their oral bioavailability. Annu Rev Food Sci Technol 6(1):299–327
95. Meroni E, Raikos V (2018) Physicochemical stability, antioxidant properties and bioaccessibility of beta-carotene in orange oil-in-water beverage emulsions: influence of carrier oil types. Food Funct 9(1):320–330
96. Michalski MC (2009) Specific molecular and colloidal structures of milk fat affecting lipolysis, absorption and postprandial lipemia. Eur J Lipid Sci Technol 111(5):413–431
97. Monsen ER (2000) Dietary reference intakes for the antioxidant nutrients: vitamin C, vitamin E, selenium, and carotenoids. J Acad Nutr Diet 100(6):637–640
98. Nguyen HT, Marquis M, Anton M, Marze S (2019) Studying the real-time interplay between triglyceride digestion and lipophilic micronutrient bioaccessibility using droplet microfluidics. 2. Application to various oils and (pro)vitamins. Food Chem 275:661–667
99. Nidhi B, Baskaran V (2011) Influence of vegetable oils on micellization of lutein in a simulated digestion model. J Am Oil Chem Soc 88(3):367–372
100. Norman AW (2008) From vitamin D to hormone D: fundamentals of the vitamin D endocrine system essential for good health. Am J Clin Nutr 88(2):491S–499S
101. Ochiai A, Itagaki S, Kurokawa T, Kobayashi M, Hirano T, Iseki K (2007) Improvement in intestinal coenzyme Q10 absorption by food intake. Yakugaku Zasshi 127(8):1251–1254
102. Ozawa Y, Mizushima Y, Koyama I, Akimoto M, Yamagata Y, Hayashi H, Murayama H (1986) Intestinal absorption enhancement of coenzyme Q10 with a lipid microsphere. Arzneimittelforschung 36(4):689–690
103. Porter CJ, Trevaskis NL, Charman WN (2007) Lipids and lipid-based formulations: optimizing the oral delivery of lipophilic drugs. Nat Rev Drug Discov 6(3):231–248
104. Qu R, Huang A (1990) Oleosin KD 18 on the surface of oil bodies in maize. Genomic and cDNA sequences and the deduced protein structure. J Biol Chem 265(4):2238–2243
105. Quideau S, Deffieux D, Douat-Casassus C, Pouységu L (2011) Plant polyphenols: chemical properties, biological activities, and synthesis. Angew Chem Int Ed 50(3):586–621
106. Quinzii CM, DiMauro S, Hirano M (2007) Human coenzyme Q10 deficiency. Neurochem Res 32(4):723–727

107. Ravisankar P, Reddy AA, Nagalakshmi B, Koushik O, Kumar B, Anvith P (2015) The comprehensive review on fat soluble vitamins. IOSR J Pharm 5(11):12–28
108. Reboul E (2015) Intestinal absorption of vitamin D: from the meal to the enterocyte. Food Funct 6(2):356–362
109. Rechner AR, Kuhnle G, Bremner P, Hubbard GP, Moore KP, Rice-Evans CA (2002) The metabolic fate of dietary polyphenols in humans. Free Radic Biol Med 33(2):220–235
110. Rein MJ, Renouf M, Cruz-Hernandez C, Actis-Goretta L, Thakkar SK, da Silva Pinto M (2013) Bioavailability of bioactive food compounds: a challenging journey to bioefficacy. Br J Clin Pharmacol 75(3):588–602
111. Riedl J, Linseisen J, Hoffmann J, Wolfram G (1999) Some dietary fibers reduce the absorption of carotenoids in women. J Nutr 129(12):2170–2176
112. Rock CL, Swendseid ME (1992) Plasma beta-carotene response in humans after meals supplemented with dietary pectin. Am J Clin Nutr 55(1):96–99
113. Salvia-Trujillo L, Verkempinck S, Rijal SK, Van Loey A, Grauwet T, Hendrickx M (2019) Lipid nanoparticles with fats or oils containing beta-carotene: storage stability and in vitro digestibility kinetics. Food Chem 278:396–405
114. Sankar MJ, Chandrasekaran A, Kumar P, Thukral A, Agarwal R, Paul VK (2016) Vitamin K prophylaxis for prevention of vitamin K deficiency bleeding: a systematic review. J Perinatol 36(S1):S29–S35
115. Sauvant P, Mekki N, Charbonnier M, Portugal H, Lairon D, Borel P (2003) Amounts and types of fatty acids in meals affect the pattern of retinoids secreted in human chylomicrons after a high-dose preformed vitamin a intake. Metab Clin Exp 52(4):514–519
116. Scalbert A, Morand C, Manach C, Rémésy C (2002) Absorption and metabolism of polyphenols in the gut and impact on health. Biomed Pharmacother 56(6):276–282
117. Segura R, Javierre C, Lizarraga MA, Ros E (2006) Other relevant components of nuts: phytosterols, folate and minerals. Br J Nutr 96(Suppl 2):S36–S44
118. Shah BR, Zhang CL, Li Y, Li B (2016) Bioaccessibility and antioxidant activity of curcumin after encapsulated by nano and Pickering emulsion based on chitosan-tripolyphosphate nanoparticles. Food Res Int 89:399–407
119. Soukoulis C, Tsevdou M, Andre CM, Cambier S, Yonekura L, Taoukis PS, Hoffmann L (2017) Modulation of chemical stability and in vitro bioaccessibility of beta-carotene loaded in kappa-carrageenan oil-in-gel emulsions. Food Chem 220:208–218
120. Sporn MB, Roberts AB, Goodman DS (eds) (1994) The Retinoids. Raven Press, New York
121. Stalmach A, Williamson G, Crozier A (2014) Impact of dose on the bioavailability of coffee chlorogenic acids in humans. Food Funct 5(8):1727–1737
122. Story JA, Kritchevsky D (1976) Comparison of binding of various bile-acids and bile-salts invitro by several types of fiber. J Nutr 106(9):1292–1294
123. Strobbe S, De Lepeleire J, Van Der Straeten D (2018) From in planta function to vitamin-rich food crops: the ACE of biofortification. Front Plant Sci 9(1862). https://doi.org/10.3389/fpls.2018.01862
124. Tanumihardjo SA (2011) Vitamin A: biomarkers of nutrition for development. Am J Clin Nutr 94(2):658S–665S
125. Tarko T, Duda-Chodak A, Zajac N (2013) Digestion and absorption of phenolic compounds assessed by in vitro simulation methods. A review. Rocz Panstw Zakl Hig 64(2):79–84
126. Turley SD, Dietschy JM (2003) Sterol absorption by the small intestine. Curr Opin Lipidol 14(3):233–240
127. Turner DC, Yin FC, Kindt JT, Zhang HL (2010) Molecular dynamics simulations of glycocholate-oleic acid mixed micelle assembly. Langmuir 26(7):4687–4692
128. Tyssandier V, Lyan B, Borel P (2001) Main factors governing the transfer of carotenoids from emulsion lipid droplets to micelles. Biochim Biophys Acta 1533(3):285–292
129. Tzen J, Huang A (1992) Surface structure and properties of plant seed oil bodies. J Cell Biol 117(2):327–335

130. USDA (2017) Rank order of vitamin A content in foods per 100 g. USDA National Nutrient Database, March 2017
131. Vahouny GV, Tombes R, Cassidy MM, Kritchevsky D, Gallo LL (1980) Dietary-fibers.5. Binding of bile-salts, phospholipids and cholesterol from mixed micelles by bile-acid sequestrants and dietary-fibers. Lipids 15(12):1012–1018
132. van Bennekum AM, Wei S, Gamble MV, Vogel S, Piantedosi R, Gottesman M, Blaner WS (2001) Biochemical basis for depressed serum retinol levels in transthyretin-deficient mice. J Biol Chem 276(2):1107–1113
133. Verrijssen TAJ, Balduyck LG, Christiaens S, Van Loey AM, Van Buggenhout S, Hendrickx ME (2014) The effect of pectin concentration and degree of methyl-esterification on the in vitro bioaccessibility of beta-carotene-enriched emulsions. Food Res Int 57:71–78
134. Verrijssen TAJ, Verkempinck SHE, Christiaens S, Van Loey AM, Hendrickx ME (2015) The effect of pectin on in vitro beta-carotene bioaccessibility and lipid digestion in low fat emulsions. Food Hydrocoll 49:73–81
135. Verrijssen TAJ, Christiaens S, Verkempinck SHE, Boeve J, Grauwet T, Van Loey AM, Salvia-Trujillo L, Hendrickx ME (2016) In vitro β-carotene bioaccessibility and lipid digestion in emulsions: influence of pectin type and degree of methyl-esterification. J Food Sci 81(10):C2327–C2336
136. Weber C, Bysted A, Hølmer G (1997) Intestinal absorption of coenzyme Q10 administered in a meal or as capsules to healthy subjects. Nutr Res 17(6):941–945
137. Wei JH, Yin X, Welander PV (2016) Sterol synthesis in diverse bacteria. Front Microbiol 7(990). https://doi.org/10.3389/fmicb.2016.00990
138. West CE (1996) Strategies to control nutritional anemia. Am J Clin Nutr 64(5):789–790
139. White WS, Zhou Y, Crane A, Dixon P, Quadt F, Flendrig LM (2017) Modeling the dose effects of soybean oil in salad dressing on carotenoid and fat-soluble vitamin bioavailability in salad vegetables. Am J Clin Nutr 106(4):1041–1051
140. WHO (2009) Global prevalence of vitamin A deficiency in populations at risk 1995–2005. WHO Global Database on Vitamin A Deficiency. Newsletter 4(2):5–6, 8
141. Wiedmann TS, Kamel L (2002) Examination of the solubilization of drugs by bile salt micelles. J Pharm Sci 91(8):1743–1764
142. Williamson G, Day A, Plumb G, Couteau D (2000) Human metabolic pathways of dietary flavonoids and cinnamates. Biochem Soc Trans 28(2):16–21
143. Wolf G (2001) The discovery of the visual function of vitamin a. J Nutr 131(6):1647–1650
144. Wright AJ, Pietrangelo C, MacNaughton A (2008) Influence of simulated upper intestinal parameters on the efficiency of beta carotene micellarisation using an in vitro model of digestion. Food Chem 107(3):1253–1260
145. Xavier AAO, Mercadante AZ, Garrido-Fernandez J, Perez-Galvez A (2014) Fat content affects bioaccessibility and efficiency of enzymatic hydrolysis of lutein esters added to milk and yogurt. Food Res Int 65:171–176
146. Xu DX, Yuan F, Gao YX, Panya A, McClements DJ, Decker EA (2014) Influence of whey protein-beet pectin conjugate on the properties and digestibility of beta-carotene emulsion during in vitro digestion. Food Chem 156:374–379
147. Xu F, Pandya JK, Chung C, McClements DJ, Kinchla AJ (2018) Emulsions as delivery systems for gamma and delta tocotrienols: formation, properties and simulated gastrointestinal fate. Food Res Int 105:570–579
148. Ye A, Cui J, Singh H (2011) Proteolysis of milk fat globule membrane proteins during in vitro gastric digestion of milk. J Dairy Sci 94(6):2762–2770
149. Yuan X, Liu XJ, McClements DJ, Cao Y, Xiao H (2018) Enhancement of phytochemical bioaccessibility from plant-based foods using excipient emulsions: impact of lipid type on carotenoid solubilization from spinach. Food Funct 9(8):4352–4365
150. Zhang ZP, Zhang RJ, Zou LQ, Chen L, Ahmed Y, Al Bishri W, McClements DJ (2016) Encapsulation of curcumin in polysaccharide-based hydrogel beads: impact of bead type on lipid digestion and curcumin bioaccessibility. Food Hydrocoll 58:160–170

151. Zheng JK, Li Y, Song MY, Fang X, Cao Y, McClements DJ, Xiao H (2014) Improving intracellular uptake of 5-demethyltangeretin by food grade nanoemulsions. Food Res Int 62:98–103
152. Zhou XH, Wang H, Wang CN, Zhao C, Peng Q, Zhang TH, Zhao CH (2018) Stability and in vitro digestibility of beta-carotene in nanoemulsions fabricated with different carrier oils. Food Sci Nutr 6(8):2537–2544
153. Zielbauer BI, Jackson AJ, Maurer S, Waschatko G, Ghebremedhin M, Rogers SE, Vilgis TA (2018) Soybean oleosomes studied by small angle neutron scattering (SANS). J Colloid Interface Sci 529:197–204
154. Zou LQ, Zheng BJ, Zhang RJ, Zhang ZP, Liu W, Liu CM, McClements DJ (2016) Influence of lipid phase composition of excipient emulsions on curcumin solubility, stability, and bioaccessibility. Food Biophys 11(3):213–225

Chapter 11
Sterol Digestion in Plant Sterol-Enriched Foods: Bioaccessibility and Fermentation

Virginia Blanco-Morales, Guadalupe Garcia-Llatas, and Antonio Cilla

11.1 Introduction

Sterols comprise the major portion of the unsaponifiable fraction of most edible fats and oils being minor components of human dietary lipids. In general, sterols derive from squalene consisting of a tetracyclic cyclopenta[α]-phenanthrene structure with a flexible side chain with 8–10 carbons at C-17 and a hydroxyl group at C-3. They can be classified according to their origin as animal sterols or plant sterols (PS). Cholesterol is the predominant sterol in foods from animal origin followed by desmosterol, whereas plant membranes contain several types of PS. More than 200 different types of PS have been reported in plant species, where the most common are β-sitosterol, campesterol and stigmasterol, called phytosterols, that are structurally similar to cholesterol, but include a methyl or ethyl group at C-24. Structures of cholesterol and common phytosterols are shown in Fig. 11.1. In turn, saturated PS, referred as phytostanols, have no double bond in the ring structure [27, 48].

The addition of PS to food matrices is an increasingly common practice in the food industry, with the aim to prevent cardiovascular disease through their cholesterol-lowering activity at dosages of 1.5–3 g PS/day [27]. Moreover, a great deal of scientific evidence supports their additional beneficial effects on human health and, consequently, their use for the formulation of functional foods is increasing. On this basis, it is important to consider that the beneficial effects attributed to these bioactive compounds are not only related to the intake levels but also to their bioavailability. In this sense, *in vitro* methods to simulate gastrointestinal digestion allow determining sterol (cholesterol and PS) bioaccessibility and, thus recognize factors that can affect their solubility with the aim to optimize their beneficial effects improving the design of functional foods [14]. Moreover, *in vitro* gastrointestinal

V. Blanco-Morales · G. Garcia-Llatas (✉) · A. Cilla
Nutrition and Food Science Area, Faculty of Pharmacy, University of Valencia, Valencia, Spain
e-mail: virginia.blanco@uv.es; guadalupe.garcia@uv.es; antonio.cilla@uv.es

M. M.-L. Grundy, P. J. Wilde (eds.), *Bioaccessibility and Digestibility of Lipids from Food*, https://doi.org/10.1007/978-3-030-56909-9_11

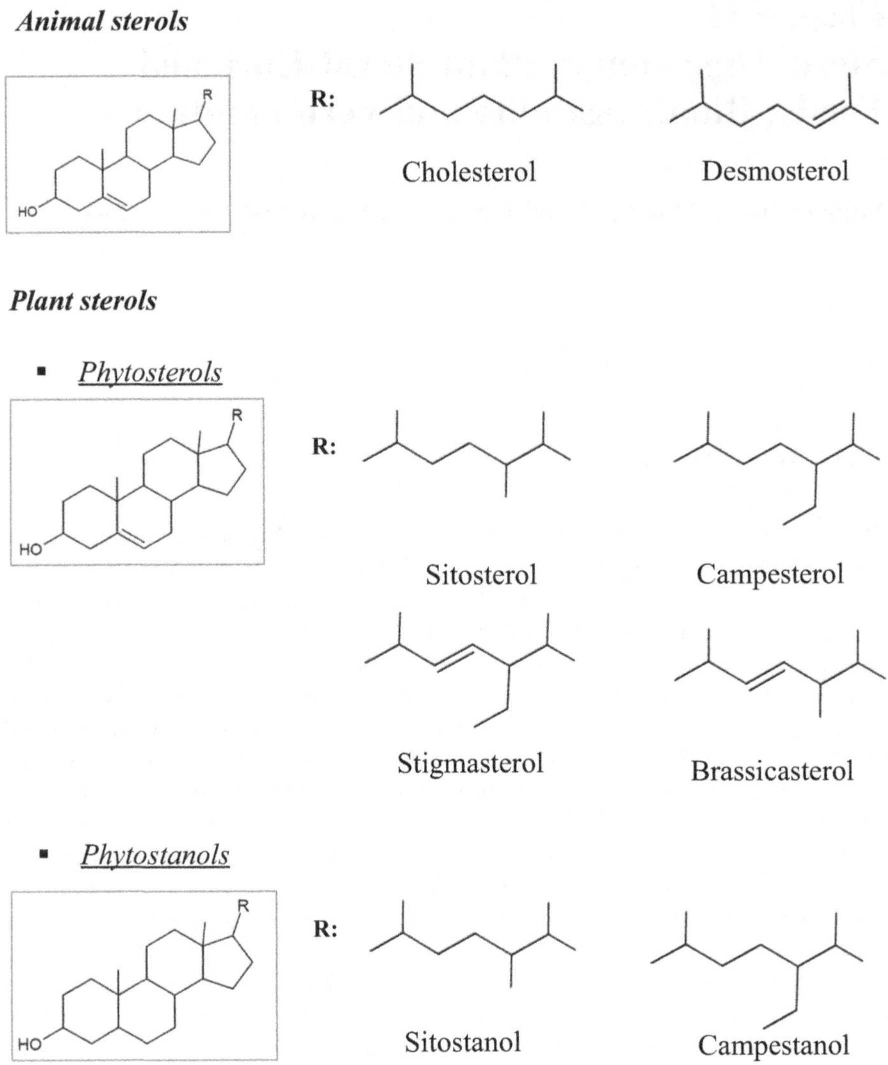

Fig. 11.1 Structure of main animal and plant sterols

digestion methods are useful tools to evaluate the solubility of sterol oxidation products (oxysterols), potentially harmful substances that can be formed during the manufacture or storage process of PS-enriched products, in order to elucidate factors that are implicated in the formation of oxysterols and minimize them [8]. In addition, sterols are also susceptible to the fermentation by gut microbiota, that lead to the formation of metabolites and, in turn, to modifications of the microbiota population. However, studies that evaluate sterols fermentation from PS-enriched products are still scarce [17].

The following chapter reviews the information available about sterol and oxysterol bioaccessibility in PS-enriched foods and the factors that could modify their solubility (contents and source of PS, food matrix composition, or digestion methodology carried out). Furthermore, current information on sterol fermentation has also been addressed, describing sterol metabolites, the microbiota species implicated in their formation and the changes in microbiota composition that can be generated by high PS concentrations (derived from PS-enriched formats).

11.2 General Concepts: Bioaccessibility and Bioavailability

In the field of nutrition research, the ambiguous application of the concepts bioavailability and bioaccessibility is common. Although used indistinctly and subject to alternative definitions, they are not synonymous terms. Thus, it is important to ascertain and be explicit about which definition is being considered [53].

On the one hand, bioavailability is defined as the fraction of an ingested component available for use in normal physiological functions and is determined by *in vivo* assays [34]. Bioavailability encompasses three main steps: (i) digestibility and solubility of the element in the gastrointestinal tract, (ii) absorption of the element by the intestinal cells and transport into the circulation, and (iii) incorporation from the circulation to the functional entity [53, 63]. In addition, in the literature, there are numerous examples of researchers who only measure the uptake or absorption with cell cultures (i.e. Caco-2 cells) in *in vitro* assays, yet refer to their analysis as bioavailability [24].

On the other hand, bioaccessibility has been given two alternative definitions, as reviewed by Cardoso et al. [13]: (a) the fraction of a compound that is released from its food matrix in the gastrointestinal tract and thus becomes available for intestinal absorption (typically based on *in vitro* procedures). Hence, this definition does not include absorption across the intestinal wall or any metabolic processing and depends only on digestion and release from the food matrix [24]; or a more stringent and much less widely used definition: (b) the fraction of a compound that is released from its food matrix in the gastrointestinal tract and thus becomes available for intestinal absorption (i.e. it enters the bloodstream), including the entire sequence of events that takes place during the digestive transformation of food into material that can be assimilated by the body, absorption/assimilation into the cells of the intestinal epithelium, and, finally, pre-systemic metabolism (both intestinal and hepatic) [26]. Thus, from this last standpoint, bioavailability includes bioaccessibility.

Bioaccessibility, as a first step and prerequisite of bioavailability, is an important concept when nutritional guidance is planned or for tailor-made food products to meet explicit requirements (i.e. high bioaccessibility and bioavailability of plant sterols and stanols to decrease plasmatic total cholesterol and low-density lipoprotein cholesterol levels in mildly hypercholesterolemic subjects). In this sense, people wishing to decrease their energy intake would be aligned with foods with a reduced macronutrient digestion and absorption. On the contrary, for individuals who suffer from malnutrition or present higher energy requirements, nutrient-rich foods are recommended for their high bioaccessibility, subsequent bioavailability and potential biological effects. For this reason, fully understanding how the food matrix acts within the gastrointestinal tract during digestion and how this affects nutrient bioaccessibility is critical [33]. The determination of the bioaccessibility of sterols enables the comparison of individual sterols in the same and different samples and the identification of which sterols are more available in the intestinal tract for later absorption [35].

Based on these considerations, authors would like to make clear that for this book chapter review, the term bioaccessibility will only be used for assays accounting for *in vitro* digestion (excluding absorption and pre-systemic metabolism), and bioavailability considering both, the *in vitro* (uptake and transport in cell cultures) and *in vivo* (animal and human) perspectives. Therefore, it is convenient to stick to the terminology mainly used by many of the authors referenced.

11.3 Fate of Sterols and Oxysterols During Digestion

Dietary sterols from animal or plant origin, and oxidized or not, are minor components present in fat of the human diet which belong to the unsaponifiable fraction [56]. They are assumed to follow the fate of lipids in the gastrointestinal tract, consisting of several sequential steps, including physicochemical and enzymatic events, such as: their release by breakdown from the food matrix, then their dispersion/emulsification into lipid droplets, followed by their enzymatic hydrolysis by specific lipases at the emulsion-water interface, ensued by their solubilization into intestinal mixed micelles formed by bile salts and phospholipids, then their diffusion across the unstirred water layer, and finally their permeation/absorption by passive diffusion or active transport involving specific proteins in the enterocyte brush border membrane [1, 11, 44, 57]. In the particular research field of lipophilic molecules, the bioaccessibility endpoint is considered after their incorporation into an absorbable form (intestinal mixed micelles), which delivers digested lipids to the aqueous-enterocyte membrane for subsequent absorption [44]. Bioavailability, in turn, can be ascribed to their absorption into the enterocytes and/or their location in the vascular channels of lymph or blood [53].

Regarding the specific fate of sterols in the gastrointestinal tract, first they are released from the food matrix, a fact that mainly depends on two key issues such as type/form of sterol (which affect their hydrophobicity and extraction efficiency) and type of food matrix [31]. Next, in the intestinal lumen, esterified sterols locate into the core of emulsified fat droplets whereas free sterols partition between fat droplets and micelles. Then, in the duodenum, sterol esters are mainly hydrolyzed to free sterols by pancreatic cholesterol esterase, although the implication of pancreatic lipase could not be ruled out [50], and subsequently transferred to mixed micelles (transfer efficiency dependent on sterols form and degree of enzymatic hydrolysis), which provide a vehicle for absorption into enterocytes [31]. The intestinal absorption of sterols is mediated by the Niemann-Pick C1-like 1 (NPC1L1) protein from the brush border membrane into the intestinal mucosa [65] although a slight portion is absorbed by passive diffusion [59]. This enzyme would exhibit a different response to sterols and stanols of different shape and size (i.e. cholesterol uptake being 4 times higher than that of sitosterol since cholesterol has a 7% lower molecular weight) [64]. Once within the enterocyte, sterols are re-esterified by Acyl CoA cholesterol Acyl Transferase (ACAT) before their incorporation in chylomicrons for transport through the lymph due to the action of the microsomal triglyceride transfer protein (MTP). Nevertheless, due to the low affinity of ACAT for plant sterols, only a small part of them is esterified, which explains their low absorption (<5%) compared to cholesterol (40%). In addition, two ATP-binding cassette proteins (ABCG5/ABCG8) have been shown to reduce absorption of plant sterols by pumping them back into the intestinal lumen [36, 56]. The main route for sterols excretion is the bile pathway; however, a small fraction of plant sterols can be excreted through the skin [59], and more recently, it has been described that cholesterol can also be excreted from the bloodstream to the intestinal lumen via a transintestinal efflux pathway (TICE) [20]. Subsequently, all sterols effluxed from enterocytes to the lumen reach the large intestine, where they are subjected to microbial transformation with the successive kinetic reaction steps: sterol → stenone → stanone → stanol. Then, this "chemically-inert" stanol product would ultimately be excreted in feces [17, 64]. This topic is be expanded below (see Sect. 11.6).

All together the complex gastrointestinal process of lipids, and specifically of dietary sterols, is influenced by an array of factors which can promote or hamper their bioaccessibility and bioavailability as previously reported [11, 31, 64]. Such factors mainly include: (i) molecular size and type of the sterols (i.e. sterol/stanol, oxidized or not), (ii) the food matrix in which the sterol is present or incorporated (i.e. presence and nature of dietary lipids, dietary fiber), (iii) competitive interaction with other compounds (i.e phytostanols apparently inhibit phytosterols absorption), and (iv) genetic factors (i.e. phytosterolemia, where mutations in ABCG5/ABCG8 proteins cause an increase in the net absorption of phytosterols by decreasing their excretion back to the intestinal lumen). Some of these factors are related in the following sections.

11.4 Bioaccessibility of Sterols

The effectiveness of PS-enriched foods is closely related to their bioavailability, which depends on the food matrix and the composition of the ingredients used as PS source. *In vitro* gastrointestinal digestion methods can be used as predictive models to evaluate PS bioaccessibility in order to improve the design of functional foods and to evaluate the factors that enhance the incorporation of PS into mixed micelles, and thus leading to a greater displacement of cholesterol from them [2, 28].

Although they are scarce, different studies have evaluated sterol bioaccessibility in PS-enriched products such as commercially available fermented milk beverages [61], milk beverages [3], fruit beverages with or without milk [3, 4, 9], dark chocolates [58], granola bars and puddings [60], and fermented maize yogurt-like products [30]. The studies on beverages were carried out by our research group and the results of PS bioaccessibility obtained are summarized in Table 11.1. In that

Table 11.1 Sterol bioaccessibility (%) in plant sterol-enriched beverages

Fermented milk beverages[a]					**Skimmed milk-based beverage**[b]	**Fruit beverage***[b]
PS content	**2.9**	**2.0**	**1.6**	**1.5**	**0.8**	**0.8**
Campesterol	–	18.94 ± 1.48	19.46 ± 0.38	9.01 ± 0.44	3.75 ± 0.15	4.88 ± 0.16
Campestanol	12.54 ± 0.59	–	–	–	3.47 ± 0.35	4.27 ± 0.35
Stigmasterol	–	–	–	8.74 ± 0.52	2.95 ± 0.44	3.04 ± 0.17
β-Sitosterol	6.31 ± 0.30	17.17 ± 1.49	16.88 ± 0.64	8.97 ± 0.48	2.51 ± 0.07	2.93 ± 0.10
Sitostanol	11.35 ± 0.96	13.57 ± 1.24	13.42 ± 1.18	–	2.71 ± 0.12	3.22 ± 0.16
Cholesterol	–	–	–	–	79.05 ± 1.09	–
Total PS	11.28 ± 0.85	16.84 ± 1.35	16.64 ± 0.64	8.90 ± 0.40	2.62 ± 0.08	3.10 ± 0.11

PS-enriched milk-based fruit beverages								
PS content	**0.8**[b]	**0.8***[b]	**0.8***[c]	**+ Olive oil and soy lecithin*** **0.8**[c]	**+ Milk fat and whey protein concentrate enriched with MFGM*** **0.8**[c]	**1**[d]	**1****[d]	**1*****[d]
Campesterol	8.31 ± 0.65	6.66 ± 0.27	9.03 ± 0.59	26.37 ± 3.94	33.86 ± 3.83	39.55 ± 1.36	39.75 ± 0.70	38.90 ± 0.63
Campestanol	6.82 ± 0.57	5.79 ± 0.55	27.75 ± 2.76	31.41 ± 4.74	36.92 ± 3.84	43.68 ± 0.78	43.16 ± 0.83	41.53 ± 0.97
Stigmasterol	5.96 ± 0.60	4.02 ± 0.11	5.72 ± 0.60	19.37 ± 0.98	32.64 ± 3.28	33.55 ± 1.55	33.52 ± 1.40	33.96 ± 0.66
β-Sitosterol	6.42 ± 0.50	4.15 ± 0.04	8.45 ± 0.81	28.14 ± 3.53	30.72 ± 2.61	36.68 ± 1.60	37.19 ± 0.65	36.80 ± 0.62
Sitostanol	6.05 ± 0.71	4.41 ± 0.10	9.79 ± 1.02	29.94 ± 3.45	34.37 ± 2.90	39.28 ± 1.37	39.74 ± 0.64	38.22 ± 0.68
Cholesterol	53.34 ± 4.85	98.96 ± 3.3	57.60 ± 3.37	53.50 ± 4.23	64.18 ± 6.64	80.61 ± 3.44	78.45 ± 0.39	84.87 ± 0.88
Total PS	6.48 ± 0.52	4.35 ± 0.06	8.73 ± 0.81	28.19 ± 3.58	31.39 ± 2.70	37.20 ± 1.54	37.67 ± 0.65	37.12 ± 0.63

Bioaccessibility (%) calculated as (sterol content in bioaccessible fraction × 100)/(sterol content in beverage). Content of PS expressed per 100 mL of beverage except for [61] (g/100 g)

GOS galactooligosaccharides, *MFGM* milk fat globule membrane, *PS* plant sterols

[a]Vaghini et al. [61]

[b]Alemany et al. [3]

[c]Alvarez-Sala et al. [4]

[d]Blanco-Morales et al. [9]

*With mandarin juice from concentrate as source of β-cryptoxanthin; **With 1 g GOS/100 mL; ***With 2.5 g GOS/100 mL research, the same methodology for *in vitro* gastrointestinal digestion was used and, thus, allows the comparison of results, making the evaluation of food matrix effects possible. This gastrointestinal digestion, named micellar digestion, includes three-steps (salivary, gastric and intestinal digestion), and was specifically developed for lipophilic compounds by the addition of key lipid metabolism enzymes during the intestinal phase (pancreatic lipase, colipase, cholesterol esterase and phospholipase A2) and sodium taurocholate to improve generation of a micellar medium in which sterols are incorporated [32].

In general, a larger and higher bioaccessibility range for total PS is observed in milk-based beverages with higher contents of PS (8.9–16.8% containing 1.5–2.9 g PS/100 g fermented milk) [61] versus lower contents (2.6% with 0.8 g PS/100 mL milk-based beverage) [3] (Table 11.1). In addition, to the different level of PS enrichment, it should be pointed out that the alteration of milk fat globule membranes (MFGM) by acid or microbial action in fermented products may result in different adsorption of sterols compared to native membranes, thus becoming an effective delivery system for PS [7]. However, other components of the food matrix may reduce PS solubility, since lower bioaccessibility is observed in presence of higher carbohydrates and fibre contents (8.9 versus 16.6%) in beverages with similar PS enrichment (1.5–1.6 g/100 g) [61].

Other factors such as the source of the PS ingredient can also influence bioaccessibility. Determination of PS contents in beverages shows that samples with 1.6 and 2.0 g PS/100 g beverage report the same or similar source of PS (β-sitosterol (80%) > sitostanol (12%) > campesterol (5–7%)). After the application of the simulated digestion, similar solubility of individual sterols (campesterol > β-sitosterol > sitostanol) and of total PS was obtained (~17%). In turn, beverages with a similar or higher enrichment in PS (1.5 or 2.9 g PS/100 g) but with a different profile of the PS ingredient (β-sitosterol (45%) > stigmasterol (33%) > campesterol (22%), or sitostanol (85%) > campestanol (11%) > β-sitosterol (4%)) show changes in individual PS solubility (campesterol > β-sitosterol > stigmasterol, or campestanol > sitostanol > β-sitosterol) and a lower total PS bioaccessibility (9 and 11%, respectively) [61] (see Table 11.1). In fact, when comparing the fruit beverage (containing banana puree, grape juice from concentrate and mandarin fruit juice from concentrate rich in β-cryptoxanthin) with the skimmed milk-based beverage, both enriched with the same content and source of PS (0.8 g/100 mL from tall oil), similar bioaccessibility of total (approximately 3%) and individual PS (campesterol > campestanol > β-sitosterol ≈ sitostanol ≈ stigmasterol) was observed [3].

Currently, incorporating PS into low-fat foods such as milk-based fruit beverages, which can be enriched with PS [15], is a convenient way to obtain the daily recommended amount of PS in subjects with moderate hypercholesterolemia given the presence of antioxidant compounds, such as vitamin C, polyphenols, and carotenoids. In this context, our research group has evaluated PS bioaccessibility in milk-based fruit beverages with different matrix modifications [3, 4, 9] using the same PS-ingredient from tall oil (Table 11.1). As it can be observed, beverages containing milk and fruit juice showed an increase in total PS bioaccessibility (4.4–8.7%) with respect to those containing only fruit juice (2.6%) at the same level of PS-enrichment (0.8 g/100 mL). The presence of milk fat, that has been demonstrated to be an effective delivery system for lipophilic microconstituents [7], could favor the extraction and stability of PS in the food matrix, improving their dispersion in the aqueous phase of the digestion. Moreover, the incorporation of mandarin fruit juice source of β-cryptoxanthin to milk-based fruit beverages slightly reduces total PS bioaccessibility (6.5 versus 4.4%), probably due to a competition between PS and β-cryptoxanthin for its incorporation into mixed micelles [3]. Regarding solubility of individual PS in this kind of beverages (with or without mandarin juice), a larger value for campesterol/campestanol with respect to β-sitosterol/sitostanol is observed. Authors suggest that the shorter side-chain group at position 24 of campesterol/campestanol (methyl) (versus sitosterol/sitostanol (ethyl)) and the lower affinity of the first for the bile salt micelles favours their aqueous solubility. It is remarkable that in beverages containing mandarin juice, Alvarez-Sala et al. [4] reported a twofold higher total PS bioaccessibility than the one observed by Alemany et al. [3]. The larger contribution of campesterol to the bioaccessibility (27.8% versus 5.8%) explains the highest total PS bioaccessibility (Table 11.1).

One step forward in our research line was the addition of fat and emulsifiers to the PS-enriched milk-based fruit beverages (containing mandarin juice from concentrate), such as olive oil and soy lecithin [4] or milk fat and whey protein concentrate enriched with MFGM [4, 9] in order to assess the influence on PS bioaccessibility (see Table 11.1). At doses of PS enrichment of 0.8 g/100 mL, the increase in fat content (2.4% versus 1.1%) from vegetable or animal source, leads to a 6.5- or 7.2-fold increase in PS bioaccessibility, respectively [4]. Thus, the addition of fat is likely to promote the formation of mixed micelles during gastrointestinal digestion, increasing thereby PS solubility, with no influence of the origin of the lipid source. However, the bioaccessibility of individual PS is influenced by the fat origin, since β-sitosterol and sitostanol showed higher solubility than campesterol in presence of olive oil and lecithin. At doses of 1.0 g PS/100 mL [9], the increment in PS bioaccessibility observed with the presence of milk fat and MFGM in the beverages with respect to the non-fat enriched milk-based fruit beverages is even higher (8.6-fold), showing them to be suitable ingredients for improving the bioaccessibility of PS.

Regarding cholesterol, it shows a high solubility in the mixed micellar aqueous phase, and its bioaccessibility is, therefore, significantly greater than that of PS in all of the above-mentioned PS-enriched products (between 1.9- and 30.2-fold) (Table 11.1). Nevertheless, the higher hydrophobicity of PS with saturated lateral chains (campesterol, campestanol, β-sitosterol, sitostanol) with respect to cholesterol, implies a better transference to the mixed micelles, thus decreasing cholesterol incorporation when considering PS-enriched foods [47]. Additionally, cholesterol bioaccessibility seems to decrease in the presence of mandarin juice, whereas PS increase probably due to molecular competition with cholesterol incorporation into the micelles [14].

Finally, the impact of the addition of galactooligosaccharides (GOS), at different concentrations (1 and 2.5 g/100 mL), to PS-enriched beverages with milk fat and MFGM on sterol bioaccessibility has been evaluated [9]. The addition of GOS does not modify individual or total PS bioaccessibility (approximately 37%), although a slight increase on cholesterol bioaccessibility was observed in the beverage with higher GOS content (from 81% to 85%). In the same study, for the first time, the static harmonized *in vitro* gastrointestinal digestion methodology of the COST Action INFOGEST [46] was applied in order to evaluate sterol bioaccessibility. In a first trial, the Infogest digestion method carried out with porcine bile salts at a concentration of 10 mM for the intestinal digestion phase, did not allow cholesterol quantification. Probably, the extra contribution of cholesterol from the bile salts led to saturation of micelles, and therefore reduced cholesterol solubilization. In contrast, PS bioaccessibility could be determined, with percentages for total PS 2.6-fold lower than the one observed for the same beverage with micellar digestion (14 versus 37z%). Also, the profile of PS solubility was altered with respect to those obtained with our micellar digestion, with sitostanol (16%) and β-sitosterol (14%) showing the greatest bioaccessibility, followed by campesterol, campestanol and stigmasterol (13%). These results, thus, demonstrate that PS bioaccessibility is also influenced by the conditions of the applied gastrointestinal digestion method. Specifically, the higher content of bile salts proposed by Infogest with respect to micellar digestion (10 mM versus 1.4 mM), could inhibit key enzymes of lipid metabolism such as pancreatic lipase, thus diminished PS solubility. Moreover, the different origin of the bile salts (porcine versus bovine), could influence sterol micellar solubility due to their different polarity.

Therefore, in order to allow cholesterol bioaccessibility determination, we assayed two modifications of the Infogest digestion conditions [9]: the reduction of bile salts concentration from 10 mM to 1.4 mM and the change of bile salts origin from porcine to bovine, but maintaining the 10 mM concentration). Both modifications made cholesterol (62 and 38%) and PS bioaccessibility determination (22 and 14%) possible, being lower with respect to the same beverage following a/the micellar digestion protocol (79–80% for cholesterol and approximately 37% for PS). This is probably due to the presence of key enzymes in the micellar digestion protocol that improve lipophilic solubilization such as cholesterol esterase (which mediates cholesterol re-esterification promoting their incorporation in mixed micelles), phospholipase A2 (which favours the release of free fatty acids and lysophospholipid with surfactant properties and could improve cholesterol absorption efficiency) and lipase and co-lipase (that could enhance droplets lipid digestion, increasing sterol micellarization). The modifications proposed for the Infogest method also affected the PS individual solubility profile. In the case of the reduction of porcine bile salts concentration to 1.4 mM, the order of solubility was campestanol (25%) > campesterol (25%) > sitostanol (24%) > β-sitosterol (22%) > stigmasterol (18%), similar to that observed in micellar digestion using bovine bile salts at the same concentration. By contrast, when the bile salts origin was changed, the use of bovine bile salts resulted in the following order of solubility: sitostanol (16%) > campesterol (15%) > β-sitosterol (14%) > campestanol (13%) > stigmasterol (10%), similar to that observed in Infogest digestion using porcine bile salts at the same concentration (10 mM). This indicates that sterol solubility is closely related to the bile salts content beyond of the gastrointestinal *in vitro* digestion method or the bile salt origin used.

Recently, the Infogest digestion method has been applied in order to evaluate PS bioaccessibility in other PS-enriched food matrices such as dark chocolate [58] and granola bars and pudding [60]. PS bioaccessibility in dark chocolate with different cocoa (64, 72, and 85%) and microencapsulated PS contents (5, 10, and 15%) ranged from 6.5% to 8.4%. This bioaccessibility is in the range of those observed in commercially available fermented milk beverages [61] although much lower than those obtained in optimized matrix composed by the milk-based fruit beverage with milk fat and MFGM [4, 9]. On the other hand, the bioaccessibility of PS from granola bars enriched with low-crystallinity phytosterol nanoparticles impregnated in nanoporous starch aerogels (PS-NSA) or crude PS (1 g PS per 2 bars) was evaluated in different formulations, including non- (0%), low- (7%), and regular-fat content (24%) [60]. PS-NSA improved PS bioaccessibility compared to crude PS by threefold in all granola bar formulations. Low- and regular-fat granola bars showed similar PS bioaccessibility compared to crude PS (~30%) or PS-NSA (~90%). However, when non-fat granola bars were digested, bioaccessibility decreased to 16% for crude PS and to 52.7% for PS-NSA. Similar to the commented above for PS-enriched milk-based fruit beverages [4], the presence of fat in granola bars seems to improve PS bioaccessibility, although no differences were observed in PS bioaccessibility between low- and regular-fat granola bars. Authors suggest that there is a threshold lipid content required to maximize the solubilization capacity of the mixed micelles. Additionally, in the same study, the effect of the addition of crude PS versus PS-NSA on bioaccessibility in a non-fat pudding formulation was evaluated. Similar to the granola bars, the inclusion of PS in nanoparticles improved PS solubility (19% versus 2%). However, the bioaccessibility observed in pudding was lower than in granola bars, probably due to the more complex ingredients (oats, almonds, coconut, egg whites and gum arabic), including a larger content of proteins and emulsifiers that might improve the solubilization of PS in the aqueous phase during the *in vitro* digestion.

Finally, fermented maize yogurt-like products containing dispersible PS (1.7 g/100 g) and carotenoids (1.6 g/100 g) showed the lowest β-sitosterol bioaccessibility (1.6%) observed in all the aforementioned PS-enriched products [30]. The *in vitro* gastrointestinal digestion carried out, based on Reboul et al. [54] and Dhuique-Mayer et al. [21] differs from the micellar and Infogest ones since it is composed of only two digestion steps (gastric and intestinal phases). It was validated against human studies and considered to be a reliable model for carotenoid behaviour during *in vitro* digestion [24]. It included pepsin in the gastric digestion phase, and porcine pancreatin, porcine bile extract and cholesterol esterase in the intestinal step. Authors suggest that, similar to what occurred in milk-based fruit beverages containing β-cryptoxanthin mentioned above [3], the presence of carotenoids limits the incorporation of β-sitosterol into mixed micelles. With respect to cholesterol, the incorporation of PS resulted in a reduction in its solubility of 50%.

11.5 Bioaccessibility of Oxysterols from Foods

Sterols are susceptible to oxidation generating different compounds know as oxysterols, which comprise cholesterol and phytosterol oxidation products (COPs and POPs, respectively). Although information available about POPs formation is scarce compared to COPs, autoxidation is considered the main mechanism of sterol oxidation in foods [42, 51]. The process of autoxidation comprises an initial abstraction of the C-7 hydrogen from the sterol steroid ring, followed by a several reactions with reactive oxygen species which finally leads to the formation of 3β-hydroxy-5-en-7-peroxyls. After the abstraction of a hydrogen, more stable 7-hydroperoxydes molecules are formed. During heating and/or storage conditions, 7-hydroperoxydes can decompose into 7α/7β-hydroxysterols or 7-ketosterols by reduction or oxidation reactions, respectively. Other secondary products from sterol oxidation are 5α,6α/5β,6β-epoxysterols, formed from the interaction between non-sterol oxidized molecules and a hydroperoxy radical. Hydration of epoxysterols in an acidic environment leads to triol derivate formation [12, 27]. All these reactions that make up the autoxidation of sterols are summarized in Fig. 11.2. Autoxidation can be initiated, respectively, influenced by several factors such as heat, light exposure, oxygen or presence of lipids and water, which means that food composition, industrial

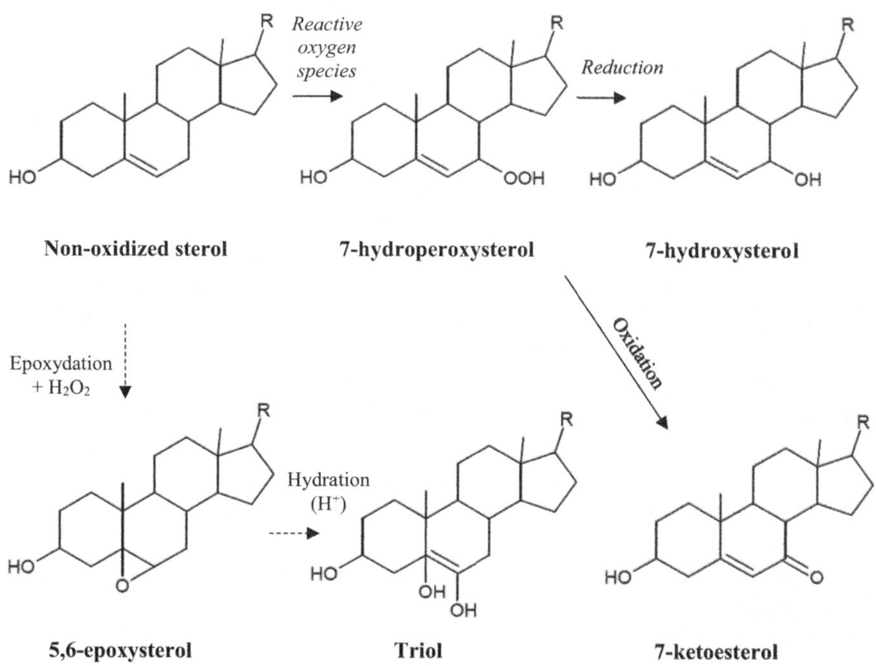

Fig. 11.2 Formation of oxysterols by autoxidation

processing, storage conditions and culinary procedures are factors to be taken into account in oxysterol formation in foods [8].

Very few studies addressing the effect of gastrointestinal digestion on oxysterols have been performed on foods. Hur et al. [37] assessed the influence of different cooking methods (oven, pan-frying, boiling, microwave heating) on COP formation in pork patties during *in vitro* digestion. Although total COP contents in the cooked samples before digestion were not significantly different, the simulated gastrointestinal digestion exerted a significant increase in microwave-cooked patties, together with a higher lipid oxidation (measured by TBARS) compared to the other cooking methods. Moreover, the COP formation during an *in vitro* digestion of biopolymer (chitosan, pectin, onion powder, or green tea powder) encapsulated-beef patties cooked in microwave showed that the encapsulation produced less cholesterol oxidation after the intestinal step of the digestion versus control (non-encapsulated) [38]. In both studies, total COP contents included 7α- and 7β-hydroxy, 7-keto-, triol, 20α-hydroxy, and 25-hydroxycholesterol, although individual concentrations were not detailed, not even initial (raw) amounts in the samples. However, results indicated that cholesterol from meat samples can be oxidized when exposed to digestive enzymes and bile salts, increasing COP production during digestion, which is promoted by cooking methods such as microwave heating and can be inhibited by biopolymer encapsulation of the food products.

The bioaccessibility of COPs and POPs from PS-enriched milk and/or fruit beverages evidenced a lower bioaccessibility of COPs relative to cholesterol (2.2–17% versus 53–99%), whereas those for POPs were much higher than those of their corresponding phytosterols (19.1–49.3% versus 2.2–6.5%) [3], in concordance with the absorption trend in animals. Also, a matrix effect regarding bioaccessibility of oxysterols was detected, since bioaccessibility of POPs was significantly lower in the PS-enriched milk beverage, and that of COPs in fruit and milk-based beverages. In all samples, 7β-hydroxycholesterol (6.8–20.7%) for COPs and sitostanetriol (62.7–176.6%) for POPs were the most bioaccessible oxides, the latter probably due to the transformation of epoxy derivatives into triol derivatives, because of the acidic conditions of a simulated gastric juice, reported by Maerker et al. [41].

The bioaccessibility of oxysterols in PS-enriched milk-based fruit beverages with milk fat and MFGM does not increase during storage (0, 3 and 6 months) at room temperature, which demonstrates that sterols are stable in beverages of this kind [5]. In this study, total POP bioaccessibility (45.4–49%) was lower than that of COPs (58.1–80%), which is in contrast with values reported by Alemany et al. [3]. However, this behaviour is comparable to the one observed for the corresponding non-oxidized sterols determined in the beverages (β-sitosterol presents lower bioaccessibility than cholesterol) [3, 4, 9]. Regarding individual oxysterols, triol (56.3–62.0%), α-epoxy (49.3–60.6%) and 7β-hydroxysitosterol (56.2–59.0%) and 7α/7β-hydroxycholesterol (80.5–121.0 and 82.0–93.3%) show the highest bioaccessibility.

11.6 Fermentation of Sterols

Intestinal microbiota is able to act upon the sterol molecules that reach the colon, generating sterol metabolites by means of hydrolyzation, hydrogenation and dehydrogenation reactions. In this sense, microbial cholesterol degradation has been widely studied by *in vitro* assays of fermentation carried out with microbiota (from human, rat, pigeon or chicken faeces) or by pure culture experiments of enteric microorganisms (*Eubacterium spp., Clostridium perfringens, Bifidobacterium spp., Enterobacter aerogenes, Escherichia coli, Bacteroides spp.*) [17]. Information available indicates that cholesterol is mainly converted into coprostanol and, in to a lesser extent into cholestanol. Coprostanol formation has been suggested to follow two routes: a direct pathway through cholesterol reduction or an indirect by means of intermediate metabolite formation (cholestenone, cholesterone, coprostanone, coprostenol). Moreover, it has been suggested that coprostanol may transform slightly into epicoprostanol [17, 29, 39] (Fig. 11.3).

Cholesterol degradation into metabolites by the microbiota is related mainly to the abundance of cholesterol-reducing bacteria. *Eubacterium* and *Bacteroides* genus have been reported to be responsible for the direct as well as indirect routes for cholesterol metabolism, whereas *Bifidobacterium* is only implicated in the direct pathway [17, 43, 49, 52, 55]. Other bacteria groups such as genus *Clostridium* and *Lactobacillus* and *Lachnospiraceae* and *Rumunococcaceae* families have been suggested as coprostanoligenic bacteria with no association with the aforementioned routes [6, 16, 39, 40].

Similar to cholesterol, the biotransformation of PS has been described through the formation of different metabolites such as phytostanone and phytostenone intermediates, and ultimately phytostanols [64]. Sitosterol and campesterol are mainly transformed into ethyl- and methyl-coprostanol and subsequently into ethyl- and methyl-coprostanone, respectively. It also been suggested that stigmasterol could be converted into ethylcoprostenol and ethylcoprostanone. However, PS biotransformation routes have not been fully elucidated, and *Eubacterium spp.* are, thus far, the only species associated with PS degradation [17, 23, 25, 45].

In order to shed light on the sterol biotransformation by gut microbiota, a fermentation study by batch culture of a beverage residue obtained after the simulated gastrointestinal digestion of a PS-enriched milk-based fruit beverage that provided ~1.5 mg PS was carried out by our research group [18]. Results showed a decrease of neutral sterol contents of β-sitosterol (8 and 14%), sitostanol (21 and 16%), campesterol (19 and 39%), campestanol (17 and 15%), stigmasterol (17 and 19%) and brassicasterol (10 and 1%) at 24 and 48 h of fermentation, respectively. Correlated with these diminutions, increments of ethylcoprostanol (18 and 50%), ethylcoprostanone (16 and 39%), methylcoprostanone (63 and 21%) and stigmastenol (13 and 16%) were observed. The absence of cholesterol metabolites suggests a preference of PS as substrate by the action of the gut microbiota. Moreover, PS fermentation diminished the abundance of *Flavonifractor plautii* and *Allobaculum spp.* (from the *Erysipelotrichaceae* family), while it promoted *Eubacterium hallii* and some members of the uncharacterized S-24 family of *Bacteroidetes*. However,

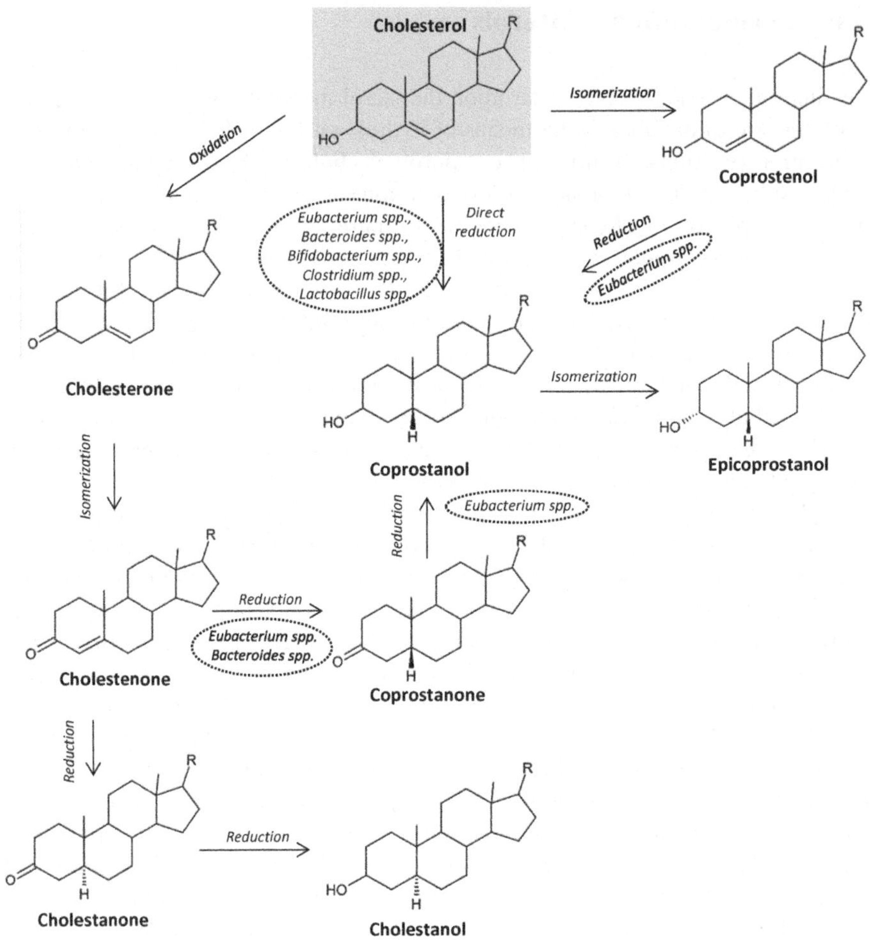

Fig. 11.3 Biotransformation pathways of cholesterol by gut microbiota [17, 39]

it has been noted that differences in microbiota composition were more strongly associated to time than PS presence. The accumulation of anaerobic fermentation products could result in a lowering of the pH that probably leads to microbial changes [22, 62].

As far as we are aware, there are only two studies, carried out by our research group, conducted by means of *in vitro* dynamic fermentation systems in order to evaluate PS biotransformation by gut microbiota [10, 19]. Cuevas-Tena et al. [19] evaluated interactions between PS and microbiota using a TIM-2 system with a PS-enriched dose (~2 g/day) from two sources (an ingredient source of PS or a commercial standard mixture of PS) using microbiota from lean and obese populations. The authors observed a lower concentration of animal and PS metabolites (coprostanol, coprostanone, cholestanol, ethylcoprostanol and methylcoprostanone) with microbiota from lean than obese subjects, suggesting a different microbial activity. The increase of cholestanol with the concomitant decrease of coprostanol and coprostanone observed during the fermentation in presence of PS for microbiota from lean subjects suggest a modification of the main pathway of cholesterol biotransformation into coprostanol. In the case of experiments carried out with microbiota from obese population, the increase of both metabolites, coprostanol and cholesterol, indicates that cholesterol is biotransformed by two different routes. However, the fact that the diet was uncontrolled (real sterol intake) or the conversion patterns of the donors (high or low sterol converters) make it difficult to only correlate this effect to the presence of PS. An increase of β-sitosterol, sitostanol, campesterol and campestanol was observed for lean and obese microbiota assays in presence of PS. Also, a drop of ethylcoprostanol contents (32–35% for lean and 87–88 for obese) with increments of ethylcoprostanone (21- and 46-fold for lean and 33- and 35-fold for obese) and methylcoprostanone (8- and 25-fold for lean) were observed, with respect to control. We linked the absence of β-sitosterol metabolism to ethylcoprostanol by a saturation of the microbiota activity in presence of a PS-enriched dose. Similar to what was observed in a batch culture PS fermentation study [18], gut microbiota seems to prefer PS as a substrate when they are present in a higher proportion with respect to cholesterol. Furthermore, microbiota analysis showed that *Catenibacterium* and *Coprococcus* genera increase in the presence of PS in experiments with lean microbiota. Both of them belong to Erysipelotrichaceae family, and thus, these results are in disagreement with those obtained in the batch culture fermentation study [18]. Regarding obese microbiota experiments, *Clostridium* genus increase when PS was added.

Recently, we published novel results related to the study of PS biotransformation using a food matrix (PS- and GOS-enriched milk-based fruit beverage) by means of an *in vitro* gastrointestinal and colonic fermentation system [10]. In general, neutral sterols determined in the assay (cholesterol, β-sitosterol, sitostanol, campesterol, campestanol and stigmasterol) showed a trend of accumulation in all colon compartments (corresponding to the ascending, transverse and descending colon). Formation of coprostanol, ethylcoprostanol, sitostenone, methylcoprostanol and ethylcoprostenol was observed only in the distal colon compartments during the fermentation of the beverage, demonstrating that sterol metabolism is colon-compartment-dependent. Indeed, higher conversion ratios of the main sterols were detected in the descending versus transverse colon for cholesterol (0.9–1.3 versus 0.2–0.7), β-sitosterol (0.02–0.7 versus 0.01–0.46), campesterol (0.2–0.5 versus 0.08–0.26) and stigmasterol (0.8–1.0 versus 0.5–0.6). In contrast with our previous assays abovementioned [18, 19], higher ratios for cholesterol biotransformation than PS were observed, suggesting that the presence of GOS enhances its metabolism. Moreover, the absence of sterol intermediate metabolites suggests their rapid bioconversion or, in the case of cholesterol, that the direct pathway of conversion to coprostanol was predominant. In contrast, the digestion and fermentation of the PS-enriched milk-based fruit beverage (without addition of GOS) showed an absence of biotransformation of both PS and cholesterol (Blanco-Morales et al., unpublished results), thus evidencing an important role of GOS in this process.

Furthermore, results of the analysis of microbiota composition in the distal colon compartments turned out to be similar and, also, different from the one present in the ascending compartment, supporting the similarities regarding sterol metabolism observed in these compartments. We identified *Bacteroides* as the only coprostanoligenic bacteria present in the colon compartments with cholesterol metabolism and also suggested the involvement of other bacteria species in the PS metabolization aside from *Eubacterium* spp., due to the PS metabolism observed in absence of these bacteria [10]. In the case of the non-GOS-added beverage, the fermentation also modulated certain members of the microbiota composition: in general, *Bifidobacterium* (within 24 h) and Megasphaera genera were stimulated, and *Mitsuokella* was reduced in the ascending colon, while the growth of *Klebsiella* genus was promoted, *Akkermansia* was maintained and Bacteroidetes phylum decreased in the distal compartments.

It has to be pointed out that discrepancies between the results obtained in these above-mentioned fermentation studies could be due to the different conditions applied. In this sense, the use of static methods (batch culture) [18] could have improved the PS metabolism since the absence of a continuous flow, such as in dynamic systems [10, 19], could imply a faster biotransformation. When comparing dynamic fermentation studies, the TIM-2 model [19] includes a dialysis membrane that simulates the absorption of water and metabolites preventing their accumulation and, therefore, a possible inhibition of microbial activity. Moreover, the use of PS included in a food matrix [10] versus provided by a PS-ingredient source or commercial standard mixture [19] could hamper their biotransformation due to the protection provided by the food matrix against the microbiota activity.

11.7 Conclusions

Dietary sterols are minor components present in fat of the human diet. Similar to other lipophilic compounds, in the gastrointestinal tract they are subjected to several sequential reactions (physicochemical and enzymatic) to obtain an absorbable form for subsequent absorption. Bioaccessibility studies, in which sterol incorporation into intestinal mixed micelles are determined, are useful to identify the factors that can promote or hamper sterol solubility and, subsequently, design optimal formulations in the case of functional foods.

Considering the studies related to the bioaccessibility of PS in PS-enriched foods, it can be concluded that their solubility is influenced by several conditions derived not only from the PS enrichment (dosage or source of ingredient) but also from food matrix composition. Furthermore, the presence of compounds such as carbohydrates or carotenoids can limit PS bioaccessibility, while fat and emulsifiers can improve it, with PS-enriched milk-based fruit beverages containing milk fat and MGFM, representing the matrix that offers the highest PS bioaccessibility among the ones reported in literature. However, the use of different *in vitro* gastrointestinal digestion methodologies results in a limiting factor when comparing PS solubility between different studies, therefore it is of interest to standardize and optimize digestion methods focused on lipophilic compounds such as PS.

Furthermore, studies on bioaccessibility of oxysterols could be used as a predictive model of their rate of absorption, although information related to the behaviour of oxysterols during *in vitro* digestion is rather limited, to the effect of the food matrix on their bioaccessibility, and taking into account that an *in vivo* oxide formation can occur as well.

Finally, the non-absorbed sterols are susceptible to subsequent transformations once they reach the colon for their excretion. Although there are studies that determine sterol excretion, information about the microbiota population related to the degradation of sterols and the pathways involved are scarce, in particular, concerning PS biotransformation from enriched products. In studies contemplating *in vitro* fermentation assays in the presence of high concentrations of PS, it has been demonstrated that: (i) microbiota origin (from lean or obese subjects) can influence sterol biotransformation, (ii) other bacterial species are implicated in PS biotransformation, aside from *Eubacterium spp.*, (iii) the presence of PS can modulate microbiota composition, (iv) the sterol metabolism is higher in the distal parts of the colon, and that (v) the food matrix in which PS are included can modulate the behaviour of sterol fermentation (the presence of fibre improves the metabolism of cholesterol versus PS).

Considering all the above facts on the fate of sterols during digestion and fermentation, more studies are needed in this field to provide further information and evidence to improve functional food products and to take advantage of their beneficial effects on health.

Acknowledgments Authors would like to thank the Spanish Ministry of Science and Innovation and the European Regional Development Fund (FEDER) (PID2019-104167RB-I00 project). Thanks are due to Prof. Amparo Alegría for her critical remarks of this chapter.

References

1. Arroyo-Maya IJ, McClements DJ (2016) Application of ITC in foods: a powerful tool for understanding the gastrointestinal fate of lipophilic compounds. Biochim Biophys Acta 1860:1026–1035
2. Alegría A, Garcia-Llatas G, Cilla A (2015) Static digestion models: general introduction. In: Verhoeckx K, Cotter P, López-Expósito I, Kleiveland C, Lea T, Mackie A, Requena T, Swiatecka D, Wichers H (eds) The impact of food bioactives on health: in vitro and ex vivo models. Springer International Publishing, Cham, pp 3–12
3. Alemany L, Cilla A, Garcia-Llatas G, Rodriguez-Estrada MT, Cardenia V, Alegría A (2013) Effect of a simulated gastrointestinal digestion on plant sterols and their oxides in enriched beverages. Food Res Int 52:1–7
4. Alvarez-Sala A, Garcia-Llatas G, Cilla A, Barberá R, Sánchez-Siles LM, Lagarda MJ (2016) Impact of lipid components and emulsifiers on plant sterols bioaccessibility from milk-based fruit beverages. J Agric Food Chem 64:5686–5691
5. Alvarez-Sala A, Blanco-Morales V, Cilla A, Garcia-Llatas G, Sánchez-Siles LM, Barberá R, Lagarda MJ (2018) Safe intake of a plant sterol-enriched beverage with milk fat globule membrane: bioaccessibility of sterol oxides during storage. J Food Compos Anal 68:111–117

6. Antharam VC, McEwen DC, Garrett TJ, Dossey AT, Li EC, Kozlov AN, Mesbah Z, Wang GP (2016) An integrated metabolomic and microbiome analysis identified specific gut microbiota associated with fecal cholesterol and coprostanol in *Clostridium difficile* infection. PLoS One 11:e0148824
7. Baldi A, Pinotti L (2008) Lipophilic microconstituents of milk. In: Bösze Z (ed) Bioactive components of milk. Advances in experimental medicine and biology. Springer, Cham, pp 109–125
8. Barriuso B, Ansorena D, Astiasarán I (2017) Oxysterols formation: a review of a multifactorial process. J Steroid Biochem Mol Biol 169:39–45
9. Blanco-Morales V, López-García G, Cilla A, Garcia-Llatas G, Barberá R, Lagarda MJ, Sánchez-Siles LM, Alegría A (2018) The impact of galactooligosaccharides on the bioaccessibility of sterols in a plant sterol-enriched beverage: adaptation of the harmonized INFOGEST digestion method. Food Funct 9:2080–2089
10. Blanco-Morales V, Garcia-Llatas G, Yebra MJ, Sentandreu V, Lagarda MJ, Alegría A (2020) Impact of a plant sterol- and galactooligosaccharide-enriched beverage on colonic metabolism and gut microbiota composition using an *in vitro* dynamic model. J Agric Food Chem 68:1884–1895
11. Borel P (2003) Factors affecting intestinal absorption of highly lipophilic food microconstituents (fat-soluble vitamins, carotenoids and phytosterols). Clin Chem Lab Med 41:979–994
12. Brzeska M, Szymczyk K, Szterk A (2016) Current knowledge about oxysterols: a review. J Food Sci 81:R2299–R2308
13. Cardoso C, Afonso C, Lourenço H, Costa S, Nunes ML (2015) Bioaccessibility assessment methodologies and their consequences for the risk benefit evaluation of food. Trends Food Sci Tech 41:5–23
14. Cilla A, García-Llatas G, Lagarda MJ, Barberá R, Alegría A (2019) Development of functional beverages: the case of plant sterol-enriched milk-based fruit beverages. In: Grumezescu AM, Holban AM (eds) Functional and medicinal beverages. Elsevier, Amsterdam, pp 285–312
15. Commission Decision 2004/336/EC (2004) Authorizing the placing on the market of yellow fat spreads, milk based fruit drinks, yoghurt type products and cheese type products with added phytosterols/phytostanols as novel foods or novel food ingredients under Regulation (EC) No 258/97 of the European Parliament and of the Council. Off J Eur Comm L105:49–51
16. Crowther JS, Drasar BS, Goddard P, Hill MJ, Johnson K (1973) The effect of a chemically defined diet on the faecal flora and faecal steroid concentration. Gut 14:790–793
17. Cuevas-Tena M, Alegría A, Lagarda MJ (2018) Relationship between dietary sterols and gut microbiota: a review. Eur J Lipid Sci Technol 120:1800054
18. Cuevas-Tena M, Gómez del Pulgar EM, Benítez-Páez A, Sanz Y, Alegría A, Lagarda MJ (2018) Plant sterols and human gut microbiota relationship: an *in vitro* colonic fermentation study. J Funct Foods 44:322–329
19. Cuevas-Tena M, Alegria A, Lagarda MJ, Venema K (2019) Impact of plant sterols enrichment dose on gut microbiota from lean and obese subjects using TIM-2 *in vitro* fermentation model. J Funct Foods 54:164–174
20. De Boer JF, Kuipers F, Groen AK (2018) Cholesterol transport revisited: a new turbo mechanism to drive cholesterol excretion. Trends Endocrinol Metab 29:123–133
21. Dhuique-Mayer C, Servent A, Descalzo A, Mouquet-Rivier C, Amiot MJ, Achir N (2016) Culinary practices mimicking a polysaccharide-rich recipe enhance the bioaccessibility of fat-soluble micronutrients. Food Chem 210:182–188
22. Duncan SH, Louis P, Thomson JM, Flint HJ (2009) The role of pH in determining the species composition of the human colonic microbiota. Environ Microbiol 11:2112–2122
23. Eneroth P, Hellström K, Ryhage R (1964) Identification and quantification of neutral fecal steroids by gas-liquid chromatography and mass spectrometry: studies of human excretion during two dietary regimens. J Lipid Res 5:245–262
24. Etcheverry P, Grusak MA, Fleige LE (2012) Application of *in vitro* bioaccessibility and bioavailability methods for calcium, carotenoids, folate, iron, magnesium, polyphenols, zinc, and vitamins B6, B12, D, and E. Front Physiol 3:1–21

25. Eyssen HJ, Parmentier GG, Compernolle FC, De Pauw G, Piessens-Denef M (1973) Biohydrogenation of sterols by *Eubacterium* ATCC 21,408 -Nova species. Eur J Biochem 36:411–421
26. Fernández-García E, Carvajal-Lérida I, Pérez-Gálvez A (2009) *In vitro* bioaccessibility assessment as a prediction tool of nutritional efficiency. Nutr Res 29:751–760
27. García-Llatas G, Rodríguez-Estrada MT (2011) Current and new insights on phytosterol oxides in plant sterol-enriched food. Chem Phys Lipids 164:607–624
28. Garcia-Llatas G, Cilla A, Alegría A, Lagarda MJ (2015) Bioavailability of plant sterol-enriched milk-based fruit beverages: *in vivo* and *in vitro* studies. J Funct Foods 14:44–50
29. Gérard P (2014) Metabolism of cholesterol and bile acids by the gut microbiota. Pathogens 3:14–24
30. Gies M, Servent A, Borel P, Dhuique-Mayer C (2020) Phytosterol vehicles used in a functional product modify carotenoid/cholesterol bioaccessibility and uptake by Caco-2 cells. J Funct Foods 68:103920
31. Gleize B, Nowicki M, Daval C, Koutnikova H, Borel P (2016) Form of phytosterols and food matrix in which they are incorporated modulate their incorporation into mixed micelles and impact cholesterol micellarization. Mol Nutr Food Res 60:749–759
32. Granado-Lorencio F, Olmedilla-Alonso B, Herrero-Barbudo C, Blanco-Navarro I, Pérez-Sacristán B, Blázquez-García S (2007) Comparative *in vitro* bioaccessibility of carotenoids from relevant contributors to carotenoid intake. Food Chem 102:641–648
33. Grundy MML, Edwards CH, Mackie AR, Gidley MJ, Butterworth PJ, Ellis PR (2016) Re-evaluation of the mechanisms of dietary fibre and implications for macronutrient bioaccessibility, digestion and postprandial metabolism. Br J Nutr 116:816–833
34. Guerra A, Etienne-Mesmin L, Livrelli V, Denis S, Blanquet-Diot S, Alric M (2012) Relevance and challenges in modeling human gastric and small intestinal digestion. Trends Biotechnol 30:591–600
35. Hamdan IJA, Sanchez-Siles LM, Matencio E, Bermúdez JD, Garcia-Llatas G, Lagarda MJ (2018) Sterols in human milk during lactation: bioaccessibility and estimated intakes. Food Funct 9:6566–6576
36. Hunter PM, Hegele RA (2017) Functional foods and dietary supplements for the management of dyslipidaemia. Nat Rev Endocrinol 13:278–288
37. Hur SJ, Lee SY, Moon SS, Lee SJ (2014) *In vitro* effects of cooking methods on digestibility of lipids and formation of cholesterol oxidation products in pork. Korean J Food Sci An 34:280–286
38. Hur SJ, Lee SY, Lee SJ (2015) Effect of biopolymer encapsulation on the digestibility of lipid and cholesterol oxidation products in beef during *in vitro* human digestion. Food Chem 166:254–260
39. Kriaa A, Bourgin M, Mkaouar H, Jablaoui A, Akermi N, Soussou S, Maguin E, Rhimi M (2019) Microbial reduction of cholesterol to coprostanol: an old concept and new insights. Catalysts 9:167. https://doi.org/10.3390/catal9020167
40. Lye HS, Rusul G, Liong MT (2010) Removal of cholesterol by lactobacilli via incorporation and conversion to coprostanol. J Dairy Sci 93:1383–1392
41. Maerker G, Nungesser EH, Bunick FJ (1988) Reaction of cholesterol 5,6-epoxides with simulated gastric juice. Lipids 23:761–765
42. Maldonado-Pereira L, Schweiss M, Barnaba C, Medina-Meza IG (2018) The role of cholesterol oxidation products in food toxicity. Food Chem Toxicol 118:908–939
43. Martin WJ, Ravi Subbiah MT, Kottke BA, Birk CC, Naylor MC (1973) Nature of fecal sterols and intestinal bacterial flora. Lipids 8:208–215
44. Marze S (2015) Bioaccessibility of lipophilic micro-constituents from a lipid emulsion. Food Funct 6:3218–3227
45. Miettinen TA (1982) Gas-liquid chromatographic determination of fecal neutral sterols using a capillary column. Clin Chim Acta 124:245–248
46. Minekus M, Alminger M, Alvito P, Ballance S, Bohn T, Bourlieu C, Carrière F, Boutrou R, Corredig M, Dupont D, Dufour C, Egger L, Golding M, Karakaya S, Kirkhus B, Le Feunteun S, Lesmes U, Macierzanka A, Mackie A, Marze S, McClements DJ, Ménard O, Recio I, Santos

CN, Singh RP, Vegarud GE, Wickham MSJ, Weitschies W, Brodkorb A (2014) A standardised static *in vitro* digestion method suitable for food – an international consensus. Food Funct 5:1113–1124

47. Moran-Valero MI, Martin D, Torrelo G, Reglero G, Torres CF (2012) Phytosterols esterified with conjugated linoleic acid. *In vitro* intestinal digestion and interaction on cholesterol bioaccessibility. J Agric Food Chem 60:11323–11330
48. Moreau RA, Nyström L, Whitaker BD, Winkler-Moser JK, Baer DJ, Gebauer SK, Hicks KB (2018) Phytosterols and their derivatives: structural diversity, distribution, metabolism, analysis, and health-promoting uses. Prog Lipid Res 70:35–61
49. Mott GE, Brinkley AW, Mersinger CL (1980) Biochemical characterization of cholesterol-reducing *Eubacterium*. Appl Environ Microbiol 40:1017–1022
50. Nissinen MJ, Vuoristo M, Gylling H, Miettinen TA (2007) Respective hydrolysis and esterification of esterified and free plant stanols occur rapidly in human intestine after their duodenal infusion in triacyl- or diacylglycerol. Lipids 42:603–612
51. O'Callaghan Y, McCarthy FO, O'Brien NM (2014) Recent advances in phytosterol oxidation products. Biochem Biophys Res Commun 446:786–791
52. Parmentier G, Eyssen H (1974) Mechanism of biohydrogenation of cholesterol to coprostanol by *Eubacterium* ATCC 21408. Biochim Biophys Acta-Lipids Lipid Metab 348:279–284
53. Pérez-Gálvez A (2019) *In vitro* digestion protocols. The benchmark for estimation of *in vivo* data. In: Mercadante AZ (ed) Carotenoid esters in foods: physical, chemical and biological properties. Royal Society of Chemistry (RSC), pp 421–458. https://doi.org/10.1039/9781788015851-00421
54. Reboul E, Richelle M, Perrot E, Desmoulins-Malezet C, Pirisi V, Borel P (2006) Bioaccessibility of carotenoids and vitamin E from their main dietary sources. J Agric Food Chem 54:8749–8755
55. Ren D, Li L, Schwabacher AW, Young JW, Beitz DC (1996) Mechanism of cholesterol reduction to coprostanol by *Eubacterium coprostanoligenes* ATCC 51222. Steroids 61:33–40
56. Rodriguez-Estrada MT, Cilla A, Cardenia V, Garcia-Llatas G (2014) Fats and sterols. In: Caplan M (ed) Reference module in biomedical sciences. Elsevier, pp 1–17. https://doi.org/10.1016/B978-0-12-801238-3.00228-2
57. Singh H, Ye A, Horne D (2009) Structuring food emulsions in the gastrointestinal tract to modify lipid digestion. Prog Lipid Res 48:92–100
58. Tolve R, Condelli N, Caruso MC, Barletta D, Favati F, Galgano F (2018) Fortification of dark chocolate with microencapsulated phytosterols: chemical and sensory evaluation. Food Funct 9:1265–1273
59. Trautwein EA, Duchateau GSMJE, Lin Y, Mel'nikov SM, Molhuizen HOF, Ntanios FY (2003) Proposed mechanisms of cholesterol-lowering action of plant sterols. Eur J Lipid Sci Technol 105:171–185
60. Ubeyitogullari A, Ciftci ON (2019) *In vitro* bioaccessibility of novel low-crystallinity phytosterol nanoparticles in non-fat and regular-fat foods. Food Res Int 123:27–35
61. Vaghini S, Cilla A, Garcia-Llatas G, Lagarda MJ (2016) Bioaccessibility study of plant sterol-enriched fermented milks. Food Funct 7:110–117
62. Walker AW, Duncan SH, Leitch ECM, Child MW, Flint HJ (2005) pH and peptide supply can radically alter bacterial populations and short-chain fatty acid ratios within microbial communities from the human colon. Appl Environ Microbiol 71:3692–3700
63. Wienk KJH, Marx JJM, Beynen AC (1999) The concept of iron bioavailability and its assessment. Eur J Nutr 38:51–75
64. Wong A (2014) Chemical and microbiological considerations of phytosterols and their relative efficacies in functional foods for the lowering of serum cholesterol levels in humans: a review. J Funct Foods 6:60–72
65. Yu L (2008) The structure and function of Niemann-Pick-C1-like 1 protein. Curr Opin Lipidol 19:263–269

Conclusion

As described in this book, lipids, as essential nutrients, are present in a wide range of food matrices, under various forms, and often associated to other food compounds. These parameters determine the way they are delivered to the GI tract which is paramount to their bioaccessibility and subsequent digestibility. Progress is being made in this relatively new research topic, relating food structure to macronutrient bioaccessibility/digestibility. Overall, more and more scientists and researchers are considering not only the nutritional composition of the food but also its structure and the organisation of nutrients within the food matrix. Indeed, some techniques are now commonly used to characterise the food being studied to further understand nutritional impact, e.g. rheology, microscopy, and optical spectroscopy.

This book has demonstrated the wide range of matrix effects in foods that influence lipid digestion, from indigestible plant-based structures to the gastric coagulation in dairy foods, as well as important molecular interactions such as the impact of dietary fibre. Nonetheless, further work is still needed, and future studies should tend to focus on macronutrient digestibility in more realistic conditions, for instance where other ingredients and structures are included. However, before being able to do this it is necessary to obtain a greater understanding of degradation of individual food components and matrices. As discussed, not all plants, and the lipids they contain, behave in a similar manner during digestion, and even within a single variety differences in digestibility can be observed. The same apply to animal sources of lipids, such as dairy products, where each processing step has a significant effect on the final food structure and therefore its fate in the GI tract.

More precise knowledge of the interactions of lipids with other nutrients and food compounds is also required, although some research already demonstrated the critical role played notably by calcium and protein on lipid digestion. This will permit to better predict and eventually control certain health outcomes, in particular cardiovascular diseases.

Finally, with the current global challenges faced in terms of climate change and food security, new sources of lipids and/or plant varieties, such as algae, should be investigated to target the multiple forms of malnutrition.

M. M.-L. Grundy, P. J. Wilde (eds.), *Bioaccessibility and Digestibility of Lipids from Food*, https://doi.org/10.1007/978-3-030-56909-9

Index

M. M.-L. Grundy, P. J. Wilde (eds.), *Bioaccessibility and Digestibility of Lipids from Food*, https://doi.org/10.1007/978-3-030-56909-9

The manufacturer's authorised representative in the EU is Springer Nature Customer Service Centre GmbH, Europaplatz 3, 69115 Heidelberg, Germany. If you have any concerns regarding our products, please contact ProductSafety@springernature.com

Printed and bound by CPI Group (UK) Ltd, Croydon, CR0 4YY
28/09/2026
02224090-0005